Cracked Coverage

Cracked Coverage

TELEVISION NEWS,

THE ANTI-COCAINE CRUSADE, AND

THE REAGAN LEGACY

JIMMIE L. REEVES AND

RICHARD CAMPBELL

Duke University Press, Durham and London   1994

© **1994 Duke University Press**
All rights reserved   Printed in the United States of America
on acid-free paper ∞
Designed by Cherie Holma Westmoreland
Typeset in Sabon with Frutiger display by Tseng Information Systems, Inc.

Library of Congress Cataloging-in-Publication Data

Reeves, Jimmie Lynn.
Cracked coverage : television news, the anti-cocaine crusade, and the Reagan legacy /
Jimmie L. Reeves and Richard Campbell.
p.   cm.
Includes bibliographical references and index.
ISBN 0-8223-1449-5 (alk. paper). — ISBN 0-8223-1491-6 (pbk. : alk. paper)
1. Television broadcasting of news—United States.   2. Narcotics, Control of—United
States.   3. Cocaine industry—United States.   I. Campbell, Richard, 1949–   .   II. Title.
PN4888.T4R44   1994
070.1′95—dc20   93-41411   CIP

To Gay D. Reeves and Dianna Campbell,

Our Partners in Life, For Keeping

the Faith

Contents

Acknowledgments

This is, perhaps, the best place for us to briefly acknowledge our own distinct roles in this collaboration. As in any extensive project, this book progressed through several developmental stages. Campbell took the leadership in the early stages. Although we both worked on the design of the study and the initial analysis, Campbell is primarily responsible for generating funding, coordinating research assistants, and securing the book contract. However, Reeves is listed as the first author because he took the leadership in writing and organizing the manuscript. To be more specific about the authorship of each segment: Reeves is the primary author of the introduction, and chapters 4, 6, and 8; chapters 1, 2, 3, 7, and 9 are jointly authored; and Campbell is the primary author of chapter 5 and the epilogue. We are happy to report that, at least so far, our friendship has managed to survive this intellectual association.

Although undertaking this project has been complicated by the fact that not much funding is available for work that challenges federal drug policies, we were able to generate some modest financial support for this study through the University of Michigan: the Warner G. Rice Humanities Award provided us with seed money; a Rackham Graduate School Faculty Fellowship partly sustained us during the middle stages of the project; and support from the Department of Communication's Howard R. Marsh Center for the Study of Journalistic Performance helped us wrap up the study. We are grateful to the administrators of these funding sources for underwriting our work.

We also want to express our gratitude to the staff and administration of the Vanderbilt Television News Archive in Nashville. We hope that this study further demonstrates the critical value of this archive. A repository for the metanarratives of the day, this archive preserves key moments in American media and culture during the age of television. Obviously, without this resource, our study would not have been possible.

We would like to acknowledge the people who helped in the final stages of producing this book: D. C. Goings, our patient photographer; Bob Mirandon, our scrupulous copy editor; and Pam Morrison, our managing editor. However, we are especially indebted to Larry Malley at Duke for having confidence in this project and encouraging us along the way.

While we, admittedly, risk overlooking someone in dispensing our thanks, we believe it is important to name the good people who contributed their time and their ideas to this work. We want to start by thanking the University of Michigan students who worked on this project as research assistants: Matt Katzive, Ingrid Kelley, Barb Matusak, Nora McLaughlin, Kristina Smith, and Yolanda Taylor. We are also grateful to Michigan's College of Literature, Science, and the Arts Undergraduate Research Opportunity Program (UROP), the Department of Communication, Rackham's Summer Research Opportunity Program (SROP), and the Office of the Vice President for Research (OVPR). Each of these programs and offices provided some of the financial support for these students.

While we have benefited from the good will of most of our colleagues and students in the Department of Communication, we are especially indebted to the following people: Richard Allen, Debra Dobbs, Bettina Fabos, Melissa Goldberg, Lamia Karim, Don Kubit, Szu-ping Lin, Ronan Lynch, Christopher Martin, Barbra Morris, Laura Moseley, Brian Nienhaus, Tasha Oren, Hayg Oshagan, Andrea Press, Terri Sarris, Frank Ukadike, and Nabeel Zuberi. An even larger share of our gratitude, though, is reserved for our colleagues and students in the University of Michigan's Program in American Culture. Much more than a "home away from home" during our time at Michigan, the American Culture Program was for us—thanks largely to June Howard's leadership—a place of refuge, nurturing, and growth. Words cannot capture how much we appreciate the following people associated with the program: Marsha Ackermann, Frances Aparicio, Liz Brent, Sean Buffington, Shuqin Cui, Corey Dolgon, Herb Eagle, Linda Eggert, Michael Epstein, Kristin Hass, Carol Karlsen, Howard Kimeldorf, Michael Kline, Jim Morris-Knower, Frank Mitchell, Mike Niklaus, Gail Nomura, Sylvia Pedraza, Eric Porter, Mark C. Rogers, Roger Rouse, Lisa Saaf, David Scobey, Timothy Shuker-Haines, Steve Sumida, Jeanne Theoharis, Alan Wald, and Joe Won.

Several people not directly associated with the University of Michigan have also provided either intellectual contributions to or moral support for this work. These people include friends and family scattered from North Carolina to southern California: Susan Brinson, Mary Helen Brown, Stanton Brown, Jackie Byars, James Byers, Jim Byers, LaVerne Byers, Terry Byers, Christopher P. Campbell, Molly Campbell, Richard E. Campbell, David Eason, Steve Fore, Herman Gray, Larry Grossberg, Dan Hahn, Bob Kasmer, Cathy Kasmer, Osamu Kusatsu, Wayne Lemon, Cindy Meason, Danny Meason, Carl Michels, Madeline Moore, Stanley L. Moore, Horace Newcomb, Sarah Newcomb, Lisa O'Neil, Clinton Reeves, Joe B. Reeves, Martha Smith Reeves, Steven Stack, Matt Tomasz, Sasha Torres, Ron Weyand, and Steve Yount.

Four people, though, have had such a profound impact on this project that we feel obligated to single them out for special recognition and thanks: Rosanne Freed, Robin D. G. Kelley, Samuel Craig Watkins, and Bruce Williams. These four comrades patiently waded through multiple drafts of the manuscript, suggested important changes in the structure of our arguments, and inspired us to explore new areas of inquiry. We only hope that, some day, we have an opportunity to repay these extraordinary folks for the time, care, and passion that they expended on improving this work.

A Note to Our Readers

Since this *is* a book about visual representations, we had originally planned to include forty-four still photos from the three networks. These stills were taken from videotapes obtained from the Vanderbilt Television News Archive in Nashville. The archive referred us to the networks for copyright permission. In our dealing with news archivists at the various networks, we ran into numerous obstacles. Two of the networks charged extremely high fees for permission to reprint single images ($100 per still at CBS and $200 per still at NBC). And after we reduced the number of images we wanted to reprint, both CBS and NBC finally refused us permission to reprint a single image. ABC did not even consider negotiation. A letter from Joel Kanoff, manager of news archives at ABC, states that network's position: "Please be advised that it is not a policy of ABC News to grant permission to reprint still images from our programs for publication. Accordingly, the photographs in question are not available for your use." Kanoff did not respond to our letters and phone calls questioning this policy.

In our own minds we regard the printing and crediting of a single image from a network news program as "fair use" under copyright law, similar to the way scholars have always quoted and credited small representative passages from various printed texts. However, we do not have the resources to contest the networks and have reluctantly withdrawn the still photos. We suspect from the kinds of questions that the archivists asked us that the networks would have been less obstructionist in this matter if the tone of the book had been more laudatory. We remain genuinely disturbed that the three major television networks—networks that depend on the First Amendment and access to public images daily in their news operations—advance policies that prevent researchers and others from reprinting a tiny portion of those images for educational and noncommercial uses.

Jimmie Reeves and Richard Campbell

Introduction

American society, which seems consumed with the notion of helping the deviant, whether it be the drug abuser or the criminal, has become obsessed with addiction. We have sanitized the problem by suggesting that "hard" addictions can happen to anybody exposed to powerful drugs. Save our children from the devil's grasp, we proclaim, and say no to drugs. Yet most of what we hear is hyperbole, and it is encouraged by those who profit from a continuance of the drug hysteria. It is hyped by medical researchers who get paid to study drugs. It is hyped by the social service industry that gets paid to help rehabilitate the addict. It is hyped by politicians who get elected by showing they have a social conscience.—Michael S. Gazzaniga [1]

I cannot help feeling that what we are now doing in the name of stopping the drug problem *is* the drug problem.—Andrew Weil [2]

The Indictment

An Obligatory Disclaimer   We feel obligated to begin with a disclaimer meant to ease our loving parents' troubled minds and to undermine other less-loving misunderstandings of our motives, our politics, or our indictment. We are not close and personal friends of Dr. Feelgood, devoted and stoned members of the National Organization for the Reform of Marijuana Laws (NORML), or cloak-and-dagger operatives of the Medellin cartel. While this book might accurately be characterized as "anti-antidrug," let us be perfectly clear at the outset that, in this instance, two "antis" do not make a "pro." In other words, this indictment of the Reagan era anti-cocaine crusade is not, in any way, intended to be interpreted as a pro-drug manifesto. As will become clear, we are contemptuous of the anything-goes hedonism associated with the mythical drug culture, but (in contrast to the sanctimonious William Bennett and other professional moral crusaders) we do not ground our contempt on the bogus proposition that America's drug problem springs from some kind of sinister "alien value system." Nothing could be further from the truth. The American infatuation with getting high is not "epidemic" but "endemic": endemic to a socioeconomic system based on the ethic of consumption, on the instant gratifications of short-term profit, on turning a buck no matter who gets hurt. It only stands to reason that a society addicted to waste would be overpopulated by "the wasted." For us, then, the self-indulgent, live-for-today values of consumerism and the predatory, dog-eat-dog imperatives of entrepreneurialism are what have driven both Reaganomics and the drug underworld—and, unfortunately, they are as genuinely American as Coca-Cola and Lee Iacocca.

In challenging the counterfeit claims underwriting Reagan's war on drugs, however, we have conscientiously avoided endorsing some alternative policy position on the so-called drug problem. Instead, we believe it is much more productive to focus on how the "behaviorist" orientation of drug control experts and the "individualist" orientation of mainstream journalists were co-opted by the "moralist" crusaders of the New Right during the Reagan era. Ultimately, this is the central charge that we set out to prove in this book—a charge that connects news coverage of the war on drugs to the "new racism," the "family values," and the "orthodoxy of nostalgia" of the Reagan counterrevolution. This, for us, is a bold indictment—and one that we think we have adequately supported. To go much beyond this indictment and take up a policy position based on an alternative interpretation of "the facts" runs counter to the spirit of the larger project. We leave it to others who are better suited to dealing with the methodological flaws and ideological contamination of enterprising drug research to transform this critique into a policy statement. Scholars like Craig Reinarman and Harry Levine, who we cite heavily in our study, are actively involved in this kind of project—and we do not feel at all qualified to move onto their turf. Our qualifications as media analysts, though, do give us a license to arraign television journalism for its role in legitimating a reactionary political agenda in its knee-jerk support of the war on drugs.

Partners in Hysteria Our indictment, then, involves three parties: the New Right, the drug control establishment, and mainstream journalism. The primary culprit, as we see it, is the New Right operating under the banner of Reaganism. The war on drugs in the 1980s was, first and foremost, a project of the New Right and one that helped it sustain support from many of its single-issue constituencies. The drug control establishment and mainstream journalism, though, also are culpable for aiding and abetting the New Right in producing and warranting this reactionary sociopolitical spectacle. Indeed, in authorizing and advocating the New Right's antidrug activism, drug experts and network journalists operated as moral entrepreneurs in the political economy of Reagan's America—entrepreneurs who benefited personally and professionally from coproducing a series of moral panics that centered on controlling this stuff called cocaine and disciplining the people who used it. In other words, in adopting a "support-the-troops" mentality in their promotions of the war on drugs, drug experts and journalists were not simply involved in disseminating the disciplinary "wisdom" of just saying "no." Instead, they also were deeply implicated in advancing, even mainstreaming, the backlash politics of the New Right in a way that helped mask the economic devastation of deindustrialization, aggravated white-black tensions in the electorate, and, ultimately, helped solidify middle-class support for policies that favored the rich over the poor.

According to our indictment, then, *the journalistic recruitment in the anti-cocaine crusade was absolutely crucial to converting the war on drugs into a political spectacle that depicted social problems grounded on economic transformations as individual moral or behaviorial problems that could be remedied by simply embracing family values, modifying bad habits, policing mean streets, and incarcerating the fiendish "enemies within."* While we do not expect that our charges will be at all convincing to drug warriors, we do think they are compelling enough to make most thoughtful readers reevaluate journalistic performance during the 1980s. But, perhaps even more important, we believe our arguments are incisive, comprehensive, and accessible enough to force thoughtful journalists (at least those who have the courage to confront our critique) to reconsider: (1) how reporters deal with government officials and enterprising experts who have vested interests in cultivating drug hysteria; and (2) how reporters mark off certain segments of the population as deviants who are "beyond rehabilitation."

Pledging Allegiances

Border Crossings This study is purposely designed to speak to a number of reading publics. Drawing on such diverse fields as cultural anthropology, cultural history, political economy, narrative theory, trans-linguistics, and the sociologies of race, ethnicity, gender, deviance, and knowledge, we set out to expose many of the stigmatizing visions associated with the 1980s anti-cocaine crusade. In striving to speak to a multi-disciplinary readership, we have attempted to eliminate much of the jargon that, unfortunately, pollutes so much of what is overvalued in the academy as "serious" scholarship. We agree with Steve Fore:

> Most academic research does not circulate beyond this closed circle of peer specialists, primarily because of the wall of specialized language that usually shuts out even the most interested non-specialist. Moreover, the educational establishment in general does little to encourage researchers to communicate their ideas with the public at large: in fact such contact is institutionally discouraged through pressure placed on academics to publish in the "right" journals (a category that specifically excludes publications with a broad general readership).[3]

In general, we lend our voices to the chorus that decries certain academic discourses that are inscrutable to anyone outside an exclusionary clique. Rewarded for retreating from engagement with the radical inequities of our age, such scholarship is often guilty of indulging in a kind of posturing that confuses masturbatory play in the ozone of high theoretical specialization with intellectual activism.

In other words, we take seriously the distinction that Stuart Hall makes between intellectual work and academic work. According to Hall, although "they overlap, they abut with one another, they feed off one another, the one provides you with the means to do the other," intellectual work and academic work are "not the same thing":

> I come back to the difficulty of instituting a genuine cultural and critical practice, which is intended to produce some kind of organic intellectual political work, which does not try to inscribe itself in the overarching metanarratives of achieved knowledges, within the institutions. I come back to theory and politics, the politics of theory. Not theory as the will to truth, but theory as a set of contested, localized, conjunctural knowledges, which have to be debated in a dialogic way. But also as a practice which always thinks about

its intervention in a world in which it would make a difference, in which it would have some effect. Finally, a practice which understands the need for intellectual modesty. I do think there is all the difference in the world between understanding the politics of intellectual work and substituting intellectual work for politics.[4]

Under the influence of Hall's model of intellectual work, we have structured this study as *both* a political *and* a theoretical intervention that strives for *both* accessibility *and* complexity. Edward Said asks for just such intervention on the part of academics: "Instead of noninterference and specialization, there must be interference, crossing of borders and obstacles, a determined attempt to generalize exactly at those points where generalizations seem impossible to make." Said demands that academics self-consciously carve out spaces in that symbolic terrain intended for larger public audiences "now covered by journalism and the production of information, that employ representation but are supposed to be objective and powerful."[5]

Although this is an interdisciplinary work aimed at a multidisciplinary audience, and we hope a larger public audience, certain portions are written with a more specific intellectual community in mind: the growing number of scholars involved in critical/cultural studies in mass communication. Since this book is meant to advance and broaden this relatively new movement in the academy, we strategically deploy certain terms and concepts that might not be immediately accessible to every reader attracted to this book. To make this book more user-friendly, we have attempted to explain these terms throughout the end notes. Furthermore, as a service to readers outside critical media studies, we also have marked two "insider" passages that might not be of much interest to a general readership. Here, for example, we suggest that people outside media studies might want to skim through the rest of this section on "Border Crossings" and resume a more careful reading at the section marked "Outlining the Dialogue."

Because our primary audience is communication scholars, we think it is important to briefly address the background of cultural theory and critical scholarship that has served as inspiration for this collaboration. This study is aligned with three interdisciplinary traditions in communication research—humanistic television studies, meaning-oriented journalism scholarship, and British Cultural Studies. Since we see our work as more "radical" and less "aesthetic" than the first camp, more "textual" and less "sociological" than the second, and more "Bakhtinian" and less "Grams-

cian" than the third, we want to acknowledge their combined influences
without holding any of the three liable for problems in the conceptualization or the execution of this project.

Humanistic Television Criticism Each member of a collaborative endeavor brings along a separate set of baggage. In this particular intellectual
partnership, the Reeves side of the collaboration is more closely linked to
the relatively young field of humanist television criticism. Emerging as a
distinct field of inquiry in the early 1970s and still only marginally legitimate in the American academy, this camp is personified by the pioneering
work and iconoclastic careers of Horace Newcomb and David Thorburn.
Both Newcomb and Thorburn received their graduate training in literature
departments—and both, in taking television seriously as a narrative art
form, defied the elitism that still characterizes most scholarly appraisals
of the medium. Extending John Cawelti's work on the analysis of popular fictional genres to television programming, Newcomb's *TV: The Most
Popular Art*[6] is considered by many to be the first rigorously academic
analysis of television's familiar story formulas. And Thorburn's "Television
Melodrama"[7] and "Television as an Aesthetic Medium"[8] rank as two of
the most influential articles in the humanistic tradition.

Although both of these scholars are often summarily and unfairly dismissed with the appellation of "liberal pluralist," Newcomb and Thorburn
deserve recognition for their courage and leadership in opening up spaces
in the academy for the critical study of television, of journalism, of popular
culture in general. While Newcomb and Thorburn might not entirely approve of how we appropriate their work, their ideas activate many features
of this project. Our orientation on narrative, our criticism of mainstream
social science, our attempt to make interpretive analysis more "empirical,"
our dialogic view of meaning production, our resistance to conspiracy theories, and our rejection of nostalgic elitism—all of these critical stances have
been heavily informed by the learning and the mentoring of Newcomb and
Thorburn.

Meaning-Oriented Journalistic Scholarship The Campbell side of this
partnership is more closely connected to meaning-oriented journalistic
scholarship. This camp incorporates a body of scholarship—all of it informed by the sociology of knowledge—that treats news as a vital social
force in the construction of reality, the distribution of expert knowledge,
the maintenance of common sense, the enforcement of norms, and the
production of deviancy. Of the scholars who fit into this camp, we have

alcohol-related death of athlete David Overstreet, and Reaganism's orthodoxy of nostalgia (chapter 6);

–the "Big Lie" campaign of the National Institute of Drug Abuse (NIDA), the University of Michigan's "Monitoring the Future" survey, the public opinion polling sponsored by news organizations, and the "Jar Wars" of the 1986 congressional campaigns (chapter 7);

–the history of the nuclear family ideal, the profamily agenda of the New Right, the "image problems" of Nancy Reagan, and the "labeling" of crack babies as "tomorrow's delinquents" (chapter 8);

–the mainstream criticism of journalism's "hyping" of the crack crisis, the "participatory journalism" of Geraldo Rivera, and the discourse of doom invoked by voices from the grave (chapter 9).

The study concludes by considering how mainstream journalism's treatment of the cocaine problem influenced other forms of American popular culture—from the quasijournalistic discourse of "reality" programming like "Cops" to the counter-journalistic discourse of the gangster rap of Ice-T. Finally, we project a possible future for the war on drugs that accentuates the positive by exploring cracks in the reactionary, top-heavy coalition that dominated presidential politics during the 1980s.

Part I

RE-COVERING THE WAR ON DRUGS

The Cocaine Narrative:

A Thoroughly Modern Morality Tale

[William] Bennett, who has been secretary of education without solving the problems of education and drug czar without solving the problems of drugs, now wants to write books about how to solve the problems of both. In America, this is what we call "expertise."—Roger Simon, *Baltimore Sun*

Re-Covering the Cocaine Narrative

In documenting and analyzing the continuities and disruptions of the latest struggle over the meaning of cocaine, this study examines the representations of drug culture in 270 television news reports broadcast between 1981 and 1988. Collectively, these reports form a kind of grand mosaic that we call *the cocaine narrative*—a narrative that took shape around social conflicts and cultural distinctions related to the contemporary politics of race, class, gender, sexuality, region, religion, age, and taste. At various moments during the narrative the meaning of cocaine would be inflected by gender issues, it would take on racial overtones, and it would even animate myths about the sanctity of small-town life in middle America. Therefore, like other major cultural controversies associated with AIDS, abortion rights, television evangelism, and environmental protection, the cocaine narrative expressed many of the prominent themes and antagonisms of the Reagan era. Ultimately, we demonstrate how, in *constructing* and *reaffirming* cocaine use as a moral disease or a criminal pathology, the network news also facilitated the *staging* and *legitimating* of Reagan's war on drugs as a major political spectacle.

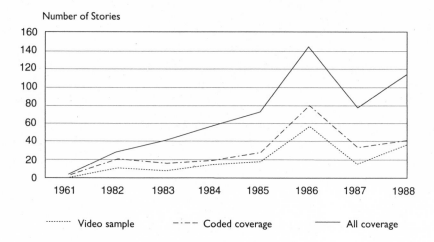

Number of Stories

FIGURE I   Network TV Coverage of Cocaine

Slicing Through the Coverage   To help make our study more man-
ageable, we systematically identified a group of stories appearing on the
evening newscasts of ABC, CBS, and NBC that captures the main thrust and
illustrates the main trends of drug coverage during the 1980s. First, we
screened all the news items between 1981 and 1988 (the Reagan era) listed
under the subject heading of "Cocaine" in the *Vanderbilt Television News
Index:* 528 items listed.[1] As figure 1 indicates, the yearly number of cocaine
stories rose dramatically during the decade before peaking in 1986.[2] In 1987
the coverage dwindled a bit, only to rise again in the following presidential
election year.

In this screening we eliminated most news items that were not reporter
packages (especially 15–20-second studio "readers" presented by anchors).
Because our study focuses on drugs as a domestic social problem, we eliminated all the "foreign intrigue" stories that dealt chiefly with international
drug trafficking.[3] Figure 1 also shows how the remaining 228 stories were
distributed over time, and Appendix A lists these stories chronologically.

Using abstracts published in the Vanderbilt Index, we coded each of
the stories for sourcing and genre. After eliminating routine arrest, criminal
investigation, and prosecution stories, we were left with 147 news stories
that specifically treated cocaine as a domestic social problem. Videotapes
of this collection of stories were then ordered from Vanderbilt University's

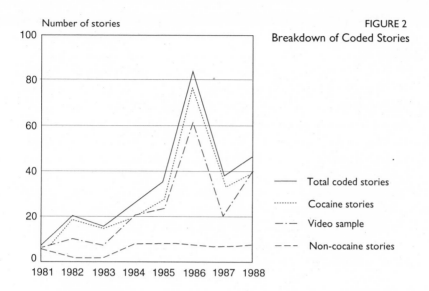

FIGURE 2
Breakdown of Coded Stories

—— Total coded stories

········ Cocaine stories

—·— Video sample

——— Non-cocaine stories

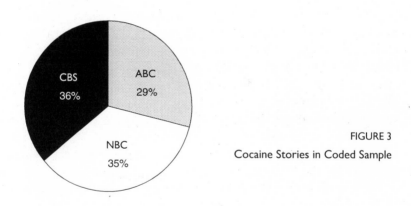

FIGURE 3
Cocaine Stories in Coded Sample

News Archive for close critical analysis. In Appendix A these stories are listed.

In addition to the cocaine stories we also ordered and analyzed video-tapes of forty-two other reporter packages that a general screening of the news index identified as relevant to our study. This smaller and highly selective sample (listed in Appendix B) includes stories on the dangers of other drugs (marijuana, heroin, PCP, "Ts and Blues," and "Wack"), stories on drug use by youngsters, stories on the link between AIDS and heroin addiction, and stories on Nancy Reagan's role in the war on drugs.

Figures 2 and 3 illustrate how the 270 stories are distributed over time and how they break down according to analytic categories. We are confident that these reports afford a comprehensive portrait of the network news treatment of cocaine as a social problem during the Reagan administration. Thanks to the forty-two noncocaine stories included in the study, we have at least a rough sketch of how other drug issues were related to the anti-cocaine crusade.

Phases of Coverage Our frame analysis[4] of these 270 news stories suggests that the cocaine crusade of the 1980s took shape in three distinct phases. In the initial stage (Phase I), running roughly from January 1981 to November 1985, cocaine would be defined according to a "trickle-down" paradigm in which what was once a decadent "glamor" drug became increasingly available and abused by middle-class Americans. As we discuss in chapter 5, this trickle-down framing of the drug news was largely governed by discourses of recovery in which white offenders were able to be purified of the shame of cocaine transgression by way of therapeutic intercession. Soon after the journalistic discovery of crack, the coverage reached a crisis level (Phase II) that lasted from December 1985 to November 1986. During this crisis, race and class dimensions of the cocaine problem established during Phase I coverage are almost completely inverted as the trickle-down formulation is displaced by a siege paradigm accentuated by disturbing images of chaos in the streets. Often featuring an Us vs. Them orientation toward a color-coded "enemy within," this siege paradigm activated discourses of discrimination that were distinct from recovery operations prominent during Phase I coverage. The final "aftermath" or "postcrisis" period (Phase III), November 1986 to December 1988, would culminate in the war on drugs emerging as the major issue of the 1988 presidential campaign. Although the siege paradigm would continue to prevail during this Phase III coverage, the discourse of discrimination was somewhat softened by a growing sense of doom as reporters reacted to criticism from without about the crusading Phase II coverage and to self-doubt from within about the continued failures of the war on drugs to solve the drug problem.

In drawing on narrative theory to mark out these phases we treat the chain of events that appear in the 1980s' cocaine story as a "change from one state of affairs to another."[5] Arranged in chronological order according to the logic of cause and effect, story events conventionally move from an initial "equilibrium state" (which typically represents normalcy) through a period of disequilibrium (which generally represents a disruption of normalcy) to a new equilibrium (which often represents the restoration of normalcy).[6] In the cocaine narrative the disturbance of normalcy is generally defined in terms of pollution.[7] Cocaine, the pollutant, threatens the purity of some vulnerable group (children, teenagers, mothers), some important segment of the economy (the stock market, professional sports, the trucking industry), or some culturally charged setting (Hollywood, small-town America, the college campus). The period of disequilibrium generally foregrounds a struggle for purification that almost always involves the operation of some agency affiliated with the drug control establishment. While Phase I featured therapeutic intervention as the approved solution to the drug pollution (see chapter 5), Phases II and III would place more emphasis on surveillance mechanisms, punitive disciplinary operations, and partisan political maneuvers (see chapters 6, 7, and 8). For various reasons associated with modern journalism's orientation toward "bad news," the cocaine reports dealt almost exclusively with disruption and disequilibrium. The restoration of normalcy, according to most journalists' view of the reported world, is simply not "newsworthy."

But there is an even more insidious reason that the restoration of normalcy in drug news is an event that is only anticipated and never realized—the journalistic reliance on experts who make their living off the cocaine problem. Although journalism rarely explores this conflict of interest, a substantial part of the therapeutic, surveillance, disciplinary, enforcement, judicial, penal, and political components of the drug control establishment is devoted to and sustained by the production and reproduction of drug delinquency. As Howard S. Becker observes in his classic study of what he terms "moral entrepreneurs," the agent of control—especially the "rule enforcer"—is often caught in a double bind:

> On the one hand, he must demonstrate to others that the problem still exists; the rules he is supposed to enforce have some point, because infractions occur. On the other hand, he must show that his attempts at enforcement are effective and worthwhile, that the evil he is supposed to deal with is in fact being dealt with adequately. Therefore, enforcement organizations, particularly when they are seeking funds, typically oscillate between two kinds of claims. First, they say that by reason of their efforts the problem they deal with is approaching solution. But, in the same breath, they say the problem is

perhaps worse than ever (though through no fault of their own) and requires renewed and increased effort to keep it under control. Enforcement officials can be more vehement than anyone else in their insistence that the problem they are supposed to deal with is still with us, in fact is more with us than ever before. In making these claims, enforcement officials provide good reason for continuing the existence of the position that they occupy.[8]

Consequently, with nothing to gain and everything to lose from declaring a victory in the war on drugs, the drug control establishment's networks of power, knowledge, and discipline have a vested interest in maintaining a perpetual sense of urgency, even a sense of hysteria, about cocaine pollution.

Kernel Events   In approaching the cocaine news drama and hysteria through the lens of narrative theory, we make a distinction between two kinds of events: *Kernels* and *Satellites*. Kernels are those crucial events that actively contribute to the story's progression.[9] Appendix C presents a chronology of the events that we have identified as crucial to the development of the cocaine narrative. In Phase I, for instance, the most important kernel event occurred on October 14, 1982, when President Reagan, in a speech delivered at the Justice Department, declared war on crime and pledged an "unshakable" commitment to do "what is necessary to end the drug menace." Many point to this moment as the official commencement of the latest war on drugs.[10]

The overdose deaths of actor-comedian John Belushi and Robert Kennedy's son David Kennedy, the arrest of failed auto tycoon John DeLorean, two cycles of professional sports scandals (the first in 1983, the second in 1985), and Hollywood scandals involving the highly publicized drug trials of musician John Phillips, and actors Richard Dreyfuss and Stacy Keach—all of these events also constitute defining moments in Phase I that situate cocaine as a decadent taboo and a dangerous substance. But these troubling developments would pale in symbolic magnitude when compared to the cultural trauma of the back-to-back deaths of athletes Len Bias and Don Rogers in the summer of 1986. In chapter 6 we devote a considerable amount of attention to the role these last two deaths played in authorizing the drug hysteria that swept the nation during 1986.

Satellite Events   In contrast to the gravity of kernel events, satellite events are less significant moments that are more routine and not as central to the cause-effect chain of the unfolding story. The bulk of the cocaine stories fit into the satellite category. These satellite stories steadily increased in number from 1981 through 1986 before sharply declining in 1987 during the beginning of Phase III. In Phase I the satellite stories often took shape in the form of "status reports" couched as warnings about the consequences of cocaine abuse and its spread from the privileged to the middle classes (see chapter 5). With the rising sense of urgency and hysteria of Phase II, the satellite stories took the form of "crisis updates" that often featured the campaigning voices of crusading journalists (see chapters 6 and 7). In Phase III many satellite stories expressed second thoughts as news organizations responded to criticism of their "hyping" the crack crisis (see chapter 9).

Reflections on the Cocaine Narrative

The Bounty Hunter's Angle   In focusing on the meaning of the drug news, in examining the linkages between expert knowledge and state power, and in interrogating the hypocrisies of the journalistic ideal of objectivity, this study represents a drastic departure from much traditional quantitative news research that is not without risks. As Steve Fore maintains: "As long as the realm of common sense belief holds that there is a more or less direct correlation between media content and audience behavior, and that this relationship may be verified and usefully scrutinized through the comforting solidity of hard numbers and controlled scientific experiments, other modes of explanation will not be accepted or sought."[11] Based on an information-oriented view of the communication process, such modern orthodox news research, like orthodox news, is often informed by the kind of no-nonsense positivism summed up in Jack Webb's laconic dictum, "Just the facts, ma'am." According to this view, reality is something concrete, unitary, and natural—a maze of facts, figures, opinions, confidential sources, and eyewitnesses that can be discovered, perceived, surveyed, quantified, and reported "objectively." Generally speaking, research inspired by this view of communication attempts to evaluate the news in terms of its fidelity to the facts.

In framing journalistic practice in terms of a "reflection" or "mirror" analogy, traditional news research conceives of the deviation between the journalistic account and this mythical reality as an index of something bad or negative called "bias." Take, for instance, John Merriam's analysis of national media coverage of drug issues between 1983 and 1987. Dusting off the mirror model, Merriam concludes that in this coverage "the media have reflected the public's attitude at large": "In the business of producing and selling news, large media organizations are generally in the business of reporting back to the public what it needs to know and is willing to pay for. The management skill of news organizations is to be able to recognize stories that fit this test. Drug issues have generally ranked fairly well, but they have not proved as durable as many people, knowing the seriousness of the problem, would like them to be." [12] Notice how Reaganesque assumptions about the virtues of the so-called free market are naturalized in Merriam's assessment of the news industry's performance in reflecting what the public "needs to know and is willing to pay for."

Such social-scientific research tends to provide what is best described as a "top-down," administrator's picture of the news. Framed and composed from the point of view of a power bloc vying for the attention of a sluggish public, this research is often inspired by the idea of using the news media to achieve "civic mobilization." Indeed, Donald Shaw and Maxwell McCombs, leading practitioners of agenda-setting research, are surprisingly straightforward in pledging their allegiances to this power bloc. In their article, "Dealing with Illicit Drugs: The Power—and Limits—of Mass Media Agenda Setting," they argue that understanding the dynamics of agenda-building is important "both to communication scholars and to policymakers." [13] It is important to communication scholars because it provides a theoretical way of explaining "how ideas and information are disseminated." But Shaw and McCombs are careful to suggest that this research is equally worthwhile because of its "practical implications for policymakers concerned with arousing public concern over drugs and with maintaining public support for government programs designed to combat the problem." [14]

Clearly, so-called value-neutral agenda-setting research is far from neutral in its application to the media's construction of the drug problem. In the tradition of mercenary social scientific expertise, the concluding paragraph of Shaw and McCombs's article reads like promotional literature for a consulting operation:

> In the research presented here, we have learned a great deal about the press and public agendas in regard to drugs. Most importantly, we have gained additional insights into the dynamics of agenda building. *There is excellent guidance here for those leaders and interest groups who wish to maintain public and governmental attention on drugs. There is an opportunity to be seized,* but only a thoughtful policy extending beyond the publicity of the moment will ameliorate the drug problem. The press and the public agendas can be

shaped by transitory events. But if the drug problem is to be reduced, neither the press nor the public must be diverted by news events at the periphery of the issue. *Using the knowledge presented here, a thoughtful leadership—in the press, government, and public—must frame the drug issue for the national agenda in such a way that ameliorating actions are possible.* [Our emphasis] [15]

There is an *opportunity* to be *seized?* For us, these words express the prevailing mentality of a group of modern specialists who operate as "moral entrepreneurs" in the cocaine narrative. [16] As hired guns in the war on drugs, Shaw and McCombs understand that a lot of money is being thrown at the drug problem and that, because of its utility to "leaders and interest groups," agenda-setting researchers deserve to cash in on at least some of the bounty.

The Modern Context While our study is also concerned with understanding how the framing of the drug issue in the news media makes "ameliorating actions" possible, we have very different allegiances from those embraced by mainstream social science. As proposed in chapter 2, our mole's eye view of the war on drugs identifies more with "the policed" than with elite "policymakers." In keeping with Foucault's work, we challenge the rhetorics of liberal reform and humanitarian rehabilitation that have traditionally played an important role in justifying:

1. the scientific management of populations;
2. the expanding authority and power of expertise;
3. the ongoing production, "verminization," and punishment of drug delinquency. [17]

Or, put another way, our analysis, unlike that of orthodox news research, re-frames the 1980s drug news in the context of the warp and woof of modernity.

But what do we mean by modernity? Unfortunately, like such nebulous concepts as "art" and "ideology," modernity is a word that, depending on the angle of the sun, the inflection of the speaker, or the prejudices of the audience, can stand for almost anything or approximately nothing. Because of this dubious distinction, modernity inspires both great thought and hopeless confusion. For clarity's sake, then, we want to situate "Modernity" (with a capital "M") in relation to the two epochs in Western thought that immediately preceded it: although the seeds of the "modern attitude," the "modern perspective," and the "modern condition" were planted during the radically new historical self-consciousness of the

Renaissance (ca. 1400–1650), and modern tendencies would continue to develop during the Age of Reason (ca. 1650–1790), for us (as for Foucault) the Enlightenment represents the true arrival of the Modern Age.

A European intellectual movement of the late eighteenth century that included such noteworthy thinkers as Diderot, Hume, Adam Smith, Voltaire, and Bentham, the Enlightenment would provoke the birth of the human sciences—sciences devoted to the study of human "life, labor, and language."[18] Unlike the taxonomies of classical epistemologies that were preoccupied with discovering the divine order of things, these new human sciences substituted Man for God as both the subject and object of modern knowledge. As Alexander Pope put it, "the proper study of mankind is man."[19] In promoting the study of the human condition, Enlightenment thinkers hoped to free mankind from hunger, disease, crime, war, and superstition. The sciences of life would promote health, temperance, productive bodies, and sober minds; the sciences of labor would eliminate want, poverty, laziness, and greed; and the sciences of language would facilitate democracy, equality, morality, and creativity.

In theory, Enlightenment thought represented a secular manifestation of the same lofty purpose embraced by Saint Thomas Aquinas—the quest for a just and good life. In practice, though, as Nietzsche argued, the seductive and ambiguous thing called "human progress" that propelled this intellectual movement would lead the Enlightenment project to be both "creatively destructive" and "destructively creative."[20] "To the degree that it also lauded human creativity, scientific discovery, and the pursuit of individual excellence in the name of human progress," notes David Harvey, "Enlightenment thinkers welcomed the maelstrom of change and saw the transitoriness, the fleeting, the fragmentary as a necessary condition through which the modernizing project could be achieved."[21] Of course, the genocide, the nerve gas, the gulags, the ethnic cleansing, and the other technologically advanced and systematically applied horrors of the Modern Age have caused many to voice regrets and confess second thoughts about the Enlightenment project. Some, like Harvey and Jürgen Habermas, in renouncing modernity still manage to defend the rationalism of the Enlightenment as the only way to save the planet and redeem the human race. But detractors, like Weber and Foucault, condemn the Enlightenment project as a utopian folly that promises liberation and delivers oppression. Weber, for instance, blames the instrumental rationalism of the Enlightenment project for fostering the bureaucratic domination of modern societies. According to Weber, in this "iron cage" of bureaucratic administration, "specialists without spirit, [and] sensualists without heart, [are] caught in the delusion that [they] have achieved a level of development never before attained by mankind."[22] And Foucault puts one of his many denunciations of administrative knowledge and power this way: "The relationship between rationalization and excesses of political power is evident. And we should not wait for bureaucracy and concentration camps to recognize the existence of such relations."[23] Following Foucault, this study conceives of modernity in terms of the triumph of a control culture grounded on the utopian vision of the Enlightenment project—a control culture whose instrumental rationalism is dedicated to regulating populations and disciplining bodies through an array of surveillance mechanisms, policing techniques, and normalizing procedures.

Disciplinary Metaphors   Most research based on the information paradigm and the mirror metaphor is so unself-consciously mired in the Enlightenment project that it takes, as given and natural, the vested interests of modern experts who are in the business of promoting drug hysteria "in such a way that ameliorating actions are possible." Because of this blind spot, such research does not begin to account for the active role the news media play in producing what John E. Merriam naively describes as "the public's attitude at large." For us, this productivity is animated by the very language of routine journalistic discourse on drugs. Recall for a moment how the language associated with the familiar "Disease-Crime Model" of drug use has helped define and sustain the cocaine crisis. "Plague," "epidemic," "contagion," "infestation," "cancer"—metaphors linked to this model literally saturate journalistic accounts of the drug problem. Unfortunately, disease metaphors are so commonplace that many journalists, medical experts, and governmental officials no longer seem able to recognize actual differences between disease and outlaw habit, between illness and taboo behavior. The double jeopardies associated with the Disease-Crime Model are perhaps best summed up by Lester Grinspoon and James B. Bakalar:

> The most important restraint on our treatment of sick people is the requirement that the patient feel ill and want to be cured; the most important restraint on our treatment of criminals is the requirement that they have committed some specific clearly harmful act; by regarding the status of addiction to certain drugs or the habit of using them as somehow both an illness and crime, we dispense with both restraints. There is no need to show evidence of a harmful act, because the drug use is after all in itself a disease; but there is no need for

the consent to treatment ordinarily required with disease, because it is a crime, or at best a morally wrong, immature, irresponsible habit that must be forcibly suppressed. Finally, by treating certain habits as diseases and crimes, without demanding the usual evidence of disease or crime, we produce pernicious consequences which we then falsely attribute to the habits themselves and treat as further evidence that they should be treated as diseases and crimes.[24]

In defining drug transgression as both a disease and a crime, then, the state and the medical profession make the user of illegal drugs into a *super deviant* who is subject to quarantine and imprisonment, involuntary treatment and mandatory sentencing.

Strangely enough, official and journalistic language associated with the rhetoric of the "war on drugs" is much more honest, instructive, and revealing than the metaphors of the Disease-Crime Model. As President Bush's former "drug czar" William Bennett put it, "The drug crisis is a crisis of authority—in every sense of the term 'authority' ": "What those of us in Washington, in the states, and in the localities can do is exert the political authority necessary to make a sustained commitment to the drug war. We must build more prisons. There must be more jails. There must be more judges to hear drug cases and more prosecutors to bring them to trial, including military judges and prosecutors to supplement what we already have."[25] The language of war, then, properly characterizes drug regulation as a battle over control—who controls definitions of legal and illegal, medication and intoxication, therapy and sin, prescription and addiction, morality and immorality, mainstream and marginal, order and chaos.

After 1986, the demographic profile of the drug war's dehumanized enemy that emerges on the evening news is predictable enough—predominantly black or Latino, young, male, poor, and isolated in the inner city. As Ralph Brauer put it, "anyone who watches television knows that inner cities are the real 'front lines' [of the war on drugs]": "As a result, those living there face suspicion, condescension and hate. Their neighborhoods have become contemporary versions of Southeast Asia—dangerous, exotic places where even a child might kill you. (Notice how often this scenario appears in the drug war and nourishes the same ugly emotions it fed during Vietnam.)"[26] That the Bush-Quayle campaign successfully exploited this climate of "suspicion, condescension and hate" in 1988 with its Willie Horton TV spots is only one particularly spectacular example of how the metaphors and images of the anti-cocaine crusade played out in the political arena.

Dialogic Analysis

Where our view of drug regulation relies on Foucault's critique of modernity, our concrete analysis of journalistic discourse is heavily influenced by the literary theory of Mikhail Bakhtin. To examine how the television news figured into the political uses of drug delinquency in the 1980s, we take a Bakhtinian "dialogic approach" to interpreting the process by which reporters and anchors rewrite the cocaine problem in their packaging of the news. This approach is based on the simple proposition that the television news is a dynamically open and self-regulating system of communication. In many ways this proposition is at the heart of what separates our study from those that adopt an information-oriented view of news as content or a straight ideological reading of news as propaganda. Such content analyses and ideological readings fail to recognize that (quoting Bakhtin) "we must deal with the life and behavior of discourse in a contradictory and multi-languaged world." Because of this failure, traditional content analyses and propaganda studies arrive at only a "passive understanding of discourse"— that is, "an understanding of an utterance's *neutral signification* and not its *actual meaning*." According to Bakhtin, a passive understanding "is no understanding at all":

> it is only the abstract aspect of meaning. But even a more concrete *passive* understanding of the meaning of the utterance, an understanding of the speaker's intention insofar as that understanding remains purely passive, purely receptive, contributes nothing new to the word under consideration, only mirroring it, seeking, at its most ambitious, merely the full reproduction of that which is already given in the word—even such an understanding never goes beyond the boundaries of the word's context and in no way enriches the word.[27]

What we hope to generate in this study is an active response to cocaine news. Such a response "assimilates" the news story under consideration "into a new conceptual system, that of the one striving to understand."[28]

Modern Storytelling In justifying our appropriation of Bakhtin's work on the novel to our interpretation of the journalistic treatment of the cocaine problem, it is worth noting that the proper names given to "the News" and "the Novel" are roughly synonymous. As narrative forms, the news and the novel are, after all, both devoted to what Bakhtin terms the "eternal re-thinking and re-evaluating" of contemporary reality. The story world of the news, also like that of the novel, is "a world-in-the-making" that is "stamped with the seal of inclusiveness." Unlike epic storytelling in which the present is merely a transitory station between a glorious past and predetermined destiny, in both the news and the novel the present becomes "the center of human orientation in time and in the world." In fact, the news is so fixed on the present that it, as a general rule, expands its "zone of contact" with contemporary reality to an ahistoric extreme; the news is so trained on the here and now that it is, indeed, often sealed off from the past and the future.[29] The news, therefore, might be accurately characterized as a "factualized" version of the novel—the novel with amnesia and tunnel vision. For us, then, both the news and the novel are linked by their status as distinctly *modern* narrative forms. Indeed, this study conceives of the news as one of the major accomplishments of the modern project—an accomplishment that is at the same time *both* a principal product *and* prominent producer of modernity.

Conventional journalism, obsessed with the present, with the timely, with the now or new(s), often endures critical attack for its failure to flesh out historical context and analytic frameworks in the examination of contemporary phenomena. But journalism is merely enacting an aspect of modernism in its fetishism of the present. In fact, Marshall Berman regards contradictions associated with the journalistic fetishization of the present as the very stuff of modernity: "To be modern is to find ourselves in an environment that promises adventure, power, growth, transformation of ourselves—and, at the same time, that threatens to destroy everything we have, everything we know, everything we are. . . . To be modern is to be part of a universe in which, as Marx said, 'all that is solid melts into air.' "[30] Berman has captured here the range of experience that may be found in a newspaper or a television newscast on any given day: from the grim portrait of the latest drug-related gang death to a feature story on a heroic rescue to useful tips on removing troubling carpet stains to the drama of last night's boxing match to the antics of our favorite comic-strip characters. Mike Wallace caught the spirit of modernity at work in news at the end of CBS's very first "60 Minutes" program in 1968: "And there you have our first . . . broadcast. Looking back, it had quite a range, as the problems and interests of our daily lives have quite a range. Our perception of reality roams, in a given day, from the light to the heavy, from warmth to menace, and if this broadcast does what we hope it will do, it will report reality."[31] In identifying the range from warmth to menace, light to heavy, news reporters and other modern professionals typically believe that contemporary experience divides dualistically, along a linear continuum between good and evil. Such a definition of experience is, however, a moral—not a neutral—definition. By pretending they are neutral subjects, however, journalists also claim success in capturing and identifying reality, another distinct virtue of the modern expert.

Of course, reportorial conventions that fetishize the present also satisfy day-to-day deadline pressures. The timely, practical news report responds to "particular needs," then serves those needs "directly and practically."[32] As Theodore Glasser and James Ettema contend, "Among journalists . . . news is not a theoretical construct but a practical accomplishment . . . ."[33] Modern news advocates a literal or descriptive posture, implying that "facts" are transparent and generally unambiguous. Such descriptive "information" details "what is going on here" without plunging explicitly into analytical reflection or the multitude of contradictions that saturate "the real." *Real* news stories are marked off and categorized apart from opinion pieces and political columns. Consequently, journalism's infatuation with the here and now, as well as its reduction of "the real" to "the factual," ultimately serves a modernist utilitarian ideology. And the obsession with "the new" explains why the drug narrative virtually dropped out of the 1992 election coverage after playing such a central political role in 1986 and 1988. Journalism, from its modern and insular perspective, had simply used up the cocaine narrative, even though most of the social and economic circumstances that surrounded and contextualized the first wave of cocaine stories remained in place, carried forward in the policies of the Bush administration. In this regard, journalism also often deludes itself and us, the readers and viewers, into thinking that fundamental social and economic conditions have altered or even disappeared simply because reporters have depleted the narrative during an earlier moral panic that pandered to timeliness and exhausted "the facts."

The Language of Modern Journalism   Even the common "language" of the well-informed reporter is an artifact of journalism's passage from the partisan to the modern era. This transformation was accompanied by the establishment of a new, lean, efficient, "universal" language that could pass over telegraph wires instantly—an unadorned language that was cheaper to send and transcended regional, ethnic, and class boundaries. Suggesting again the connection between the news and the novel, James Carey makes connections among the modern efficiency of the telegraph, conventional journalism, and Ernest Hemingway's modernist writing, which was influenced by journalism ("helping him to pare his prose to the bone, dispossessed of every adornment"):[34]

> The telegraph, by creating the wire services, led to a fundamental change in news. It snapped the tradition of partisan journalism by forcing the wire services to generate "objective" news, news that could be used by papers of any political stripe. Yet the issue is deeper than that. The wire services demanded a form of language stripped of the local, the regional; and colloquial. They demanded something closer to a "scientific" language, a language of strict denotation in which the connotative features of utterance were under rigid control. If the same story were to be understood in the same way from Maine to California, language had to be flattened out and standardized. The telegraph, therefore, led to the disappearance of forms of speech and styles of journalism and story telling—the tall story, the hoax, much humor, irony, and satire—that depended on a more traditional use of the symbolic. . . .
>
> . . . The spareness of the prose and the sheer volume of it allowed news—indeed, forced news—to be treated like a commodity: something that could be transported, measured, reduced, and timed.[35]

Carey further implicates the telegraph and streamlined news in turning "colonialism into imperialism" with their power to coordinate shipping and military operations, especially after the completion of the transatlantic cable in 1870. The telegraph—and by extension, the press wire services—became a conduit for the exercise of modern power.

The language of contemporary, conventional journalism is also closely allied with another modernist separation—the break from a romanticist ethos. The shift to the utilitarian *real* from the amorphous *romantic* was part of a larger turn-of-the-century set of modernist triumphs: science over religion, urbane city over provincial country, abundance over scarcity, social consumption over individual production, and self-realization over self-denial.[36] As Kenneth Gergen notes, this modernist turn attempted to obliterate "the romantic preoccupation with the deep interior" and replace it "with the rational, well-ordered, and accessible self."[37] Whereas the romantic ethos "attributes to each person characteristics of personal depth: passion, soul, creativity, and moral fiber," the chief characteristics of modernists "reside not in the domain of depth, but rather in our ability to reason—in our beliefs, opinions, and conscious attitudes."[38] Modernists in general openly reject "spiritual experiences, mysticism, and a sense of bonded unity with nature."[39]

Certainly, the romantic reverie of drug guru Timothy Leary in the

1960s came under severe attack from modernists, in part, because Leary connected drug experiences with the mystical and the pleasurable in opposition to the rational and utilitarian. Under the modernist view, then, "reason and observation" became "the central ingredients of human functioning. This latter view pervades the sciences, government, and business. . . ." [40] It also pervades the language of modern journalism—as demonstrated in Philip Bell's early 1980s' examination of Australian drug stories. Bell discovered that in this coverage "perverse pleasure" is regarded as "self-indulgent and therefore is exposed to the glare of psychiatric, legal, and other normative examination." [41] The language of the news supports values of productivity and efficiency; on the other hand, illegal drugs disturb middle-class values such as the work ethic and family unity. Bell argues that in these types of news stories from the early 1980s "the mythical hedonistic spontaneity of drug use is emphasized in media images of non-productive self-indulgence while the ingestion of alcohol and tobacco, being widely accepted as consistent with formal work values, must be understood, however contradictorily, as not dysfunctional or, alternatively, as constituting a problem within the consensus, not an attack on its central values." [42] In other words, the moral impulse toward realism and common sense in the news constructs an artificial binary opposition that cordons off illegal from legal drugs and keeps their discourses separate.[43]

Embracing Contradiction   In its most paranoid forms contemporary news studies explain journalistic "bias" in terms of conspiracy. This tendency is most vividly displayed in the clashing work of media watch groups that preach ideals of "fairness," "accuracy," and "balance" while practicing the slanted politics of partisan activism. On one extreme, Accuracy in the Media (AIM) sees secular humanism operating everywhere in the news, with Dan Rather playing the role of the Antichrist; at the other extreme, Fairness and Accuracy in Reporting (FAIR) sees reactionary propaganda everywhere in the same news, with Diane Sawyer playing the role of Leni Riefenstahl. Strangely enough, our study found evidence that supports both of these diametrically opposed positions. The coverage on CBS is, in fact, distinguished by a consistently anti-Reagan spin in its reporting of the futilities of the war on drugs. At the same time, we also can state that coverage on all of the networks, including CBS, periodically encourages such strong identification with police action that it becomes little more than propaganda for the expansion of state power in the surveillance and repression of visible and vulnerable populations in America's inner cities.

For conspiracy theorists, though, such contradictions are troubling. Generally, the stories that do not fit into their schemes are ignored altogether, or dismissed as irrelevant, or offered up as the exceptions that prove the conspiracies. But, for us, the contradictory character of the network news is not at all troubling. Instead, we embrace these contradictions because they express a much more complex—and liberating—way of thinking about news, reality, communication, and culture.

How are these contradictions liberating? Primarily because they subvert the ideology of objectivity that has informed the histories of both journalism and social science. This ideology is premised on the notion that human agents can somehow stand outside of language and identity, transcending both culture and history in grasping The Truth of objective reality. Interestingly, in the Western world, this search for "objective reality" began in earnest during the Renaissance about the same time that the powers-that-be suspended the search for the Holy Grail. Perhaps it is too simplistic to suggest that one pursuit merely replaced the other. And yet, though radically different in terms of method, both of these ventures use similar strategies in anticipation of reaching the same end. They look beyond human history and culture for a transcendent Truth—a Truth that both explains the mysteries of the human condition and enables the mastery of the material world. Both quests, then, are essentially dedicated to achieving omniscience and omnipotence. And, from our point of view, the secular realistic quest for the Objective Truth is just as doomed and, ultimately, just as bent on domination as that ancient romantic crusade for a sacred relic that would reveal, once and for all, the secrets and the powers of the universe.

This study is provoked by more modest objectives. In fact, our quest for knowledge begins by denying—or at least doubting—the existence of an essential or universal human Truth, whether economic, or technological, or psychological, or biological, that determines all of our actions in, or governs all of our experiences of, the world. Instead of searching for The Truth, we are more intent on understanding how what passes for the provisional truth about a particular substance (cocaine) and a particular transgression (drug "abuse") is produced, communicated, and revised in American culture during a particular stretch of recent history (the Reagan era). In other words, we are after the microtruths that never add up to an absolute, eternal Truth, but do speak to the shifting and uncertain commonsense knowledge that is exploited in licensing the deployment of state power.

As a regime of truth, commonsense knowledge that organizes everyday life in a culture is produced and reproduced, maintained and revised, by way of a symbolic process that is unique to the human species. For lack of a better word, we call this symbolic process "communication." But this study takes a view of communication that departs from the transmission-of-information paradigm informing most news research. In stark contrast to this paradigm, we think of communication in terms of surveillance, spectacle, narrative, and ritual. And, here, we adopt James Carey's particularly expansive perspective on the character of communication in modern life—a perspective that connects sacred ceremonies that "draw people together in fellowship and commonality" with secular practices that also act as "projections of community ideals and their embodiment in material form—dances, plays, architecture, news stories, strings of speech." [44] For us, it is by way of these communication rituals that modern cultures enact and reenact the spectacle of a never-finished struggle for meaning, order, control, and freedom.

This struggle is never finished, in part, because of in-group rivalry, out-group resistance, and opposition-in-isolation. These rivalries, resistances, and oppositions insure that modern meanings are never static. For example, the dynamic character of the struggle for meaning is readily apparent in the making, the unmaking, and the rehabilitation of famous reputations. Consider how the commonsense meanings of such human images as George Armstrong Custer, Billy the Kid, Harry Truman, Martin Luther King, Jr., Malcolm X, Muhammad Ali, Jimmy Carter, Rock Hudson, Woody Allen—or, more to the point, John Belushi, John DeLorean, Nancy Reagan, Manuel Noriega, Marion Barry, William Bennett, and Len Bias—have shifted over time according to the impromptu truths of widely accepted regrets, mundane fears, and ordinary hopes.

Embedded in our ritual view of communication, then, is a sense of culture that is less concerned with the imposition of a core curriculum than it is with the interpretation of everyday life. In other words, rather than take Matthew Arnold's "great [white] man" view of culture as a kind of top-down educational enterprise devoted to the perpetuation and appreciation of "legitimate" aesthetic values, we adopt Bakhtin's more anthropological view of culture as an ongoing dialogic process. This process involves the dynamic interplay of two competing forces: the *centripetal* forces of normalization that seek uniformity and coherence (and are often put in the service of domination, regulation, and control) versus the *centrifugal* forces of resistance that seek diversity and freedom (and are often the subject of repression, surveillance, and discipline). In the living process we call culture, the news is generally aligned with the centripetal forces—for news stories are devoted to ordering the clashing voices of culture into a coherent structure by discovering or even manufacturing a sense of consensus ("And that's the way it is . . ."), however fragile and fleeting. Yet, in translating the conflicts and contradictions of modern life into rituals that are comprehensible and compelling to a vast, heterogeneous (though predominantly masculine) audience, the network news also can, sometimes even intentionally, become a platform for dissent, an ally of civil disobedience, a whistle-blowing advocate of the disenfranchised, an enemy of tradition. This journalistic schizophrenia, of course, explains the curious coexistence of right- and left-wing conspiracy theories that are equally valid, and equally blind, to the complexities of the news, of reality, of the communication process, and of the cultural enterprise. To sum up our position, network cocaine news is full of contradictions because it attempts to account not for one reality, but for multiple realities; not for one informative "truth," but for conflicting meanings associated with competing moral agendas and contrary definitions of the situation.

Whereas Part II of this book is devoted to interrogating the defining moments of all three phases of the 1980s cocaine narrative, the remaining chapters of Part I, to enrich that analysis, place the cocaine narrative in dialogue with the tangled histories of modern drug regulation, modern journalism, and Reaganism. In the next chapter we undertake the first of the historical excursions by situating the 1980s drug hysteria in the context of the rise of the modern control culture and the parallel formation of the medical-industrial complex. In this extremely relevant context, the 1980s cocaine narrative can be seen as the latest episode in a morality tale that began as early as 1875 when the first modern drug regulations were enacted in San Francisco—a social drama that largely involved controlling and disciplining subordinated and demonized segments of the population. In our brief genealogy of drug regulation we hope to demonstrate how these struggles for control prompted not only a drastic expansion in the power and authority of modern experts, but also succeeded in masking political, economic, and racial agendas with the humanitarian rhetoric of recovery, rehabilitation, and reform.

2

Merchants of Modern Discipline:

The Drug Control Establishment

A critical aspect of the drug policy debate, often overlooked or taken for granted, is the question of culture. The drug policy debate is ultimately a cultural debate. To speak of the legality, indeed the propriety of putting any number of chemicals in the human body calls into question the structure of social reality. The economic data on potential tax money lost, like the criminal justice statistics on crime rates, and the figures on lost hours of work and lower productivity—cited to support or refute current drug policies—are grounded in a particular social perspective.—Dwight B. Heath[1]

The current situation won't do. The failure to get serious about the drug issue is, I think, a failure of civic courage—the kind of courage shown by many who have been among the victims of the drug scourge. But it betokens as well a betrayal of the self-declared mission of intellectuals as the bearer of society's conscience. There are many reasons for this reluctance, this hostility, this failure. But I would remind you that not all crusades led by the U.S. government, enjoying broad popular support, are brutish, corrupt and sinister. What is brutish, corrupt and sinister is murder and mayhem being committed on our cities' streets. One would think that a little more concern and thought would come from those who claim to care deeply about America's problems.—William J. Bennett[2]

Contextualizing Mayhem

In dialogue, the preceding quotes from Dwight Heath and William Bennett speak to the tensions, concerns, and anxieties that motivate this study. Like Heath, we see drug regulation as a question of culture. At the same time, since we count ourselves as intellectuals who "care deeply about America's problems," we cannot conscientiously take the position that the "murder and mayhem being committed on our cities' streets" is simply a media fantasy. The moral panic underlying the widespread public and journalistic support of the Reagan era war on drugs was not manufactured out of thin air. Instead, it was a response, however reactionary and misguided, to real material conditions, real violence, real murder, and real mayhem.

Hawks, Owls, and Doves Acknowledging these realities, though, does not mean that we are willing to accept the hawks' perspective on the cocaine problem embraced by Bennett, Reagan, and many mainstream journalists. According to this Us vs. Them outlook, the drug economy represents an alien system far removed from the values embraced by the rest of society. Bennett simply does not have the "civic courage" to admit that the belief system driving Reaganomics is widely shared by the people who turn to the drug trade as a way of getting ahead. As Philippe Bourgois observes in his street ethnography of the crack culture, "The assertions of the culture-of-poverty theorists that the poor have been badly socialized and do not share mainstream values are inaccurate":

> On the contrary—ambitious, energetic, inner-city youths are attracted into the underground economy precisely because they believe in Horatio Alger's version of the American dream. They are frantically trying to get their piece of the pie as fast as possible. In fact they often follow the traditional U.S. model for upward mobility to the letter, aggressively setting themselves up as private entrepreneurs. They are the ultimate rugged individualists, braving an unpredictable frontier where fortune, fame, and destruction are all just around the corner.[3]

The orientation on short-term profits, on competitive individualism, on possession-based status, on instant gratification—these are the get-rich-quick entrepreneurial values and live-for-today consumerist mentalities that have given us a $4 trillion national debt, a multibillion-dollar savings-and-loan bailout, a deteriorating environment, an upward redistribution of wealth, an army of homeless people, escalating health care costs, the highest per capita incarceration rate in the world, as well as the murder and mayhem of the cocaine economy.

In contemporary policy debates, according to Peter Reuter, two other drug war strategies currently challenge Bennett's hawkish doctrine: an "owl" and a "dove" strategy. The owl stance, which Reuter himself endorses, favors the treatment branch of the drug control establishment. Whereas hawks, like Bennett, see drug abuse as a crime problem, owls see it as a disease associated with adverse social conditions. While owls

oppose legalization, they question the amount of resources devoted to punitive enforcement measures and would put the funding priority, instead, on
expanding drug prevention and treatment programs.[4]

The dove position, in contrast, favors legalization. Doves consider the
murder and mayhem of the cocaine economy to be a direct outcome of prohibition. In justifying this position, the doves short-circuit criticism that
they condone drug abuse by pointing out that, as a public health problem, illicit drugs pale beside the dangers posed by alcohol and tobacco
(in 1985, for instance, alcohol and cigarettes accounted for 100,000 and
320,000 deaths, respectively, while deaths attributable to illegal narcotics
reportedly claimed just more than 3,500 lives).[5] For the doves, the hawks
have failed to learn the lessons of the Prohibition era. For example, Daniel
Lazare, declaring that the recent "explosion" in street crime "did not occur
despite the war on drugs, but because of it," explores similarities between
the 1980s and the 1920s. According to Lazare, as the Prohibition era progressed, bootleggers shifted from fronting beer to pushing bathtub gin
because they were increasingly unwilling "to risk their lives for something
that was 95 percent water and hops." Lazare sees this same mentality at
work in the shift from marijuana to cocaine in the 1980s, and he argues
that the same consequences associated with the failed attempt to prohibit
the use of alcohol in the 1920s are evident in contemporary America: gang
violence, immoderate use, and decline in the use of "softer substances."[6]

The Mole's Vantage Point   All of these positions, as the aviary imagery
suggests, are top-down administrative views of the war on drugs. While
our critique admittedly converges with the owl and dove positions at many
points, we are not prepared to endorse, or for that matter oppose, more
funding for the rehabilitation industry or a policy of legalization. What
we hope to do instead is develop a mole's-eye-view of the war on drugs—
a bottom-up perspective that, in contrast to the hawks, owls, and doves,
identifies more with "the policed" than with elite policymakers.[7] From the
mole's perspective, we believe the war on drugs is most properly understood as a classic example of the modern "political spectacle." According
to Murray Edelman, such spectacles tend to "perpetuate or intensify the
conditions that are defined as the problem, an outcome that typically stems
from efforts to cope with a condition by changing the consciousness or
behavior of individuals while preserving the institutions that generate consciousness and behavior."[8] Ultimately, by way of this vicious circle, the
failure of the drug control establishment to solve the cocaine problem with

technodisciplinary measures has only reinforced and extended the power and authority of its therapeutic and penal networks, facilitating the implementation of more of the same failed measures: more police raids, more jails, more prisons, more prosecutors, more drug testing, more treatment, more education, more rallies, more surveillance—more disciplining of the body and more regulating of the population.

This vicious cycle is especially evident in the meteoric rise of the national drug control budget during the Reagan-Bush administrations. In 1981, $1.5 billion was devoted to domestic enforcement, international and border control, and demand reduction. Under Reagan's leadership, the drug budget more than quadrupled ($6.6 billion spent in 1989) at a time of drastic cuts in many other government programs. Under Bush, that enforcement number nearly doubled again: the 1993 allocation was $12.7 billion, with the lion's share going to domestic enforcement programs.[9]

As moles, we see these ballooning numbers in terms of the contradictions and inequities of an era in which the ruling bloc embarked on a highly self-conscious program to loosen governmental constraints on the rich and tighten "margins of illegality" on the poor. The war on drugs was absolutely central to enabling the incongruities of this cynical political agenda: it succeeded in defining *social problems* grounded on global transformations in late capitalism (deindustrialization, job migration, the vanishing "family wage" of a vanishing manufacturing economy, the flexible exploitation of fragmented labor markets in a burgeoning service economy, the rise of transnational corporations, etc.) as *individual moral problems* that could be resolved by way of voluntary therapeutic treatment, compulsory drug testing, mandatory prison sentences, even the penalty of death.

Rites of the Reagan Order

Whereas much news coverage after the 1986 crack crisis would converge on hawkish definitions of the drug problem that were congruent with Reaganism's political, economic, racial, and moral agenda, there were other seemingly incongruent owlish moments, especially during the early 1980s. To address the shifting dialogue between the cocaine narrative and the backlash politics of the New Right, we treat television news as a spectacle of surveillance that displays a range of cultural performances—all of which articulate images of order by representing authority, publicizing common sense, and visualizing deviance (see chapter 3). Located at one end of this range are rites of inclusion, at the other end, rites of exclusion.[10] It is important to stress that these rites have both a *discursive* dimension in news narratives and a *coercive* materiality in "correctional" networks of the modern control culture. As Cohen suggests, coercive discipline in modern societies includes both "soft" and "hard" domains of control: "on the soft side, there is indefinite inclusion, on the hard side, rigid exclusion."[11]

Discourse of Recovery   Rites of inclusion are not centrally about Us vs. a marginal Them but, instead, are devoted generally to the edification and internal discipline of those who are within the fold. Rites of inclusion, in other words, are stories about Us: about what it means to be Us; about what it means to stray from Us; about what it means to betray Us; about what it means to be welcomed back to Us. In the controlling visions of modern journalism and Reaganism, rites of inclusion are often saturated with what might best be called "the discourse of recovery." In the cocaine coverage this discourse is especially prominent in stories about reformed addicts (most notably, professional football players Carl Eller and Mercury Morris, and actor Stacy Keach) who have come to "see the light" and now "just say 'no.'" We also know this discourse in the affirmations of Reaganism—the discourse of renewal, rejuvenation, restoration, revival, and rebirth. As Joanne Morreale put it, "Reagan's rebirth rhetoric offered secular salvation, a symbolic resolution of personal and public crises":

> Ronald Reagan's rhetorical appeals attempted to renew popular faith in the ideals of capitalism by affirming traditional American values and their manifestation in myths such as those of the individual, the community, and the American Dream. If successful, this merger of myth and theory would allow him to instate his version of the market economy and effectively redirect the course of American government and society. Reagan realized his aims through use of rebirth rhetoric that resonated with American populism and religious fundamentalism.[12]

To be born again in America, to be morning again in America, to stand tall again in America, all are Reaganite slogans that speak of the exuberant moral, economic, and psychic recoveries of a nation going "back to the future."

Recovery is also a central feature of the normalizing procedures of the soft sector of drug regulation. In this domain of control, *technologies of the self* devoted to restoring and enforcing normalcy are deployed in rehabilitative treatment—treatment meant to transform the drug transgressor by way of the purifying confession. According to Hubert Dreyfus and Paul

Rabinow, "the key to the technology of the self is the belief that one can, with the help of experts, tell the truth about oneself." [13] Applied primarily to transgressing members of the middle and professional classes, this procedure represents the secularization of an ancient form of psychic control and moral extortion—a form of theocratic domination that Foucault calls *pastoral power*. [14] However, in relocating the site of this technology of the self from the priest's confessional to the psychiatrist's couch, this procedure turns human desire into a quasiscientific discourse that, under the cloak of therapy and the camouflage of promoting the "good and just life," is chiefly devoted to defining mental health in terms of "the normal" and negotiating the "self-discovery" or "self-recovery" of voluntary conformity. The confession, then, provides therapeutic replication and psychic repair of the bourgeois soul—a soul that may find something like salvation in the comfort of being (or becoming) normal.

During the Reagan era, as Stanton Peele demonstrates, addiction treatment was one of the most expansive growth industries in the U.S. service economy:

> Private treatment centers have become important operations for many hospitals, and numerous specialty hospitals and chains—like CompCare—devote themselves to the treatment of alcoholism, chemical dependency, obesity, and assorted new maladies like compulsive gambling, compulsive shopping, PMS, and postpartum depression. . . . The number of such centers more than quadrupled and the number of patients treated in them for alcoholism alone quintupled between 1978 and 1984. [15]

And what Peele condemns as the "diseasing of America" would continue to advance after 1984. In a scathing (though tardy) network exposé of the questionable practices of the drug treatment industry, ABC's Sylvia Chase reported that "the number of profit-making psychiatric hospitals" doubled yet again between 1984 and 1991. [16]

Discourse of Discrimination   But the soft sector of the drug control establishment was not the only beneficiary of the Reagan order. The hard sector flourished, too. Where disciplinary modes of inclusion treat the drug *offender* as a diseased soul in need of therapeutic transformation, modes of exclusion stigmatize the transgressor as pathological Other—a *delinquent* beyond rehabilitation. As Foucault observes, the delinquent is a special category of nonconformist: "The delinquent is distinguishable from the offender by the fact that it is not so much his act as his life that is relevant in characterizing him." [17] A place of separation, of surveillance, of *in*carceration, where the delinquent is detained and delinquency is contained (and maintained), the prison is the chief manifestation of the hard sector of the "correctional" network of the medical-industrial complex. In the 1980s, in large part because of how the Reagan administration waged its war on drugs, the pathology of delinquency was literally inscribed on the body of the young, urban, poor, black male whose very life was a punishable offense requiring disciplinary modes of exclusion.[18]

This stigmatization was especially evident after the judicial and journalistic discovery of crack. As Michael Massing notes, in many cases "the courts treat crack far more severely than cocaine, the effect of which is to imprison blacks for much longer than whites": "In Minnesota, for instance, first-time crack users convicted under a 1989 law received a four-year prison term, while first-time users of cocaine powder received probation; about 95 percent of those charged for crack were black and 80 percent of those charged for cocaine powder were white."[19] Although the law was declared unconstitutional in 1991 by the Minnesota Supreme Court, it still underscores widespread differences in the treatment of white "offenders" and black "delinquents." But in Ohio where the prison system was operating at 175 percent capacity in mid-1993, with 30 percent of that population imprisoned for drug convictions, such racially coded disparities in punishment continue. Possession of ten grams or less of powdered cocaine remains a third-degree felony punishable by one year in prison, while the same amount of crack remains a second-degree felony with a two-year mandatory term.[20]

During the Reagan era, in fact, the U.S. prison population nearly doubled (from 329,821 in 1980 to 627,402 in 1989)[21] as the number of drug arrests nationwide increased from 471,000 in 1980 to 1,247,000 in 1989.[22] By 1990 the United States had the highest incarceration rate in the world (426 per 100,000 compared to 333 per 100,000 in South Africa, its closest competitor).[23] In that same year—when about half the inmates in federal prisons were there on drug offenses[24]—African Americans made up almost half of the U.S. prison population, and about one in four young black males in their twenties were either in jail, on parole, or on probation (compared to only 6 percent for white males).[25]

In the realm of the discursive, justification for the criminalization of a generation of black youth was initiated by way of journalistic rites of exclusion. As opposed to rites of inclusion, rites of exclusion are preoccupied with sustaining the central tenets of the existing moral order against threats from the margins. News reports that operate in this do phasize the reporter's role of maintaining the horizons of common sense by distinguishing between the threatened realm of Us and the threatening realm of Them. Under Reaganism, these rites were often shot through by what Roger Rouse terms the "discourse of discrimination":

> Constituting the world as one in which old boundaries have grown permeable and appearances can no longer serve to mark intention and identity, this discourse emphasizes that crucial moral differences still exist and valorizes the capacity both to recognize the differences and to reinscribe the line between them. We know this discourse in the Reagan rhetoric of evil empires and Nicaraguans invading Texas . . . in the images of popular movies such as *Something Wild, After Hours,* and *Blue Velvet,* and, most notably perhaps, in the panic idiom of the War on Drugs, an idiom given particularly vivid expression by Daryl Gates and his supporters in licensing their assaults on the poor black neighborhoods of south-central Los Angeles.[26]

Our study confirms Rouse's observations—the panic idiom was adopted as the public idiom of journalistic coverage during the 1986 crack crisis (see chapters 6 and 7).

A Genealogy of Drug Regulation and Modern Power

Perhaps the central finding of this study, then, is the disparity in news treatment of "white offenders" and "black delinquents." Borrowing Joseph R. Gusfield's terminology, whites were generally depicted as "repentant deviants," while blacks were often branded as "enemy" or "contesting deviants."[27] However, in reviewing the history of American antidrug movements, we also have discovered that the link between drug regulation and minority oppression is nothing new. In fact, we would characterize this malicious connection as a long-standing tradition that is as American as Manifest Destiny and the Ku Klux Klan. After all, the original attempts to regulate drugs coincide with the modern industrial transformation of American society in the decades following the Civil War. Urbanization, bureaucratization, professionalization, specialization, individualization, mass communication—all the hallmarks of modernity and the modern state—would accompany this transformation. During this period the initial moves to restrict drugs were explicitly concerned with governing the behavior, restraining the pleasures, and disciplining the consciousness of subordinate groups: Asian immigrants, newly emancipated slaves, American Indians, women, children, and the urban "lower classes." As David F. Musto put it, "in the nineteenth century addicts were identified with foreign groups and internal minorities who were already actively feared and objects of elaborate and massive social and legal restraints."[28]

The earliest modern drug regulations, antiopium ordinances in San Francisco (1875) and Virginia City, Nevada (1876), were inspired, not by medical considerations, but by xenophobia directed against Chinese immigrants.[29] According to John Helmer, this xenophobia was connected to a "split-labor market" that pitted white laborers against Chinese workers. The savage competition between these factions of the labor market helped inspire mean-spirited ordinances meant to punish the Chinese for, well, being Chinese. According to Helmer, the opium issue "was part of the general ideological response to labor market failure, reflecting the extent to which its secondary labor market, with its Chinese concentration, offered no 'work relief' to the unemployed, insecure, white working class": "The ideological role of the anti-opium campaign was to get rid of the Chinese. It had a practical consequence—providing a legal basis for unrestrained and arbitrary police raids and searches of Chinese premises in San Francisco. Ostensibly to identify opium dens, these raids served the same purpose as that of the vigilantes in the mine fields, against Chinese encampments in the mid-1850s."[30] As we suggest in chapter 4, during the transition from a manufacturing to service economy in the 1970s and 1980s, antagonism among racially coded fragments of competing labor markets would provide the background for both mobilizing the backlash politics of the New Right and revitalizing the war on drugs.

Racism also would play an important role in the initial campaigns to make cocaine illegal.[31] In a 1903 report to the American Pharmacological Association, a committee on the Acquirement of the Drug Habit concluded that "the negroes, the lower and criminal classes are naturally most readily influenced" by cocaine.[32] As Musto observes in an extensive note on this subject, the "association of cocaine with the southern Negro became a cliché a decade or more before the Harrison Act" was passed in 1914:

> the problem of cocaine proceeded from an association with Negroes in about 1900, when a massive repression and disenfranchisement were underway in the South, to a convenient explanation for crime waves, and eventually Northerners used it as an argument against Southern fear of infringement of states' rights. . . . In each instance there were ulterior motives to magnify the problem

of cocaine among Negroes, and it was to almost no one's interest to minimize or portray it objectively. As a result, by 1910 it was not difficult to get
legislation almost totally prohibiting the drug.[33]

And, according to Smith, print journalism contributed to the climate of
racist hysteria surrounding the regulation of cocaine: "One newspaper
called cocaine a 'potent incentive in driving the humbler negroes all over
the country to abnormal crimes.' Another warned that cocaine could keep
criminals from dying of normally mortal wounds by providing 'temporary
immunity from shock.' Newspapers reported frequently on murder and
mayhem among the inner-city poor—most frequently blacks—'caused' by
cocaine addiction."[34] In this crusading journalism "the murder and mayhem among the inner-city poor" was often "attributed to coke even when
it was unknown what drug was responsible, or when it was clear *no* drug
was involved."[35] This chemical scapegoating of social ills is a journalistic
tradition that is resurrected in the crusading reportage of the 1986 crack
crisis—especially in the controlling visions of maternity projected in the
sinister discourse of the "crack mother" (see chapter 8).

The Narco-Carceral Complex Not coincidentally, during this forma-
tive period in modern drug regulation a therapeutic orthodoxy emerged
that sought to expand medical authority by limiting the practice of self-
medication and eliminating the competition posed by quacks who peddled
patent medicines.[36] By requiring the labeling of proprietary remedies con-
taining alcohol, opiates, cocaine, and other substances, the Pure Food and
Drug Act (1906) represented the first step in attaining these objectives at
the federal level. By 1914, with the passage of the Harrison Act, the modern
medical-industrial complex was firmly entrenched.[37]

According to Craig Reinarman, the professionalization of medicine
was connected to the proliferation of drug laws in at least two ways. First,
the development of drug laws marked the start of a "medical imperialism" in which medical authority colonized several "new fields—including
moral, personal, and political spheres unrelated to medical knowledge but
also including more closely related fields like drug use." Second, prescribing
unproven drugs was considered to be a professional hazard that hampered
"the progress of scientific medicine." Quoting Reinarman:

> the AMA, unlike today, welcomed the assistance of the federal government in
> defining disciplinary boundaries, especially if it helped put into bold relief the
> difference between "legitimate" medical use and "illegitimate," nonmedical/

recreational use of drugs which they saw as both dangerous and professionally threatening. In short, the rise of drug controls and the rise of professional
medicine were symbiotic developments.[38]

The rhetoric of this orthodox medicine was a rhetoric of science and reform
that succeeded in masking not only political and racial agendas, but also
the economic self-interest of the medical profession and the pharmaceutical industry. "Narcotics laws aided the institutional development of modern professions of medicine and pharmacy," write Grinspoon and Bakalar,
"by helping to define their areas of competence and exclude surplus and
unlicensed practitioners": "The impulse to clean up society and reduce disorder worked against such practices as free self-medication and chaotic
small-scale entrepreneurial competition. New standards for medical and
pharmaceutical practice were professional hygiene, and insistence on clear
and legally enforced categories for psychoactive drugs was regarded as
intellectual hygiene."[39] Thus, modern forces of control and regulation were
advanced by the language of reform and rehabilitation—a language that
recasts the politics of race, of class, of domination, and of inequality into
the "neutral language of science." In this revision, political objectives and
economic aims are transformed into technical problems—a transformation
that, of course, reinforces and extends the power of modern experts.[40]

Today, the medical-industrial complex continues to contribute to the
disease/crime framing of drugs as a social problem—a problem whose
solution therefore involves therapy and rehabilitation for some, and punishment and imprisonment for others. Ushered in, ironically, when Reagan
administrators were preaching the gospel of deregulation out of one side of
their collective mouth, the latest war on drugs is served by a host of health,
welfare, and law enforcement organizations and agencies battling over precious and contested moral borders. The domestic and foreign fronts of this
police action are being waged by, among others, the Justice Department
and its Drug Enforcement Administration (DEA), the State Department,
the National Guard, the Department of Housing and Urban Development,
U.S. Customs, the U.S. Coast Guard, the FBI, the National Institute on
Drug Abuse, and the Alcohol, Drug Abuse and Mental Health Administration. By 1989 no fewer than eighty congressional committees would
claim "oversight" responsibilities for these interlocking bureaucracies.[41]
But this tangled web of officialdom makes up only part of the antidrug
establishment. Add to this public- and private-sector abuse and treatment
clinics, self-help books, guides to abuse and treatment centers, and guides

to self-help books—and what emerges is a "narco-carceral network" that represents a self-perpetuating system of power, authority, and moral enterprise.[42]

As a central institution in the operation of modern power, journalism, too, is implicated in the carceral network of the medical-industrial complex. For, as Philip Bell observes, drug discourses "in the press and on television consistently derive from scientific, bureaucratic and normative institutions. The 'factual,' balanced discourse of news itself hierarchically organizes these into complex texts about drugs as a social problem . . . ."[43] That implicit moral hierarchy places journalists and expert authorities somewhere near the top, victims and viewers somewhere in the middle, and the criminally diseased drug transgressor at the bottom.

The Marketplace of Credibility In reproducing this "hierarchy of credibility,"[44] the national news subsidizes the power and influence of experts among the medical and scientific community by enhancing their status and legitimacy; in other words, in the political economy of drug control, journalism is a market force that often raises the stock of moral entrepreneurs who profit from escalations in the war on drugs. Again quoting Bell:

> health (and much social welfare) news is highly "subsidized": it is usually a result of the cheap sources of routine information channelled from the bureaucracy, academic researchers or drug companies which are "rewarded" by public relations exposure when a source is taken up. . . . Of course, the police and courts also furnish much of the information on which drug stories are routinely built and dependence on all such sources is magnified by the inability of reporters to evaluate critically the literature received.[45]

Bell concludes, and we agree, that drug stories in particular "are one shop window for the display of the bureaucracy's and research institution's own activities."[46]

One drug expert, Robert L. DuPont, the original director of the National Institute on Drug Abuse (NIDA), grateful for his time in the national spotlight, offers a revealing insider's view of journalism's marketplace of expertise:

> I have never had a complaint about the coverage my ideas have gotten. Most of the problem experts have with the media is on the side of the experts: They simply do not understand the media's role. Experts usually are pompous and boring. They want to lecture the ignorant public. They rarely have an understanding of the marketplace of ideas the media actually is, or what it takes to make a sale in the marketplace. I have considered myself most fortunate to

have been included in the national marketplace (the national media) from time to time. I have always used as best I can my lifetime 15 minutes of fame. In the national media, that 15 minutes usually is doled out in 20-second sound bites, so it lasts quite a while for most of us who are fortunate enough to get any air time at all.[47]

DuPont's remarks implicitly support Bell's observations on the privileged role of medical and scientific experts in the American drug drama.

Drug dealers, then, are not the only ones profiting from the shadow economies driven by drug control. As Arnold Trebach maintains in his 1987 book *The Great Drug War,* "we devote billions to the enforcement of criminal drug laws while too many of our treatment experts, including leading medical doctors, are devoting their thinking to how they can fill beds in their chains of profit-making hospitals. . . . The same type of venality is true of too many government officials, academic scholars, and journalists: they know better but make conscious decisions to obtain money or power by playing the ever-popular and ever-profitable drug-war game."[48] Like the merchants of war devoted to perpetuating the power of the military-industrial complex, the moral entrepreneurs of the medical-industrial complex—and their journalistic comrades—are in the hysteria business. Consequently, they might accurately be characterized as "merchants of discipline."

3

Visualizing The Drug News:

Journalistic Surveillance/Spectacle

Narratives are organizations of experience. They bring order to events by making them something that can be told about; they have power because they make the world make sense. The sense they make, however, is conventional. No story is the inevitable product of the event it reports; no event dictates its own narrative form. News occurs at the conjunction of events and texts, and while events create the story, the story also creates the event. The narrative choice made by the journalist is therefore not a free choice. It is guided by the appearance which reality has assumed for him, by institutions and routines, by conventions that shape his perceptions and that provide the formal reper- tory for presenting them. It is the interaction of these forces that produces the news, and their relationship that determine its diversity or uniformity. . . .
—Robert Carl Manoff[1]

Television journalists see life as bursts of distinct, discontinuous events, how- ever much we may live it as a steady hum. TV is better at telling us "the way it is" than at threading together the several ways things might seem to be, as seen by different cultures or religions or kinds of men. American television news, like the rest of American journalism, is scrupulously "objec- tive"—which means it does not challenge the prevailing biases of a predomi- nantly white, Judeo-Christian, imperial, internationalist, capitalist society.
—William A. Henry III[2]

Overview

By far our most theoretical, this chapter presents a model of news dis- course that emphasizes journalism's entanglement in modern networks of power and knowledge. In outlining what we term the "eye/peacock model" of news discourse, we argue that mainstream television journalism is properly understood as a spectacle of surveillance that is actively engaged in representing authority, visualizing deviance, and publicizing common sense. From this perspective, we treat the nightly newscast as a cultural performance in which the journalist is actively involved in guarding the horizons of common sense by enacting the rites of inclusion and exclusion discussed in the introduction to chapter 2. To flesh out this argument we draw from and elaborate on contemporary narrative theory, on various works associated with the sociology of deviance, and, most important, on Foucault's discussion of Jeremy Bentham's panopticon. Ultimately, we develop a framework for describing and analyzing the human context of the drug news. In this framework we organize the people who appear in the cocaine narrative according to four major social types associated with the distribution of power and knowledge in modern technocracies:

1. **Primary Definers.** Generally experts and authorities associated with the drug control establishment, these news "sources" are privileged in the cocaine narrative as people "in the know." In the 1980s' drug news, these definers can be further organized into two subcategories that correlate, roughly, to the soft and hard sectors of the drug control establishment:

 a. Recovery Experts. In news coverage of drug policy debates, these definers often give a voice to an owlish orientation that treats drug transgression as a disease and, consequently, favors the thereapeutic branch of the medical-industrial complex.

 b. Law Enforcers. Defining drug transgression as a crime, these authorities often champion hawkish strategies in the war on drugs that favor punitive retribution and exclusionary incarceration.

2. **Transgressors.** These human images give a visible form to drug deviance. As suggested in chapter 2, transgressors can further be divided into two subcategories that also conform to the logic of soft and hard control:

 a. Offenders. The drug offender is often depicted as someone who, with the proper discipline, can be rehabilitated. In other words, the offender is a form of transgressor that is subject to journalistic discourses of recovery.

 b. Delinquents. The drug delinquent is often depicted as beyond redemption. Therefore, this form of transgressor is frequently subjected to journalistic discourses of discrimination.

3. **Representatives of Common Sense.** These are the private citizens who appear in "on-the-street" interviews and are often deployed to give a body and voice to consensus.

4. **Well-informed Journalists.** These include the anchors and reporters who narrate the reported world. While some of these familiar figures are so recogniz-

able that they have attained media star status, most are relatively anonymous, even invisible, in the professional performance of their institutional roles.

The chapter culminates with a section that reports two major findings of a census we conducted of these social types as they appeared in the 1980s' drug news.

Although the material in the following three major sections of this chapter should be of interest to media scholars and communication researchers, we fully recognize that this venture into critical theory might not be at all palatable to a more general readership. For those outside media studies who may not want to wrestle with this subject matter, we suggest jumping from here to the chapter's final section, "Major Characters and Characteristics."

Beyond the Looking-Glass Analogy

In treating the news as a particularly powerful form of modern storytelling, we rely on an analytic framework derived from contemporary narrative theory. Informed by the work of Sarah Ruth Kozloff and Seymour Chatman,[3] figure 4 diagrams our framework. As this map suggests, at the heart of narrative theory is the distinction between "story" and "discourse." According to this distinction, the story dimension of narrative is concerned with "what happens to whom" and the discourse dimension is concerned with "how the story is told."[4] To move beyond the mirror analogy, and make theoretical sense of the discursive productivity of television journalism (of how news stories are told), we propose a model of news discourse that at once invokes and conflates the symbolism of the corporate logos of CBS and NBC—the Eye and the Peacock. For us, the collision of these familiar modern images expresses a synthetic model for the operation of the television news as a normalizing force in American culture: as a mixed metaphor, the eye signals surveillance, and the peacock conveys spectacle. However, in conceiving of the television news as a powerful system of surveillance and spectacle—or better yet, as a spectacle of surveillance— we find ourselves somewhat at odds with Foucault's genealogy of modern power.

Panoptic Spectacle In Foucault's genealogy, Jeremy Bentham's famous panopticon is the epitome of normalizing technology. Designed, significantly, in 1791 during the dawn of Modernity, the panopticon "consists

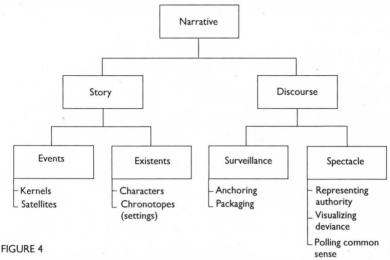

FIGURE 4
A Diagram of Narrative Structure in Television News

of a large courtyard with a tower in the center and a set of buildings, divided into levels and cells, on the periphery. In each cell there are two windows: one brings in light and the other faces the tower, where large observatory windows allow for the surveillance of cells."[5] As Thomas Mathiesen observes, Foucault treats "panopticism" as "a fundamental movement or transformation *from the situation where the many see the few to the situation where the few see the many*" [Mathiesen's emphasis].[6] Whereas the architecture of power in premodern regimes was articulated in the spectacle of the Temple, the Colosseum, and the Public Square, the architecture of modern power, according to Foucault, "is the exact reverse of the spectacle": "Our society is not one of spectacle, but of surveillance. . . . We are much less Greeks than we believe. We are neither in the amphitheater, nor on the stage, but in the panoptic machine, invested by its effects of power, which we bring to ourselves since we are part of its mechanism."[7] While our news model certainly benefits from Foucault's ideas about "the panoptic machine," we still argue that we are more Greeks than Foucault believed. Although panopticism is certainly a taken-for-granted aspect of the everyday regulation of the prison, the factory, the school, and even the city street and the department store, the disciplinary technology of the spectacle is still very much a part of the modern control culture. Indeed, the coexistence of spectacle and surveillance technology is especially apparent in the operation of modern mass media. And yet, as Mathiesen argues, the "older institutions of spectacle differ in several important respects from the modern ones":

> In the older context, people were gathered together; in the modern media context, the "audience" has increasingly become delocalized so that people have become isolated from each other. In the older context, "sender" and "receiver" were in each other's proximity; in the modern media context, distance between the two has increased. Such differences, and especially the general fragmentation which is alluded to here, may have consequences for persuasion as well as protest.[8]

As we shall argue in chapter 6, these "consequences for persuasion" are especially apparent in late summer 1986 when public opinion polls cited drugs as the number one problem facing the nation despite the fact that more than 60 percent of those surveyed also reported that drugs were not a major problem in their own hometowns.

Some of the more sophisticated mainstream communication research does, in fact, recognize the surveillance and spectacle dimensions of news discourse. Vincent Price, for instance, in a discussion of public opinion processes, uses dramaturgical terms to describe the structured interactions among "political actors, journalists, and the attentive public." The term "attentive public" refers to a subset of the general population that "pays continuing attention to political affairs, engages thoughtfully in public issues and occasionally talks about those issues with others."[9] For Price, the attentive public is basically made up of "followers":

> the differentiation between leaders and followers—between political actors and political spectators—emerges as one of the central structural features of large-scale public debate. Political actors (or "elites") are those people trying to make a difference in the behavior of the collective. Actors, both within and outside the established political system and often organized into pressure groups, create public issues in the first instance by formulating and then advocating alternative policies. "Public debate" refers primarily *to the debate between contending political actors*—one that is displayed by the mass media for people in the attentive public to observe and to contemplate (and, far less frequently, to join).[10]

The language of "political actors," "political spectators," "displayed by the mass media" provides, at least for us, implicit validation of our news-as-spectacle formulation.

However, we receive much more explicit support for the news-as-surveillance component of our model in Price's discussion of the journalist's function in the public opinion process. In fact, Price identifies two journal-istic "roles" in this process: *surveillance* and *correlation*. Following Harold Lasswell,[11] Price argues that the surveillance, or "lookout," function is "per-haps the most prominent public service attributed to, and claimed by, the news media":

> It is quite commonly reflected in the names of newspapers, for example, the "Sentinel," "Monitor," "Clarion," "Observer" or "Intelligencer." As surveil-lance agents, journalists try to alert publics to problems. They bring news about the behavior of political elites—their actions, presumed intentions, and internal disagreements—to the attention of their audiences. In so doing, re-porters provide the principal mechanism for allowing an attentive public to monitor the political environment—albeit one that may be constrained by institutional, professional and organizational factors. . . .[12]

In our model this *middle-up surveillance* of the "behavior of political elites" by Price's attentive public is conceived of as the spectacle of the many watching the few.

The second major role identified by Price, correlation, is not so obvi-ous and also not as firmly established in communication research. The role is meant to describe how the news contributes to the coordination of public responses to the political environment. In Price's words, "the news media bring together contrasting ideas and views within the attentive pub-lic, report to its members what others think, and thus help to organize the collective reaction."[13] From this point of view, the primary role of polls, as well as letters to the editor and commonsense sound bites (traditionally termed man-on-the-street interviews), is to display organized or sponta-neous public responses to the newsworthy events and issues of the day. But while Price sees these representations *of* the public *to* the public as playing a role separate from surveillance, we argue that what Price identifies as correlation is actually a *lateral form of surveillance* (the many watching the many). And yet, we want to stress that, in the news this lateral sur-veillance is *always* mediated, edited, and filtered by the elite. It is *never* simply a direct "reflection" of public response; instead, it is a *top-down administrative view* of that response.

In fairness to Price, we must note that he fully recognizes that his surveillance and correlation functions are a matter of "perspective":

> Journalism performs these same two functions for elites in the active pub-lic as well. The identical item of "news" or "commentary" can serve opposite functions, however, depending upon one's perspective—as a spectator or as

an actor in the process. Media characterizations of opinion within the attentive public (which help to *correlate* its internal responses) are simultaneously a means of surveillance for elites (by helping them monitor the attentive public's reactions). Actors pay close attention to the news to see how they are doing. The extent to which the mass media help to correlate elite responses to issues may be less apparent, but it is no less central to the process.[14]

As we argue in chapter 7, this "monitoring" of the public by elite interests is not so much a matter of being responsive to the public will as it is a matter of cultivation, colonization, and control. This is especially the case in the 1986 crack crisis and spectacle. The cultivation of public opinion was accomplished by antidrug *public*-ity including carefully crafted information campaigns sponsored by government agencies and the saturation reporting of the operations, actions, gestures, warnings, pronouncements, and prescriptions of prominent antidrug crusaders who had shared vested interests in promoting the moral panic. The news media, in *public*-izing these campaigning and crusading sentiments as the "way things are," played an important role in cultivating an apparently frenzied public response. Once this response began to register in polls, the fund-hungry agencies and crusading interests were able to then colonize the cultivated opinion for the purposes of passing tougher laws, generating new funding, and expanding the power of the controlling forces. Ultimately, this combined cultivation, colonization, and control of public opinion by elite interests explain how people can come to support economic actions (like tax cuts for the wealthy) and disciplinary practices (like random drug testing in the workplace) that are clearly not in their best interests.

The Bifocality of Control We want to complicate the either/or notion of news functioning as (using Price's terms) *either* surveillance *or* correlation "depending upon one's perspective." Our model proposes instead a both/and view of news as a hybrid of spectacle and surveillance. Or, put another way, bound up in the mixed metaphor of the Eye/Peacock model is the notion of the *bifocality of control*.[15]

The concept of bifocality is another way of thinking about what Mathiesen terms "parallels in structure" between ancient architectural and modern panoptic forms of spectacle. The fragmentation of "the many watching the few" in the modern media spectacle enforces this bifocality. From a systemic perspective, it is easy enough to conceive of this double visioning in the news as "the many" (the news audience or attentive public) watching "the few" (the elite) watching "the many" (the general population). As Mathiesen argues, both the spectacle of the many watching the few and the surveillance of the few watching the many are characteristically "fan-shaped" (like the peacock's tail in the NBC logo) in form. In the *spectacle fan* the optical vectors of the many converge on the few; in the *surveillance fan* the optical vectors radiate outward to police the many.

If we shift the perspective from the systemic to that of the "private citizen" sitting at home "attending" to the television news, then we encounter a much more complicated situation in which acts of surveillance and spectatorship merge in a bifocal moment. Indeed, for the spectator-in-privacy, watching the nightly newscast is an act of individuated surveillance (the one watching the many). In this isolated experience the news spectator is situated as the "unseen observer" of the journalistic spectacle, the addressee of journalistic narration whose "domestic gaze," whose "attention," is at the very heart of the political economy of the sponsor-supported network news.[16] For, after all, it is the "attention" of news consumers that is measured, commodified, and sold by the networks to advertisers.

Variable Visibility

But beyond entangling the news consumer in the economics of the news business, being situated as unseen and isolated observers of public spectacles is precisely what implicates the news audience in the panoptic machine. In Bentham's panopticon the guardian in the tower can observe the inmates isolated in the cells, but the inmates cannot see into the tower. Foucault argues that this disparity in visibility is a clear example of how modern power operates. Indeed, for Foucault, "Visibility is a trap."[17] Dreyfus and Rabinow offer this interpretation: "The inmate cannot see if the guardian is in the tower or not, so he must behave as if surveillance is constant, unending, and total. The architectural perfection is such that even if there is no guardian present, the apparatus of power is still operative. . . . For if the prisoner is never sure when he was being watched, he becomes his own guardian."[18] But the bifocality of the news voyeur, in fact, represents a high-tech refinement of this "architectual perfection." As a spectacle of surveillance, the news positions the attentive public *in the tower with the guardians of the general population.*

Representing Authority   In this situation, though, visibility is not necessarily a trap. On the one hand, for the guardians of the normalizing apparatus (government officials and various experts), news operates as a spectacle of legitimation. For these "news sources," being one of the "chosen few" who are seen by the many is a *mark of authority*. This marking is what Richard Ericson, Patricia Baranek, and Janet Chan mean when they argue that "News is a representation of authority":

> In contemporary knowledge society news represents *who* are the authorized knowers and *what* are their authoritative versions of reality. . . . It offers a perpetual articulation of how society is socially stratified in terms of possession and use of knowledge. It indicates who is in possession of knowledge as "cultural capital," and thereby articulates who are members of the "new class" who derive their labor and property membership from the production, distribution and administration of knowledge. . . . At the same time that it informs about who are the authorized knowers, it suggests, by relegation to minor role and by omission, who is excluded from having a say in important matters.[19]

In drug stories reporters seek out "appropriate" enforcement, medical, and academic experts who typically provide enough conflict to sustain the news narrative. Police, doctors, and social scientists contribute their *expert* voices to the pool of knowledge that the reporter then arranges and (re)presents. These news characters are from the land of specialized knowledge. And the language of this realm is frequently "jargonese," which often has to be decoded and translated into "common sense" by the reporter. The privilege of specialized knowledge includes access to information such as drug use data bases and state secrets such as the strategies of federal drug operatives. Located outside the borders of common sense, specialized knowledge is generally possessed by authorities involved in state security and surveillance (industry regulators, military and law enforcement officers, intelligence agents, etc.) or by professionals in various arts and sciences (mathematics, music, medicine, and institutes of social research, etc.). The novelist Walker Percy defines and critiques experts as "the true princes" of the modern age: "*They* know about science, *they* know about medicine, *they* know about government, *they* know about my needs, *they* know about everything in the Cosmos, even me. *They* know why I am fat and they know secrets of my soul which not even I know. There is an expert for everything that ails me, a doctor of my depression, a seer of my sadness."[20] By italicizing "they," Percy acknowledges the fundamental division in contemporary society between the haves and have-nots of special expertise.

Indeed, in representing authority, the journalist becomes something of a broker in expertise. In modern technocracies with complex divisions of labor and knowledge the journalist mediates the widening gap between "private citizens" and experts. In occupying this knowledge gap modern reporters are the exemplars of a new social type that Schutz has called "the well-informed citizen."[21] On the one hand, the well-informed type is distinguished from the lay public by an unwillingness to "unreflectively" accept the point of view or "judgment of the expert." On the other hand, unlike the expert, the well-informed generalist lacks expertise in any particular specialized field.[22] However, with privileged access to experts and their special knowledge, well-informed generalists command a critical, mediating symbolic station in modern cultures—a station located between "expertness and the lack of it," between "authorized knowers" and "private citizens."

Visualizing Deviance   For those of us who come under the scrutiny of the normalizing apparatus, visibility is still very much a trap. For the scrutinized transgressor, news operates as a spectacle of stigmatization where news visibility is a *mark of deviancy.* This view of news as stigmatization is also invoked by Ericson and his colleagues who suggest that modern news methods are primarily devoted to "visualizing deviance":

> deviance is *the* defining characteristic of what journalists regard as newsworthy and, as such, becomes inextricably linked with journalists' methods. Deviance and control are not only woven into the seamless web of news reporting, but are actually part of its fibre: they define not only the object and central character of the news stories, but also the methodological approaches of journalists as they work on their stories. They not only constitute the essence of what is deemed to be newsworthy to begin with, but also enter into choices, from assignment, through the selection and use of sources, to the final composition of the story.[23]

In cocaine coverage, deviancy is made visible in the life—and sometimes the death—of the drug transgressor. In contrast to well-informed reporters and authorized knowers, the cocaine transgressor personifies another type of knowledge that is *re*-presented as outside the bounds of both expert knowledge and common sense. This is the realm of *non-sense.* Nonsense resides in the domain of superstition, of condemned or absurd ideas, of self-destructive behavior, of obsolete knowledge that is no longer an element of consensus. The social types depicted in news images who transmit those ideas—the condemned, the deviant, the criminal, the mad—offer a marked contrast to the expert, the well-informed, or the ordinary citizen.

But the boundaries of nonsense and obsolete knowledge are also fragile and subject to the rhythms of history. Foucault, for instance, has traced the historical changes in how madness and the madman in literature— King Lear and Don Quixote serve as examples—shifted from their association with sacred forms of special insight during the Renaissance to a redefinition in direct opposition to reason during the Enlightenment. Often "center stage as the guardian of truth" during the early seventeenth century, madness is banished by the next century: "By a strange act of force, the classical age was to reduce to silence the madness whose voices the Renaissance had just liberated. . . ."[24] During the Enlightenment, as efficiency and work ethic (both central to modern notions of common sense) achieved new dominance in the social order, the mad—characterized by their inability to work and their disturbance of social order—increasingly were viewed as outside the realm of reason, and certainly outside the borders of common sense.

Publicizing Common Sense For the unseen many who watch the news spectacle, coming under the scrutiny of journalistic surveillance is a terrifying proposition that involves the "loss of privacy." Privacy, after all, is reserved for those who conform to normalcy; to stray from normalcy is to risk being "made an example of" by being visibly transformed from a privatized citizen into a publicized deviant. Even so, the nonstigmatized discourse of the private citizen does achieve some visibility in the news. In many stories the voluntary testimony of the private citizen often is constructed as "common sense." The representation of common sense is usually accomplished by way of two conventions: the man or woman on-the-street interview and the public opinion poll. Since we devote chapter 7 to a discussion of the journalistic deployment of the opinion poll as common sense, here we will limit our discussion to the street interview.

In these interviews, someone, selected not for expertise but for ordinariness, volunteers an opinion about a newsworthy event or issue. As a visible part that stands in for the whole of the unseen public's response, such interviews take on the aura of consensus. In fact, in stories presenting controversies, a lone interview or sound bite from "an ordinary private citizen" often determines which of the competing expert definitions of the situation is perceived as "correct." Consequently, in drug coverage— as in all television news and most political discourse—common sense is generally privileged as a referee for and over the contradictions of expert knowledge. As a result, we apply a hermeneutics of suspicion in interpreting the strategic use of on-the-street interviews in news stories and treat these interviews, not as overt expression of common sense, but as covert expression of the journalist's own not-so-detached perspective. At any rate, the street interview is one example of the many ways that the cultural performances of well-informed reporters represent and reproduce common sense in the news.

Journalistic Performance

Journalists today play a central role in the social construction of reason and nonsense, of normal and abnormal. In the ongoing representation of authority and visualization of deviancy, the well-informed journalist mediates two symbolic horizons of common sense.[25] As suggested earlier, on one horizon the journalist bridges the knowledge gap between "expertness and the lack of it"; on the other, the journalist guards the frontier between "normalcy and the lack of it," between "reasonable people" and "deviant nonsense." The journalistic policing of these horizons is a symbolic process that involves *both* inclusion *and* exclusion (see chapter 1)—and, in this study, we treat this symbolic processing of knowledge and people as first and foremost a "cultural performance." Like other cultural performances in both traditional and modern societies, the news operates as a ritual that, in anthropologist Sherry Ortner's words, "the people themselves see as embodying in some way the essence of their culture, as dramatizing the basic myths and visions of reality, the basic values and moral truths upon which they feel their world rests."[26]

Anchoring the Surveillance   For the journalists who enact the rites of inclusion and exclusion, news visibility operates in contradictory ways. For the anchor, who is situated as *the* guardian of the reported world, high visibility is associated with media stardom. Similar in many intriguing ways to the popular hosts of television's entertainment shows (Johnny Carson, David Letterman, Arsenio Hall, etc.), network anchors occupy one of the most highly visible posts in the flow of television. ABC's Peter Jennings, CBS's Dan Rather,[27] and NBC's Tom Brokaw (like Carson, Letterman, and Hall) are star moderators of the medium—patriarchal masters of electronic eye contact who specialize in engaging the audience with the "Hi-Mom" intimacy of direct address. Following in the footsteps of Edward R. Murrow and Walter Cronkite, they achieve star status by "individualizing" the well-informed type. In other words, the star anchor gives both a bodily form and a specific identity to the ideal of journalistic professionalism. A concept from film studies illustrates this point. Such directors as Alfred Hitchcock, Woody Allen, Steven Spielberg, or Spike Lee become "auteurs with biographies"—which is to say that they project a public identity that personalizes the anonymous category of "filmmaker."[28] In contrast to the more or less anonymous reporters who narrate the news packages, star anchors can be understood as "journalists with biographies."

The familiar presentations of the networks' star anchors command the top spot in the journalistic hierarchy of credibility. As P. H. Weaver argues: "There is hardly an aspect of the scripting, casting, and staging of a television news program that is not designed to convey an impression of authority and omniscience. This can be seen most strikingly in the role of the anchorman—Walter Cronkite is the exemplar—who is positively god-like; he summons forth men, events, and images at will; he speaks in tones of utter certainty; he is the person with whom all things begin and end."[29] Under normal conditions the anchor appears live in most parts of the country. Located in a "newsroom," this postmodern *chronotope*[30] is at the same time everywhere and nowhere. Although, admittedly, the studio facilities are physically situated in New York, or Washington, D.C., or Atlanta, the "newsroom" of the anchor subsumes the *surveillance fan* of the stories included in the telecast and the *spectacle fan* of the unseen audience tuning into the newscast. In other words, for us, the newsroom regulated by the anchor's performance includes not only his NASA-like command and control center, but also the *public*-ized "reported world" of experts and deviants as well as the *private*-ized "living rooms" of attentive spectators. In this omniscient position of surveillance, these three telegenic men supply what amounts to a *grand narration* of the news—a patriarchal narration that gives the most scattered and diverse subject matter the semblance of continuity and coherence.[31]

Packaging Surveillance Ironically, in the case of reporters news *in*visibility can actually enhance their authority. After all, the ideology of journalistic detachment militates against the reporters being perceived as individual agents, as having identities apart from their role as well-informed journalists. According to this professional ideology, reporters only present "the facts," so it follows that reporters working from the same set of facts are pretty much interchangeable. Despite the convention of identifying reporters by name in lead-ins and again in sign-offs, their presence is most often expressed in the disembodied words of voice-over narration. In this mode of address, reportorial performance generally invokes only meanings associated with the generic category "TV reporter."

But, strangely enough, the anonymity of their narration enhances the *authority* of their accounts by contributing to the illusion of journalistic distance. This distance, according to Shanto Iyengar and Donald Kinder, makes it seem as if the reporter "towers over the scene in question." [32] As "journalists without biographies," generic reporters are able to sustain the *impersonal* posture of detached observation, the seemingly omniscient pose of the dispassionate harvester and harbinger of "the facts." In the "objective" report, the *self* of the reporter disappears as attention shifts to the presentation of *facts*—"what's going on here." Deployed in various locations on the surveillance fan, reportorial agency is generally hidden by what Gaye Tuchman terms the "web of facticity," which includes the separation of hard news from opinion, the use of quotation marks, sound bites, and neutral word choices, the polarizing presentation of "two sides" of an issue (a narrative device for generating dramatic conflict), and the use of the detached third-person point of view (also a narrative device that constructs an omniscient reportorial voice). [33]

From this perspective, in Weaver's estimation, the reporter "speaks authoritatively and self-confidently about everything that comes into his field of vision: men, events, motives, intentions, meanings, significances, trends, threats, problems, solutions—all are evidently within his perfect understanding, and pronounces them without any ifs, ands, or buts." [34] Because it is practical and efficient, reportorial orientation toward facts helps establish modern journalism as an apparently neutral, legitimate institution with the power to frame events, identify reality, sell news, and make common sense of widely disparate experiences for audiences. In conventional factual news reports, however, the public is seldom privy to the embedded moral choices involved in news selection, to the relationships of reporters to their sources, communities, and corporate employer, and to the commodity aspects of news reports. Validated by the consensus of conventional wisdom, these personal and economic elements are, over time, masked in modern rationalized, institutionalized practices.

Particularly crucial in the masking of personal and moral levels of experience is the overt identification of reporters simply with their institutions in the conventional package sign-offs: "This is . . . Harold Dow, CBS News" or "Bettina Gregory, ABC News" or "Jim Cummins, NBC News." At a conventional level, such sign-offs merely signify a shorthand, closing signature. But at another level, by abandoning even the connecting "of" preposition, the reporters identify themselves exclusively as institutional extensions, not as individual agents. Such a convention reaffirms their alleged neutral institutional role and returns them to the anonymous world of detached reporting where they may escape scrutiny as individual agents implicated as both observers and shapers of experience. Steve Fore notes, "journalists working for the mainstream media outlets use their privileged access [to experts] to publicly negotiate and renegotiate the meaning of the 'normal,' guided by their allegiances to the largely intuitive world of common sense and to the often unstated but weighty expectations of the corporations that cut their paychecks." [35] From a conventional point of view, reporters and editors merely present commonsense "facts." But from our point of view, they impose a corporate/network narrative on inchoate experience where the "pressure to articulate an uncontroversial, centrist position is strong. . . ." [36]

The *secondary narration* of the newscast, the narration of the news "package," is performed by reporters whose personal identities then are only dimly perceived. As any good TV journalism student learns, the news package is made up of four elements: (1) *cover footage;* (2) *voice-over narration;* (3) *reporter stand-ups;* and (4) *sound bites.* Generally, cover footage appears on the visual track while we hear the reporter's reading of voice-over narration. In a stand-up, the reporter appears on the screen and directly addresses the audience. And in sound bites, we see and hear someone, other than the reporter, provide testimony that is relevant to the story. The broadcast sound bite, then, is the equivalent of the quotation in print journalism. Although the vast majority of sound bites are gathered in interview situations or are extracted from staged press conferences, occasionally a sound bite will be presented as *ambient sound*—the recording of environmental sounds or "unstaged" verbal interactions that are apparently not controlled by the reporter.

The package demonstrates news initiative on the part of each network as it displays reporters who have *fanned out* to cover events and issues around the nation and the world. In the journalistic ordering of expert knowledge, common sense, and deviant nonsense, the standard package translates experience, expertise, and moral conflict into what Stuart Hall calls the language of "the public idiom." In this translation, well-informed journalists construct versions of "the language of the public" based on what they agree is "the rhetoric, imagery and underlying common stock of knowledge." [37] This reportorial language—whether the inverted pyramid print news lead or the TV news package—carries shared assumptions about common sense. Ultimately, this language renders "into a public idiom the statements and viewpoints of the primary definers" or expert sources.[38] David Eason suggests that the public idiom *is* the story form; reporters, after all, have no expert jargon or "special language for reporting their findings. They make sense out of events by telling stories about them."[39] For Hall, the public idiom sets the news agenda by inserting "the language of everyday communication *back into the consensus"*—back into the commonsense world of the familiar news narrative. The definitions and interpretations of powerful sources and authoritative news anchors become part of the taken-for-granted reality of public discourse.[40] By translating experience into the taken-for-granted representations of common sense, journalism better conceals its own collusion in the spectacle of surveillance and in its own moral positions on the therapeutic recovery or punitive discrimination of social deviancy.

In concluding our discussion of the eye/peacock formulation, we want to make it clear that journalism's bifocality of control is not only about enforcing prohibitions. Clearly, the journalistic rendering of the cocaine narrative during the 1980s did much more than simply say "no" to drugs. As Foucault put it in the mid-1970s, anticipating Nancy Reagan's antidrug jingle:

> If power were never anything but repressive, if it never did anything but say no, do you really think one would be brought to obey it? What makes power hold good, what makes it accepted, is simply the fact that it doesn't only weigh on us as a force that says no, but that it traverses and produces things, it induces pleasure, forms knowledge, produces discourse. It needs to be considered as a productive network which runs through the whole social body, much more than a negative instance whose function is repression.[41]

As a "productive network" that "covers" the whole social body with fan-shaped systems of spectacle and surveillance, the news traverses and brokers the various domains of expert knowledge, it produces and reproduces common sense, it re-presents and legitimizes authority, it visualizes and stigmatizes deviance, and it induces and privatizes the voyeuristic pleasures of the unseen spectator. In other words, the cocaine narrative is as much about saying "yes" to the recovery operations and discriminatory systems of the modern control culture as it is about saying "no" to drugs.

Chief Characters and Characteristics

To comprehend shifting demographic contours in the human context of the cocaine narrative, we conducted a systematic census of the reporters, experts, private citizens, and transgressors who populated the 1980s' cocaine narrative. We are, of course, hesitant to make too much of this census that replicates the kind of content analysis in some mainstream social science research that, like journalism, reduces meaning to facts and reality to numbers. Although we fully recognize the many limitations of the *hermeneutics of counting and sorting*, we still believe that two general findings of this crude method are worth reporting. The first finding speaks to the gendering of power and knowledge in the story world of the news; the second identifies a striking correlation in the changing populations of authorized knowers and deviant transgressors as the cocaine problem was rewritten from a white, upper- and middle-class addiction tale into a black, inner-city horror story.

The Gendering of Power and Knowledge Given that all of the major networks' star anchors (*Tom Brokaw, Peter Jennings,* and *Dan Rather*) were white males during the bulk of our study, the proposition that white, patriarchal power is still the dominant discourse in the national news should come as no surprise. But women and people of color have, at least, made some inroads at the reportorial level.[42] Our census revealed some intriguing gender and racial differences among the three networks. Not surprisingly, male reporters outnumbered female on all of the networks. However, at least on the cocaine beat, NBC had by far the worst ratio of female-to-male reporters. Only four females out of a total of thirty-eight reporters turned up in our NBC sample. This 10.5 percent female representation compares poorly with CBS's 23 percent or ABC's 32 percent.

Perhaps the most disturbing census finding was the female/male ratio in the population of authorized knowers. Of the 337 expert sound bites in all phases of our video sample, 305 were gathered from men, and only thirty-two from women. More than 90 percent of the sound bites, then, were from male knowers. These numbers are consistent with another source study done during this period. Based on a forty-month analysis of guests on ABC's *Nightline* from 1985 until 1988, FAIR "found that 89 percent of the U.S. guests were men, 92 percent were white, and 80 percent were professionals, government officials, or corporate representatives."[43] Both our figures and FAIR's indicate that men in our culture not only control news discourse but are authorized to know more than women.[44] (Note that in our sample some experts contribute multiple sound bites. Therefore, this is a count of sound bites, not the number of definers.)

Chief Reporters Our screening also identified the reporters most often assigned to the domestic cocaine story. We consider those professionals who filed more than three stories in our cocaine sample to be "chief reporters." ABC had only two reporters who qualified for this category: *George Strait* with six and *Bill Greenwood* with five. Strait (who seems to us to have the perfect last name for a reporter covering the drug news) is African American. Greenwood, who is white, made this list by virtue of his being assigned to the Len Bias story. CBS had three chief reporters. Two, *Bill Plante* with five and *David Dow* with four, were white males. Plante, CBS's Washington correspondent during this period, filed most of the stories on Reagan's role in escalating the war on drugs. But by far the most significant reporter at CBS was *Harold Dow,* an African American journalist whose sixteen stories mark him as *the chief reporter* on all networks of the domestic drug beat during the 1980s. NBC's *Robert Hager* with ten cocaine stories to his credit, was the only reporter to come close to Dow's numbers. Hager was joined by two white males at NBC who qualified as chief reporters— *James Polk* with seven stories and *Brian Ross* with six. Polk mostly covered the sports/cocaine connection. Ross, whose last report in our sample was broadcast in 1984, generally covered cocaine as an elite taboo. His primary beat was Hollywood and the John DeLorean bust.

Chief Transgressors In our census of transgressors, we counted those who were either named in narration or who provided some type of sound bite. A high percentage of speaking transgressors remained anonymous. Some were referred to by pseudonyms; many had their identities concealed by techniques such as angling the camera away from the face, using silhouette lighting, electronically blacking out the face, or wearing a mask. The Phase I concentration of such veiling techniques for transgressors averaged about .40 incidents per story (concealment occurred seventeen times in forty-two stories). In Phase II this rate increased to .80 (forty-two incidents in fifty-seven stories). We suggest that, as cocaine lost much of its status-enhancing value, people were more reluctant to have their faces associated with the drug. Even so, we were continually surprised at the number of

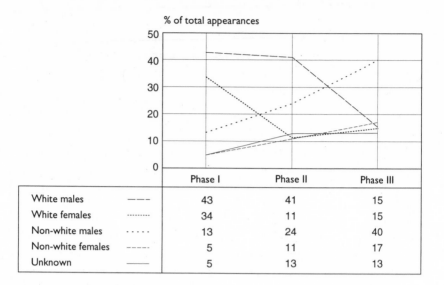

% of total appearances

		Phase I	Phase II	Phase III
White males	– – –	43	41	15
White females	··········	34	11	15
Non-white males	· · · · ·	13	24	40
Non-white females	– – – –	5	11	17
Unknown	———	5	13	13

FIGURE 5   Changing Demographic Profile of Anonymous/Pseudonymous
Drug Transgressors

transgressors who were willing to openly sniff or smoke cocaine in plain
view of a television news camera—especially in light of what happened in
Phases II and III when news cameras joined the police force.

Figure 5 illustrates how the demographic composition of the population of anonymous and pseudonymous transgressors would be drastically
transformed over the course of the Reagan era. These figures were generated from the video sampling of the cocaine stories—and the unknown
category was made up of those people whose personal identities were concealed to the point that racial background could not be coded. For this
population, the percentage of white female transgressors steeply declined
after Phase I (from 34 percent to 11 percent); the percentage of white males
took a similar decline after Phase II (from 41 percent to 15 percent). In stark
contrast to these trends, the percentage of nonwhite female transgressors
increased gradually but steadily from phase to phase to phase (from 5 to 11
to 17 percent); and the percentage of nonwhite males increased sharply in
each succeeding phase (from 13 to 24 to 40 percent). These numbers support our view that, during the Reagan era, the cocaine problem as defined
by the network news became increasingly associated with people of color.

Our analysis of the "chief transgressors" of each phase also supports
this contention. We consider chief transgressors to be those figures or

relatively well-known individuals who make multiple appearances in our sample of cocaine stories. In Phase I, for instance, an even dozen were named or appeared in four or more cocaine stories. The clear leaders were *John DeLorean* and *Curtis Strong*, who tied with nine mentions/appearances each. DeLorean is the white industrialist who was caught trying to smuggle a briefcase full of cocaine into the United States. It was reported that *William Morgan Hetrick,* another chief transgressor who appeared in six stories, cooperated with drug agents in setting DeLorean up for the bust.

Curtis Strong is the Pittsburgh caterer who was prosecuted for supplying cocaine to several major league baseball players. One-fourth of the dozen chief transgressors in this phase were athletes who received immunity to testify against Strong: *Keith Hernandez,* whose name was mentioned in eight cocaine stories; *Dale Berra* (Yogi Berra's son), in five stories; and *Enos Cabell,* in four stories.

Four other chief transgressors were athletes who got caught white-nosed in the cocaine scandal that rocked professional baseball in 1983; *Willie Aikens, Steve Howe, Jerry Martin,* and *Willie Wilson,* all appeared or were mentioned in four cocaine stories.

The final two members of this dirty dozen also fit the general thematic of this period in the cocaine narrative. Both were Hollywood luminaries. One, *John Belushi,* died of a drug overdose and was featured in four cocaine stories during Phase I. The other, *Stacy Keach,* served a prison sentence in England for possessing cocaine; Keach also appeared in four stories.

Six of the twelve, then, were white—Belushi, Berra, DeLorean, Hetrick, Howe, and Keach. However, in Phases II and III, only one of the chief transgressors would fit into this demographic category.

In Phase II, all of the six chief transgressors were associated with the life and death of *Len Bias.* Bias himself achieved posthumous stardom for the timing and circumstances of his sudden demise. From June through December 1986 Bias appeared or was mentioned in no less than twenty-three news stories—which is a record that marks Bias as *the chief transgressor of the cocaine narrative during the Reagan era.* Featured in eight stories, his coach at the University of Maryland, *Lefty Driesell,* was his closest competitor and *the only white male* who qualified as a chief transgressor in this phase. Although Driesell was never accused of abusing cocaine, he was treated as a transgressing figure in the Bias death narrative. He played the part of the shady coach who, in supposedly tolerating or ignoring drug abuse, put winning ahead of the health and welfare of his players (see chapter 5). Three of the other chief transgressors, *David Gregg, Terry Long,* and *Brian Tribble,* were friends or roommates of Bias who were implicated in the death. Tribble, who eventually would be tried for supplying Bias with the drug, appeared or was mentioned in six stories; Long and Gregg were featured in four each. The last of the chief transgressors, *Don Rogers,* was a professional athlete who played in the National Football League for the Cleveland Browns. Rogers, who died just days after Bias, was a posthumous transgressor figure in seven stories.

It seems only fitting that all of the five transgressors with four or more mentions in cocaine stories during the final phase would be caught up in political scandals. Three—*Alice Bond, Julian Bond,* and *Andrew Young*—appeared in seven stories. Allegations by Alice Bond that her estranged husband, Julian, and Atlanta's Mayor Young regularly indulged in cocaine prompted a none-too-secret investigation by Robert Barr, an African American federal prosecutor (mentioned later as a chief definer). The other scandal, involving Washington, D.C., Mayor *Marion Barry* and his mistress, *Karen Johnson,* would not peak until two years after Reagan vacated the White House. Even so, in 1987 and 1988 Barry would be featured in five stories and Johnson in four. It also should be noted that *all of the chief transgressors of Phase III were African American.*

Chief Definers   For us, one of the most compelling and provocative aspects of how television news constructed cocaine as a social problem in the 1980s is bound up with the way in which the population of chief definers shifted in relation to the changing demographics of the transgressor population. Here, we define chief definers as those authorized knowers who appeared in three or more stories in a particular phase of the cocaine narrative. In Phase I, only five people appeared in three or more cocaine stories:[45]

> 1. MARK GOLD, a medical/therapeutic expert, founder and medical director of a national cocaine hotline, logged in the most appearances with five. Gold is white.
>
> 2. ADAM RENFROE, a defense lawyer, was second with four appearances. Renfroe was the attorney for Curtis Strong, the Pittsburgh caterer cited above as a primary transgressor. Both Renfroe and Strong are African Americans.
>
> 3. RICHARD MILLER, a psychologist and the founder of COKENDERS (an up-scale treatment program), appeared in three stories. Miller is white.
>
> 4. PETER UEBERROTH, then commissioner of professional baseball, appeared in three stories. Ueberroth is white.

5. ARNOLD WASHTON, a medical/therapeutic expert who was the spokesperson for a national cocaine hotline, also appeared in three stories. Washton is white.

Notice that none of the chief definers are women, that three out of the five were involved in the treatment industry, and that the other two were connected to sports scandals.

Profiles of the nine chief definers featured in Phase II underscore how much the cocaine narrative changed during the crack crisis:

1. RONALD REAGAN made the most appearances with six. He is white.

2. ROBERT STUTMAN, chief of the DEA's New York office, was a close second with five appearances. Stutman is white.

3. ARTHUR MARSHALL, a Maryland state prosecutor in charge of the Len Bias case, appeared four times. Because we did not order videos of stories featuring Marshall, we have not been able to code his age or his racial background.

4. BENJAMIN WARD, New York City police commissioner rated four appearances. Ward is African American.

5. CHARLES RANGEL, a Democratic member of the House of Representatives who is very active in overseeing the drug problem and sponsoring drug legislation, appeared in three stories. Rangel is also African American.

6. MITCHELL ROSENTHAL made three appearances. Dr. Rosenthal is a therapeutic expert and director of Phoenix House, a treatment facility that services poor, inner-city youth in New York. Rosenthal is white.

7. JOHN SLAUGHTER, chancellor of the University of Maryland, made three appearances responding to the death of Len Bias. No video.

8. JOHN SMIALEK appeared in three stories as the medical examiner who determined that Len Bias died from smoking crack cocaine. No video.

9. ARNOLD WASHTON's three appearances as director of the cocaine hotline qualified him as the only expert to rank as a chief definer in both Phase I and Phase II.

Clearly, law enforcement officials hold a more prominent place in the ranks of chief definers of the cocaine problem during this period. The fact that three out of the nine are involved in enforcement and prosecution suggests a shift in purification measures from therapeutic solutions to armed interventions. Although therapeutic experts still appear on the list, they are not as prominent as the enforcement and judicial agents. Furthermore, the most visible treatment center, Phoenix House, represents a stark contrast to the kind of privileged detoxification (which includes saunas and aerobics) of an expensive private program like COKENDERS. The appearances of Marshall, Slaughter, and Smialek speak to the cultural significance of the death of Len Bias.

Reagan's and Rangel's appearances, of course, indicate that the cocaine crusade is gaining currency as a political spectacle. This trend continues. Consider the profile of the chief definers of Phase III:

> 1. ROBERT STUTMAN of the DEA displaced Reagan at the top of the list with five appearances.
>
> 2. RONALD REAGAN tied for second with four appearances.
>
> 3. ROBERT BARR, a federal prosecutor investigating allegations of cocaine use by Georgia politicians Julian Bond and Andrew Young, also scored four appearances. Barr, Bond, and Young are all African American.
>
> 4. GEORGE BUSH, as vice president and Republican presidential candidate, appeared in three cocaine stories. Bush is white.
>
> 5. FRANK CASHEN, general manager of the New York Mets, appeared three times to comment on pitcher Dwight Gooden's cocaine difficulties. Cashen is white.

Notice that none of the chief definers in Phase III are associated with the treatment branch of the medical-industrial complex. Stutman and Reagan are the only two authorities who qualify as chief definers in both Phase II and Phase III. And the fact that three of the five are either politicians or investigating politicians evidences the rising political currency of the war on drugs.

Stutman's prominence in the crisis and postcrisis coverage deserves recognition. For us, Stutman is to crack what Harry Jacob Anslinger was to marijuana in the 1930s (and what Donald Trump was to real estate in the 1980s). Anslinger, as director of the federal Bureau of Narcotics, launched his own "public awareness" campaign that promoted a moral hysteria based on white fear of the dangers of marijuana use by people of color, especially Mexican migrant workers in the Southwest.[46] As Becker documents, this campaign succeeded in fostering a drastic increase in the number of marijuana "atrocity" stories in the popular press that often "either explicitly acknowledged the help of the Bureau in furnishing facts and figures or gave implicit evidence of having received help by using facts and figures that had appeared earlier. . . ."[47] Anslinger himself related the following atrocity in an article published in *American Magazine:*

> An entire family was murdered by a youthful [marijuana] addict in Florida. When officers arrived at the home they found the youth staggering about in a human slaughterhouse. With an ax he had killed his father, mother,

two brothers, and a sister. He seemed to be in a daze. . . . The officers knew him ordinarily as a sane, rather quiet young man; now he was pitifully crazed. They sought the reason. The boy said he had been in the habit of smoking something which youthful friends called "muggles," a childish name for marijuana.[48]

Becker discovered that this same horror story appeared in five of the seventeen alarmist magazine articles published during this period. Becker also discovered that this press coverage, once cultivated by Anslinger, was then turned around and *used as evidence of public concern* about the marijuana problem in congressional hearings considering regulation of the substance. "The leading newspapers of the United States," said the assistant general counsel of the Treasury Department, "have recognized the seriousness of this problem and many of them have advocated Federal legislation to control the traffic in marijuana."[49] An early example of law enforcement adopting a *proactive strategy* in dealing with the press, Anslinger's moral enterprise would succeed in securing the passage of the Marijuana Tax Act of 1937, the law that made the use and sale of marijuana a felony offense.[50]

Stutman's promotion of the crack crisis also would contribute to the passage of tough new antidrug legislation in 1986; but what sets Stutman apart from Anslinger (and links him to Donald Trump) is that he also was actively involved in self-promotion. "Stutman acquired a reputation," writes Michael Massing, "as both a maverick and a self-promoter. His fondness for talking to the press earned him the nickname 'Video Bob.' "[51] It also earned him a lucrative consulting job with CBS News. And, like Trump, Stutman has cashed in on his entrepreneurial exploits with a book deal, writing *Dead on Delivery: Inside the Drug War, Straight from the Street*—a work that (according to Massing's review) "in its own unwitting way . . . provides some remarkable insights on how the 'drug war' is conducted."[52] Therefore, Stutman, along with Mark Gold and Arnold Washton of the cocaine hotlines (see chapters 5 and 9), exemplifies the enterprising drug crusader who profits professionally and personally from media exposure in drug stories—and consequently, has a vested interest in cultivating, authorizing, and maintaining a sense of public hysteria about drug pollution.

By using italics in the text to introduce the names of the chief anchors, reporters, definers, and transgressors, we are modifying a convention of film scriptwriting (in which such names are rendered in capital letters). In the usual film script, the appearance of a major character is spotlighted by way of capital letters. For us, the preceding names flagged by italics are, indeed, the major characters of the cocaine narrative. However, because of the design of our study there is at least one major character missing from our roster—*Nancy Reagan*. Why this glaring omission? Although Nancy Reagan was, indeed, a prominent figure in drug news of the 1980s, she appeared only in three stories specifically coded in the Vanderbilt News Index as dealing with cocaine. Recognizing this flaw in our sampling technique, we also gathered and analyzed stories that featured Nancy Reagan's work as a moral entrepreneur in the war on drugs (chapter 7 includes a detailed analysis of her activism).

Together, Nancy and Ronald Reagan accounted for thirteen sound bites in our cocaine sample (Nancy with three and Ronald with ten). Therefore, along with the DEA's Stutman (who appeared in eleven stories), the Reagans rank as *chief definers of the cocaine problem in the network news during the 1980s.*

4

Reaganism:

The Packaging of Backlash Politics

> In the aftermath of the Democratic defeat in 1984, Reagan received much of
> the credit for producing the radical shift of voter allegiance reflected in the
> 1984 outcome. In fact, while Reagan's role was indisputably critical, he was
> more a principal agent within—rather than the prime mover of—a sea change
> involving forces substantially more profound and extensive than the fortunes
> of the two political parties or of their candidates. To see Ronald Reagan as
> the cause of an ascendant conservatism minimizes the significance and con-
> sequence of large-scale social and economic transformations—developments
> beyond the power of any single political player to determine.
> —Thomas Byrne Edsall and Mary D. Edsall [1]

In this final chapter of Part I we shift the focus of our discussion from
modernity and the drug control establishment to Reaganism. Although the
drug control establishment certainly played a central role in authorizing
the anticocaine crusade of the 1980s, the war on drugs was, at root, a
Reaganite project that expressed the New Right's basic response to social
problems grounded in economic distress. As we are about to demonstrate,
that response is (paraphrasing Craig Reinarman and Harry Levine) to treat
people *in* trouble as people who *make* trouble. After discussing the ideological turmoil and economic instability that set the stage for the triumph
of Reaganism in the 1980s, we conclude the chapter with an examination
of a new form of racism promoted by the New Right that, ironically, is
grounded on the belief that white prejudice is no longer a factor in the
continued economic disadvantages experienced by people of color. In this
examination we propose that the war on drugs, especially after the 1986
crack crisis, was consistent with the antiwelfare and antiaffirmative action
backlash that Reaganism exploited to gain popular support for economic

policies that favored the rich. Ultimately, we suggest that the moral panic generated by the crack crisis provided the New Right with an opportunity to mask the devastating impact of Reaganomics on America's inner cities by reframing economic distress in terms of the wages of racially coded sin.

Reagan As Auteur?

Along with the revival of the war on drugs, one of the ideological hallmarks of 1980s America was the resurrection of celebratory tales of the heroic entrepreneur. Lee Iacocca, Peter Ueberroth, Steve Jobs, Donald Trump, and Ted Turner were all canonized as folk heroes who gave visible form to the American ethic of capitalist success. Their recognition and fame not only inspired a generation of young business school graduates (as well as drug dealers), but such regard lent credibility to an individualist and elitist historiography that was the basis of Reagan's economic policies: the "great man" theory of history. This view of history is founded on "the belief that history is made by the inspired acts of outstanding individuals, whose genius transcends the normal constraints of historical context."[2] Although generally discredited by contemporary historians, the great man view is still the way that common sense—and journalists as custodians of common sense—comprehend the world and its often incomprehensible complexity.[3]

Reaganism's great man theory of economic progress is expressed in its starkest terms by the conservative economist George Gilder in these passages from *The Spirit of Enterprise:*

> All of us are dependent for our livelihood and progress not on a vast and predictable machine, but on the creativity and courage of the particular men who accept the risks which generate our riches.[4]
> The key to growth is quite simple: creative men with money. The cause of stagnation is similarly clear: depriving creative individuals of financial power.[5]

Implicit in this patriarchal and individualistic spin on the operation of the domestic economy are two categories of productive activity: the entrepreneur and the drone. And, according to Robert Reich (who was to become a top economic adviser to President Bill Clinton), "in the popular mind, people were not fated for one or the other category":

> [In the popular mind] the distinction had nothing to do with class. Almost anyone could become an entrepreneur with enough drive and daring. The

economy needed both, of course—creative entrepreneurs to formulate the Big Ideas that would find their way into new products and production techniques, and drones to undertake the routine chores involved in realizing these ideas. But for the economy to grow and prosper, it was presumed necessary to reward people who opted to become entrepreneurs and discipline those who remained drones.[6]

In Reaganomics, this presumption that economic growth was simply a matter of stimulating the entrepreneurial realm with economic rewards would justify the reverse Robin Hoodism of slashing taxes on the wealthy while eliminating social programs aiding the poor.

In challenging the view of history that informs the trickle-down policies of the Reagan administration, we find ourselves running the risk of unwittingly substantiating the very great man historiography that we endeavor to condemn. In other words, we take care not to treat Reagan, the individual, as *the* great man (or villain) of the 1980s who managed against all odds to impose his vision and will on America. At the same time, we do not want to let Reagan entirely off the hook; instead, we want to endorse a view of history that still allows for enough of a sense of agency to hold people accountable for their actions and inactions, for the damage they inflict, for the hate they condone, and the misery they tolerate.

In this regard, the first step to understanding Reaganism is to take seriously its status as an "ism." As such, the proper name "Reagan" performs what Foucault terms the "author function" for a top-heavy economic, electoral, and moral coalition that solidified in the 1970s around the politics of resentment and the worship of mammon.[7] In packaging backlash and sanctioning greed, Reaganism is most properly understood as a highly versatile political-moral-ideological-economic-coercive-and-discursive formation made up of a multiplicity of forces and voices, not all of which are compatible. Expansive and multifaceted, Reaganism is the chief expression of a political "war of position" that was undertaken by entrepreneurial interests during the chaotic economic transformations of the 1970s.[8]

The centripetal core of this formation was the New Right (its political cadre) and the Religious Right (its army of moral crusaders). According to Rosalind Pollack Petchesky, the combined forces of the New Right and the Religious Right mounted a coordinated political offensive around "two interlocking themes":

> The first is the *antifeminist backlash*, aimed initially at abortion, but extending from abortion to all aspects of sexual freedom and alternatives to traditional

(patriarchal) family life. The second is the *anti-social welfare backlash,* aimed at the principle (given a certain amount of legitimacy during the New Deal and the 1960s) that the state has an obligation to provide for economic and social needs through positive government sponsored programs.[9]

Because the antisocial welfare backlash was directed primarily at out-of-work people of color concentrated in the inner city, this two-pronged attack was both masculinist *and* racist. The driving force behind this mobilization was white-male status anxiety that transcended class location—anxiety that, as we shall argue, was aggravated by the collapse of the Fordist/Keynesian economic order.

The New Right

Technically, the New Right[10] was officially formed in 1974 in the wake of the Watergate scandal by Richard Viguerie and other young, white, disaffected conservatives in response to Gerald Ford's selection of an "eastern establishment liberal," Nelson Rockefeller, for the vice presidency. By raising vast sums of money and networking with other conservative groups, the New Right was quickly able to become a formidable political force. Sponsored by grants from the Heritage and Robert M. Schuchman foundations and support from wealthy businessmen such as Joseph Coors, the New Right took the technological lead in expansive direct mail solicitation for such political action committees (PACs) as the National Conservative Political Action Committee, the National Congressional Club, Citizens for the Republic, and Committee for the Survival of a Free Congress. To combat what it termed the "special interest politics" of the Democratic party, it perfected the art of "single-issue politics" that cut across class lines. By exploiting anticivil rights, antiwelfare, anti-affirmative action, antitax, antibusing, and antigun control sentiments in Middle and Corporate America, the New Right was able to forge alliances with Phyllis Schlafly's American Conservative Union, the Young Americans for Freedom, the National Rifle Association, and various other pro-business lobbies and law-and-order groups.[11]

The New Right, of course, did not invent backlash politics. Nor is backlash politics confined to the American context. To grasp the historical roots and the global relevance of Reaganism and the New Right, both must be placed in dialogue with similar reactionary commotion in the nation's past and the transnational present. Along the national axis, the political backlash of the 1970s and 1980s is the progeny of what Michael Rogin calls the American "countersubversive tradition" of "political demonology"; along the transnational axis, Reaganism operates as a prime coordinate of larger geopolitical transformations of capitalism that have spawned parallel movements in other postindustrial Western societies (most notably, Thatcherism in Great Britain).[12]

While the American countersubversive tradition can be traced back to the Puritan witch-hunts, which, in Carol Karlsen's phrasing, constructed "the devil in the shape of a woman," our truncated genealogy starts with a more recent wave of witch-hunts undertaken during the dawn of the cold war and aimed at the devil in the shape of a Communist.[13] As a "friendly witness" for the House Committee on Un-American Activities and an FBI informer (given the code number "T-10"), Ronald Reagan played both a public and clandestine role in efforts to rid the movie colony of Communist influences.[14] But a far more significant figure would achieve prominence during this period of national paranoia—Senator Joseph McCarthy, a 1950s prophet of the coming Reagan counterrevolution. "It is not the less fortunate, or members of minority groups who have been selling this nation out, but rather those who have had all the benefits the wealthiest nation on earth has had to offer," fumed McCarthy in his famous Wheeling, West Virginia, speech. "This is glaringly true of the State Department. There the bright young men who are born with silver spoons in their mouth are the ones that have been the worst." As Jonathan Reider suggests, in such imaginative diatribes "laced with seeming class resentment," McCarthy pioneered the conservative post-New Deal strategy of mixing anticommunism with antielitism in a "pseudo-populist" appeal to Catholic, ethnic, and blue-collar Democrats.[15]

This strategy would be further refined in 1964 when Republicans— for the first time in a presidential campaign—took advantage of white, southern opposition to the civil rights movement. As Reider observes, "The effort to dismantle the South's caste system catalyzed a politics of massive resistance that redounded to the Republicans' benefit": "Barry Goldwater, the Republican candidate in 1964, seemed a good bet to capitalize on white resistance. His defense of states rights came from a Jeffersonian fear of federal power, not from heartfelt racialist ideology, but the South did not mind. One of eight senators who had voted against the Civil Rights

RE-COVERING THE WAR ON DRUGS 78

Act of 1964, he appealed directly to segregationists alienated by Johnson's civil rights efforts." [16] Although Goldwater lost by a landslide, he was the first Republican since Reconstruction to carry Mississippi, Alabama, South Carolina, Louisiana, and Georgia.[17]

Whereas McCarthy and Goldwater served as the Old Testament prophets of the New Right, Alabama Governor George Wallace was its John the Baptist. Wallace, who as governor had defied federal court orders to integrate the University of Alabama, ran as a third-party presidential candidate in 1968 on a conservative populist platform that had broad appeal for lower- and middle-class voters outside the South. Although his economic rhetoric affirmed the classical New Deal position on social security, health care, and organized labor, his hawkish support of the Vietnam war, his "My Country, Right or Wrong" brand of patriotism, his threats to do "whatever is necessary" to restore law and order, and his devotion to preserving the social arrangements of the apartheid South marked him as an American political underdog in the countersubversive tradition. While the New Right, obviously, was not enamored of Wallace's economic program (termed "Country and Western Marxism" in the *National Review*),[18] it did welcome with open arms his reactionary populism. "To budding New Right theorists," writes Reider, "the Wallace voters were the key to a transformed Republicanism centered on populist themes and lower-middle-class resentment. As they refigured their genealogy, the lineage ran from Andrew Jackson and William Jennings Bryan through Joseph McCarthy and George Wallace to Ronald Reagan and Jerry Falwell." [19]

The Religious Right

> Beware of false prophets, which come to you in sheep's clothing, but inwardly they are ravening wolves.—Matt. 7:15

The inclusion of Jerry Falwell in this family tree speaks powerfully of the place of religiosity in forming the Reaganite coalition. Perhaps more than any other factor, this is what distinguishes Reaganism from its British counterpart; and it is consequently a factor that severely qualifies and modifies our application of Stuart Hall's analysis of Thatcherism to the American context. As Cornel West argues, the place of religion in the cultural dynamic is one of the "blindnesses" of work adopting the conventional race/class/gender framework so prevalent in contemporary critical scholarship:

> We could trot out facts: 97 percent of Americans believe in God, 75 percent believe Jesus Christ is the Son of God, and we say, "Oh, God have mercy, I've struggled up here and we've got all these fundamentalists out here." Why is that so? What is it about religion as a cultural phenomenon in this distinctive capitalist nation that makes it so crucial? This is a crucial question, especially if you're an activist, because you'll be bumping up against all these folks who have all these strange beliefs, from your point of view. . . . We have got to attempt to understand what goes on in a complex phenomenon in this country: Namely, all of the various religious sects and groups and cults and denominations and temples and synagogues and the other forms of association through which folk come together. How do we understand them, especially as potential for broad progressive and oppositional praxis? These are fundamental questions. It's very different in Britain, where the working class rejected religion a long time ago.[20]

Clearly, in the 1970s and 1980s, while Fordism and the New Deal coalition were crumbling—and the American left was fragmenting—the New Right was able to tap at least a portion of this "potential" for broad regressive and reactionary "praxis."

Heterodoxy   Indeed, according to Thomas Byrne Edsall, the Religious Right is "perhaps the single most important source of new support for the Republican party":

> Voters identifying themselves as born-again Christians radically shifted their voting in presidential elections. Between 1976 and 1984, these voters went from casting a 56-to-44 margin for the Democratic candidate, Jimmy Carter, to one of the highest levels of support of any group for the reelection of President Reagan in 1984: 81-to-19, according to New York Times/CBS exit polls. This shift represents, in effect, a gain of eight million voters for the GOP.[21]

A loose confederation that includes born-again Southern Baptists, charismatic Pentecostals, Word-of-God Catholics, and a host of prominent evangelical broadcasters, the Religious Right represents the "radical" politicization of Christian fundamentalism. Like Islamic fundamentalism, the rival factions in this movement share a belief in the "absolute inerrancy" of their sacred texts and, consequently, an absolute commitment to patriarchal authority of biblical proportions. In affirmative terms this commitment is expressed as pro-life and pro-family activism; in antagonistic terms, it is profoundly antifeminist, anti-ERA, antiabortion, and antigay.

Energized by a controversial series of "un-Christian" Supreme Court decisions that banned prayer in public schools (*Engel* v. *Vitale*, 1962), made obscenity cases harder to prosecute (*Manual Enterprises* v. *Day*, 1962; *Jacobellis* v. *Ohio*, 1965), recognized a right to sexual privacy (*Griswold* v. *Connecticut*, 1965), and curbed state prohibitions against abortions (*Roe* v. *Wade*, 1973),[22] the Religious Right would not emerge as a unified, nationally organized political front until 1978. Up until that time the many sectarian rivalries and theological differences that have traditionally splintered the American religious landscape had prevented the formation of such an ecumenical conservative movement. The Catholic/Protestant rivalry, above all, had always undermined any solidarity of right-wing—or left-wing—American religious response. Indeed, as Mike Davis has shown, this rivalry, compounded by related ethnic and racial acrimony, is one of the prime reasons that the U.S. working class is different from its European counterparts:

> The increasing proletarianization of the American social structure has not been matched by an equal tendency toward homogenization of the working class as a cultural or political collectivity. Stratifications rooted in differential positions in the social labor process have been reinforced by deep-seated ethnic, religious, racial, and sexual antagonisms within the working class. In different periods these divisions have fused together as definite intra-class hierarchies (for example, "native + skilled + Protestant" versus "immigrant + unskilled + Catholic") representing unequal access to employment, consumption, legal rights, and trade-union organization.[23]

From the temperance movement of the early twentieth century through the revival of the Ku Klux Klan in the 1920s to the "Stop Kennedy" movement in the 1960 presidential campaign, the political manifestations of Protestant fundamentalism have customarily been fiercely anti-Catholic.[24] This rivalry also has served to severely limit Protestant support for the politics of ultraconservative priests (like the anti-Semitic Father Charles Coughlin) as well as restrain Protestant engagement in the more moderate moral crusades endorsed by the rank-and-file clergy of the Catholic church (like the Catholic Legion of Decency [see chapter 5]).

Beyond the Catholic/Protestant divide, theological controversies that seem inscrutable to the outsider had traditionally fractured the world of Christian fundamentalism into a political field marked more by anarchy than commonality. Some of the divisive issues involved macrostructural disputes about the divine plan for missionary work, congregational governance, and the formation of church hierarchies; other issues involved microprocedural disputes about contradictory biblical justifications for and against faith healing, speaking in tongues, choir music, instrumental music, icons, Sunday school, Seventh-Day Adventism, full-immersion baptism, weekly communion, single-cup communion, head covering, and the special observance of Christmas and Easter.[25] Despite such divisions, most of these factions still share a conservative moral agenda that features scripturally enforced aversions to "the works of the flesh": sexual "lasciviousness" (fornication, adultery, and promiscuity), sexual "abominations" (masturbation, sodomy, and homosexuality), and "hedonism" in general (drunkenness, drug abuse, even dancing).[26] At the same time, it is simply inaccurate to suggest that this conservative moral agenda necessarily translates into a conservative political agenda that would favor the haves over the have-nots. For, historically, fundamentalism has identified more with the needy than with the greedy, more with the persecuted than with the privileged. Therefore, for a good many fundamentalists (former President Jimmy Carter, among them) the collusion of the New Right and the Religious Right is, indeed, an unholy alliance.

A Common Enemy   The precipitating event that allowed right-leaning members of this religious heterodoxy to overcome their macro- and micro-differences and band together with their conservative counterparts in the Catholic Church was, strangely enough, not a Supreme Court decision, but an edict from the Internal Revenue Service. This edict, issued by Jerome Kurtz (the IRS commissioner during the Carter administration), was a delayed response to suits filed as early as 1969 that challenged the tax-exempt status of segregated private schools in the South. While many of these schools were purely and simply "segregation academies," others were part of the rapidly expanding Christian school movement—a movement motivated by a tangled set of moral concerns that transcended the simplicities of white supremacy. According to Edsall and Edsall, "the Christian school movement was a reaction to social and cultural conflict that prominently included, but was by no means limited to, issues of race":

> In many cases Christian schools were set up in response to integration
> orders; in other cases they were a response to the far more complex perception
> that violence and drugs had become commonplace in the public schools, that
> the schools no longer taught sexual restraint, that the Supreme Court deci-

sions prohibiting prayer in schools amounted to an assault on the teaching of basic values, and that what had become known as "secular humanism" was corrupting the minds of the young. These concerns meshed and overlapped with issues of race and in many cases were racially driven, but the motivations involved were far too complex to be categorized as exclusively racist.[27]

After one round of litigation, private schools continued to receive tax exemptions by observing the technicality of slipping an unenforced nondiscrimination clause into their charters. A second suit, filed in 1976, sought to close this loophole. Rather than fight this new suit, Kurtz, with Carter's blessings, ordered tough new guidelines that "shifted the burden of proof of nondiscrimination to private schools seeking tax exemptions in the years immediately following a segregation order if, as was commonly the case in the South, there were no black students in the school or if the percentage of black students was suspiciously low."[28]

The Edsalls credit the IRS controversy for transforming "a loose confederation of Christian school associations" into "a deeply political movement" whose organizational resources were mobilized in the formation of the Moral Majority in 1979. In the words of a delighted Richard Viguerie, Kurtz's action was "the spark that ignited the religious right's involvement in real politics." Or, as Robert Billings, Sr., the first executive director of the Moral Majority, put it, "Jerome Kurtz has done more to bring Christians together than any man since the Apostle Paul."[29]

Old-Time, Good-Time, Hard-Times Religion   The fundamentalist support for Ronald Reagan in 1980 is, of course, full of ironies. Carter was a deacon in the Baptist church, a staunch family man, and from the Bible Belt South; Reagan was notoriously lax about church attendance, had been divorced, and came of age in Hollywood, a symbolic den of iniquity. Garry Wills's explanation of this irony points to a central feature of Reaganism that is often masked by its "just-say-no" prohibitions: that is, the pleasure principle. According to Wills, the religious difference between Carter and Reagan "is not in the *degree* of belief but in *what* they believe": Carter "is comparatively old-fashioned in his theology; too old-fashioned, in fact, in the eyes of evangelicals themselves, who deserted him for Reagan in 1980. Carter's religion is what William James called that of the 'sick soul'— a religion of man's fall, of the need for repentance, of humility."[30] It is Carter's brand of religion that traditionally linked fundamentalism to the have-nots.

Reagan's religion, in contrast, is most properly understood as a doctrine that provides comfort to the haves. Again quoting Wills:

> America has increasingly preferred the religion James called "healthy-mindedness" which replaces sin with sadness as the real enemy of human nature. The modern evangelicals, beaming and healthy successes in the communications industry, are exemplars of that religion. Their older theological ties—which would have bound them to Carter, a more assiduous member of their historical community—they hasten to downplay in favor of Reagan's personal and national assertiveness. It is actually a pledge of one's religious acceptability, in such circles, to feel one has a lucky star; that the individual, like the nation (or the football team), is favored by heaven. Humility, never a national virtue to strenuous patriots (or football players), is not even a personal virtue now, except in the Reagan sense of affability. To acknowledge limits, like Carter, is to "love misery" and have a martyr complex.[31]

Where Carter's presidential ministry was a jeremiad of long-suffering, sacrifice, and charity, Reagan's was a carnivalesque tent revival where money changers and astrologers were welcomed into the fold, indulgences were sold by the truckload, and Jesus was privatized as a "personal savior." Therefore, bound up in the negatives of the backlash politics of Reaganism are a whole host of affirmations: to enjoy the good life; to consume wholeheartedly; to take great pleasure in privilege; to indulge without guilt in the fruits of inequality; to stand tall and take pride in being "a winner"; and, above all, to see economic advantages and good fortune as the just desserts of those chosen by God to succeed in His holy marketplace.

In fusing Christianity with capitalism and individualism (in much the same way that they conflate democracy with the "free market"), the practitioners of Reagan's religion tend to underplay the communitarian thrust of the Gospels contained in the parable of the Rich Fool (Luke 12:16–21) and the parable of the Rich Man and Lazarus (Luke 16:19–31)—a communitarianism that is perhaps best captured in the Nazarene's famous interchange with a rich, young ruler documented in Matt. 19:16–24. After Jesus tells the man to "go and sell that thou hast, and give to the poor, and thou shalt have treasure in heaven; and come and follow me," the young man "went away sorrowful: for he had great possessions." This interchange clearly does not provide much comfort for the haves because it is punctuated by Jesus telling his followers that "it is easier for a camel to go through the eye of a needle, than for a rich man to enter into the kingdom of God." The eye of the needle, then, is the great loophole that religious Reaganites are counting on. In promising guilt-free prosperity to those who put their faith in the witch doctors of Reaganomics, the ministry of the haves is the kind of old-time, good-time religion that the prophet Amos repudiated during the reign of Jeroboam II (ca. 765 B.C.):

> Hear this, O ye that swallow up the needy, even to make the poor of the land to fail,
>
> Saying, When will the new moon be gone, that we may sell corn? and the sabbath, that we may set forth wheat, making the ephah small and the shekel great, and falsifying the balances by deceit?
>
> That we may buy the poor for silver, and the needy for a pair of shoes; yea, and sell the refuse of the wheat?
>
> The Lord hath sworn by the excellency of Jacob, Surely I will never forget any of their works.[32]

The hard times that Amos predicted for Israel would come to pass. Less than fifty years after Amos rebuked the decadence of the Israelites, their capital would be in ruins and their descendants carried away by a revived Assyria. The economic turmoil and civil unrest of 1992 suggest that, to borrow a phrase from Malcolm X, the "chickens have come home to roost" also for decadent Reaganites.

Voodoo Economies

> A decade after the formal end of the sixties, there would remain in the West a guilty memory of its incomplete disruptions, a fear of its unsatisfied chaos. . . . Demands for liberation and pleasure, all fragmented but still seeking the free field of autonomy and solidarity, were by then part of the fabric of life, and there was combat for and against them everywhere. But joined by then to an economy that seemed to be collapsing in all directions at once, the fear of chaos would double. Trumpeted like a horror movie by those seeking power, on the part of the voting majorities fear mandated a radical solution: a leap back over modernity, a return to a pre-modern economy, to the economic terrorism of the Thatcherist and Reaganist regimes.—Greil Marcus[33]

As the 1992 presidential campaign demonstrates, there are many risks associated with the game of single-issue politics. Although the shifting fundamentalist vote was a crucial ingredient in margins of victory in 1980 and 1984, Republican strategists fully understood that support from the Religious Right could backfire; public identification with the leaders of the Religious Right, although deep and fanatically committed, was not all that broad. In the Moral Majority's own home state of Virginia, for instance, only 8 percent of the electorate polled in 1985 said they would be more likely to vote *for* a candidate endorsed by Jerry Falwell compared to 53 percent who reported that they would be more likely to vote *against* a Falwell-backed politician.[34] To minimize the negatives associated with such endorsements, Republican strategists masterminded what Edsall and Edsall call a "tri-level strategy for victory: the 'air war' for television; the 'ground war' of high-tech direct mail, phone banks, polls, focus groups, and computerized voter lists; and the effective political use of government policy itself to build election day majorities."[35] Clearly, the war on drugs was one such policy that resonated with a number of single-issue constituencies. "The careful separation of the air and ground wars," write Edsall and Edsall, "permitted the Reagan campaign to conduct an essentially centrist campaign on television, while using the far less visible tactics of the ground war to mobilize support among groups with more divisive, and potentially threatening, agendas—including fundamentalist Christians and anti-abortion activists."[36] This is a "war of positions" that is infinitely more intricate than even that dreamed of in Antonio Gramsci's philosophy of hegemony.

In the air war, the negatives of the politics of religion and resentment became the repressed id against which the Reagan superego orchestrated an upbeat economic message that emphasized the positives of opportunity, freedom, and liberation. In this regard, the oxymoronic "conservative egalitarianism"[37] of Reaganism duplicates many of the strategies employed by the "authoritarian populism"[38] of Thatcherism in Britain. Like Thatcher, Reagan profited from a popular revolt against the Keynesian welfare state. "The missing link between the New Right and new majority," observes Mike Davis, "was provided by Howard Jarvis and the Proposition 13 movement in California. The wave of tax-cutting crusades which followed the victory of Proposition 13 in 1978 surpassed the success of all other single-issue movements."[39] In the tax revolt the Republicans found the perfect "neutral" backlash vehicle for the air war—one that was not, at least explicitly, about race, misogyny, or morality.

Rigid Expansion The escalating antitax crusades, the heightening status anxieties of white males, and the mounting moral hysteria of the Religious Right are all inextricably linked: for all are essentially panic responses to the collapse of the Fordist/Keynesian economic order during the

1970s—an order that had regulated and sustained the postwar boom in the United States. As David Harvey observes, the key to the extended success of Fordism, and its ultimate failure, was its *rigidity*. A system whereby mass production drives mass consumption—and vice versa—Fordism relied on the stabilization of reciprocal relations among big labor, big capital, and big government. Quoting Harvey:

> The state had to take on new (Keynesian) roles and build new institutional powers; corporate capital had to trim its sails in certain respects in order to move smoothly in the track of secure profitability; and organized labour had to take on new roles and functions with respect to the performance in labour markets and in production processes. The tense but nevertheless firm balance of power that prevailed between organized labour, large corporate capital, and the nation state, and which formed the power basis for the postwar boom, was not arrived at by accident. It was the outcome of years of struggle.[40]

The rigidities of this order—including long-term and large-scale fixed capital investments, collective bargaining for long-term contracts, and long-term state commitment to entitlement programs—were sustainable only as long as there was stable economic growth. From 1945 to 1973 the rigidities of the system basically succeeded in stimulating heavily regulated growth to the general benefit of most U.S. workers. However, social barriers in the workplace, the school, and the housing market still excluded many from the Fordist dream of employment security, upward mobility through education, and home ownership. Under Fordism, the labor market was divided into a heavily unionized "monopoly sector" made up of "affluent workers," and a low-wage "competitive sector" made up of what Michael Harrington termed the "other America."[41] In Harvey's estimation, the inequities of this divided labor market "produced serious tensions and strong social movements on the part of the excluded—movements that were compounded by the way in which race, gender, and ethnicity often determined who had access to privileged employment and who did not."[42] In the 1960s the government tried to contain this discontent with Great Society measures including welfare, job training, and educational programs. But by far the most divisive element of this containment strategy was court-ordered busing to achieve racial balance in public school systems. Busing, more than other government action, sowed the seeds of popular, white, suburban discontent—especially in the North—that would be harvested by the backlash politics of the New Right.

The loose monetary policy and the escalating wartime economy of the

1960s allowed the Fordist order to absorb, for a short time, the expenses of these new social programs. However, throughout the 1970s the rigidities of Fordism would be destabilized by a series of traumatic developments. The greatest shock to the system came in 1973 when the shift to a peacetime economy and the oil embargo combined to thrust the nation into a sharp recession. The hard times that followed were marked not only by deindustrialization, high inflation, high unemployment, and high taxes, but by a steady decline in the power of organized labor, steady erosion of public support for the expensive social programs instituted by Great Society containment, and a steady failure to restore enduring economic growth. As the reciprocal relations among big labor, big government, and big capital were progressively undermined, a new economic order emerged—an order that Harvey calls "flexible accumulation."[43]

Flexible Exploitation   As the name implies, flexible accumulation departs drastically from the rigidities of Fordism. Displacing the long-term temporal and spatial stabilities of Fordism with the instabilities of short-term engagement, flexible accumulation promoted a fundamental shift in collective norms and beliefs toward the values of an entrepreneurial culture based on old-fashioned competitive individualism. Oriented on windfall profits, hit-and-run marketing, and "paper entrepreneurialism," these values emphasize, in Harvey's words, "the new, the fleeting, the ephemeral, the fugitive, and the contingent in modern life, rather than the more solid values implanted under Fordism."[44] On the supply side, flexible accumulation has been marked by expansions of the service sector (most notably, in the private health-care, banking, real estate, and fast-food industries), accelerations in the pace of product innovation, explorations of highly specialized market niches, and reductions in turnover time, while on the demand side it has entailed what Davis terms "overconsumptionism."[45]

However, two forces allied with Fordism inhibited the more flexible flow of capital under the new order—big government and big labor. Reagan, the great deregulator and the slayer of the Professional Air Traffic Controllers Organization (PATCO) dragon, was the bane of both. Although deregulation was actually initiated by the Carter administration, under Reagan most of the controls on business were drastically weakened—with drastic consequences, the most visible negative outcome being the savings and loan debacle. While it is undeniable that small businesses, as well as patriarchal and artisanal enterprise, flourished in this new pro-commerce climate, deregulation also led to massive mergers, corporate diversifications, and in some sectors (airlines, energy, communications, and financial services) increased monopolization.[46]

The decline of organized labor is too complicated to treat here.[47] Suffice it to say that collective bargaining during the Reagan era was distinguished by wage concessions and employee givebacks. But the unions had plenty of company in their misery—the radical restructuring of the labor market under flexible accumulation also had a devastating impact on the nonunionized working poor, especially those isolated in America's inner cities.

The job migrations, deregulation, and deindustrialization of the new economic order allowed employers to exert stronger pressures of control on the work force—and provided much more flexibility in controlling labor costs. Characterized by a move away from regular employment, a surge in service employment, and a reliance on part-time or temporary work arrangements, the labor market of the new order is organized into a franchised "core group" and fragmented "peripheral groups." The core, a steadily shrinking job market, is the laboring aristocracy that still basically enjoys the good life of the Fordist affluent worker, with one major difference—under the new order, the core worker is expected to be "adaptable, flexible, and if necessary, geographically mobile."[48] The peripheral job market essentially represents an explosive elaboration of the Fordist "competitive sector." According to Harvey, this periphery is composed of "two rather different sub-groups":

> 1. Low paid, full-time employees who have readily available clerical or secretarial skills or who provide routine and lesser-skilled manual work. Work force reductions in this group are eased by high turnover linked to limited career opportunities.
>
> 2. Part-timers, casuals, fixed-term contract staff, temporaries, subcontractors, and public subsidy trainees have even less job securities than the low paid, full-time employees—and provide employers with "even greater numerical flexibility."[49]

Under this split-level configuration, the new labor market takes on a bottom-heavy shape—some high-wage jobs, many low-wage jobs, and a "missing middle."[50] The widespread economic insecurities, status anxieties, and competitive tensions accompanying the polarizations of this "flexible" labor market certainly contributed to the antifeminist and antisocial-welfare backlash of the 1970s and 1980s. As always, when times are tough, scapegoats abound.

Demonizing Fordism   Reagan, like Thatcher, was able to colonize these insecurities, anxieties, and tensions by demonizing the old Fordist regime while sanctifying the new entrepreneurial order of flexible accumulation. Reaganism condemned both big government and big labor as symbols of a thoroughly discredited "liberal establishment." Paraphrasing Stuart Hall's analysis of Thatcherism, the discourse of Reagan's conservative egalitarianism condensed "at the negative pole" statism, bureaucracy, "big labor," and liberal Democrats. Against this rhetorical construction of the "power bloc," Reaganism "counterposed various condensations" of possessive individualism, personal initiative, privatization, and freedom from government restraints as the positive pole. In this way, Reaganism was able to represent "big labor" as "part of the big battalions, ranged against the 'little man' (and his family)"—part of an oppressive system of high inflation, high taxes, and intrusive government regulations. Thus, in opposing this power bloc, Reagan was portrayed as "out there struggling with little people" and the labor unions were depicted as just another special interest seeking special treatment at "Our" expense.[51]

In many profound ways this sanctification of the new order is a rewriting of the late nineteenth-century utopian myth of Horatio Alger. In Alger's capitalist utopia, the ideal society is a meritocracy in which brains, initiative, and hard work earn their just rewards. According to Reich, the "Alger cosmology" "endorsed large disparities in wealth, since riches were the award for applying yourself, saving your money and trading shrewdly. The key virtue was self-reliance; the admirable man was the self-made man; the goal was to be your own boss rather than to work for someone else. . . ."[52] Indeed, part of Reaganism's symbolic power was its ability to celebrate opportunism while undermining equal opportunity—or, at the least, redefining equal opportunity in terms of reverse discrimination. Collective middle-class (and journalistic) identification with the wealthy during the Reagan era meant that prosperous entrepreneurs were once again widely perceived as like Us, only luckier, smarter, and more daring.

Interestingly, the chief demon of the Reagan order was not a person, or even a special-interest group, but a chronotope—the 1960s. This hostile orientation toward the 1960s is also prominent in the rhetoric of Thatcherism. As Marcus observes: "Thatcher and Reagan denounced the hedonistic anarchy of the sixties as a moral wasteland responsible for economic disaster while simultaneously celebrating untrammeled capitalism as a personal quest for autonomy, self-realization, adventure, fulfillment, possibility, imagination, risk, and desire, literally taking fragments of May

'68 slogans into their mouths; the key words were adventure and risk." [53] In fact, the 1960s became the *pretext* for almost all of the political actions and moral reactions of the New Right and Religious Right during the 1970s and 1980s.

For Reaganism, the decade of the 1960s was midnight in America— and the 1980s, its new morn. The "again" in the campaign slogan—"It's morning again in America"—gave voice to a conservative egalitarianism that, in Marcus's phrasing, "promised everything to anyone with the grace to leave the damned behind." [54] Although a superficially "inclusive" message, the "morning again" slogan expressed an ideology that, according to Edsall and Edsall, "used opposition to federal tax burdens to unite the rich and the working class, as opposed to the use of federal spending to unite the poor and the middle class, an ideology purged of overt bias, tinged— in the wake of twenty years of unprecedented social change—with nostalgia, and with an implicit but stern admonition to life's losers." [55] Or put another way, the nostalgic orthodoxy of Reaganism vowed to "take back America" in a double sense. On the one hand, as an orthodoxy of inclusion, it promised to take all of America back *to* the gilded age of a pre-Fordist, Horatio Alger enterprise culture; on the other hand, as an orthodoxy of exclusion, it promised to take America back *from* the color- and gender-coded "special-interest groups" of the Keynesian welfare state.

Re-Covering Racism

> All the people like us are We,
> And every one else is They.
> —Rudyard Kipling

Where the first phase of the cocaine narrative did not coalesce with the countersubversive racial agenda of the New Right, the crack crisis coverage and its images of black youth running wild in the streets, out of control, would resonate with the prevailing attitudes toward color-coded special-interest groups of the predominantly white Reagan counterrevolution. These attitudes (which were most explicitly and notoriously voiced in focus groups made up of the infamous "Reagan Democrats" of Michigan's Macomb County) [56] are characterized in the research of a number of scholars as a "new racism." Summarizing this research, Robert M. Entman identifies three components of this backlash mentality: (1) "anti-black affect—a general emotional hostility toward blacks," (2) "resistance to the political demands of blacks," and (3) "belief that racism is dead and that racial discrimination no longer inhibits black achievement." [57] As Entman also observes, the so-called new racism springs in part from the widespread suppression of old racial bigotry that stereotyped racial minorities as lazy by nature and intellectually, as well as biologically, inferior to the white race.

In practice, though, many of the barriers guarded by the new racism are the ones erected by old-fashioned white supremacy to safeguard traditional sanctuaries of white power: the segregated neighborhood; the segregated school; the segregated country club; the white-owned business directed by nepotism; the white-dominated profession regulated by discriminatory qualifications; the "core curriculum" monopolized by Eurocentric (and chiefly masculinist) knowledge; and the legislative body governed by white male incumbency. The transgressing forces (or "special interests"), in this scheme, become any group or coalition that undermines existing sexual and racial power relations or challenges the perpetuation of white male dominance in the college seminar, the workplace, and the political arena. In other words, a siege metaphor frames not only the nationalist response to a hostile international environment, but also describes the moralist panic of the ongoing crisis in white and patriarchal prestige on the "homefront."

The Enemy Within   This reinforcing dialogue between the cocaine narrative and Reagan's version of the marginal forces threatening the American Way generated, at least in part, the extraordinary consensus about the seriousness of cocaine as a social problem in 1986. During this 1986 crisis coverage the trickle-down narrative was replaced with the siege paradigm. This mentality is, in fact, vividly captured in the title of a report published in 1990 by the U.S. House Ways and Means Committee: *The Enemy Within: Crack-Cocaine and America's Families.* The framers of the report probably did not realize how its title resonates with the rhetoric of Enoch Powell, a British politician active in the 1960s and 1970s who is to Thatcherism what George Wallace is to Reaganism. In 1970 Powell delivered a famous speech titled "Enemies Within" that, according to Kobena Mercer, "marked a crucial turning point in the popularization of a New Right perspective in British politics." In this speech Powell developed a coherent theory of national identity based on what Mercer describes as "the cultural construction of Little England as a domain of ethnic homogeneity, a unified and monocultural 'imagined community.'" [58]

For Mercer, the historical significance of Powellism is twofold: (1) Powellism successfully used immigration as a wedge issue that helped broaden the appeal of "neo-liberal anti-statism"; and (2) Powellism's fusion of populism with nationalism in scapegoating immigration managed to displace "the old biologizing language of racism, whose 'morphological equation' to superiority and inferiority was associated with Nazi ideology, in favor of a culturalist vocabulary that depended on a binary system of identities and differences." In this rearticulation, according to Mercer, Powell "contributed to the authorship of the new racism" by actually speaking the language of "liberal multiculturalism"—but in a way that reaccentuated "the concept of ethnicity" so that it supported "an anti-democratic discourse of right-wing populism." [59]

For us, the "culturalist vocabulary" that Powell deploys in his "authorship of the new racism" is strikingly similar to the new racism sponsored by Reaganism—with one profound difference. In "a nation of immigrants," Powell's idealization of a "unified and monocultural" community would not do. Instead, the New Right's reformulation of racial ideology in the United States drew from a very different set of ideals associated with the melting-pot myth—a myth that, in stark contrast to Powellism, celebrates cultural pluralism and assimilation.

However, like Powellism, Reaganism also managed to "civilize" [60] the "discourse of discrimination" by eliminating the "old biologizing language of racism." As Omi and Winant argue, the New Right has "gained political currency by rearticulating a racial ideology" that "does not display *overt* racism." [61] To understand this symbolic recovery operation, the racial ideology of Reaganism must be seen as a response to two competing discourse systems: *the discourse of "special interests" who demand group rather than individual rights* (which is associated with the radicalism of the demonized 1960s), and *the older discourse of white supremacy* (which is still very much alive today in such far right fringe groups as the Ku Klux Klan, the White American Resistance, the Order, the Posse Commitatus, the Aryan Nations, and the Farmers Liberation Army). In the space between these two discourse systems, Reaganism fostered an emergent racial formation that replaced the social Darwinism of traditional white supremacy with the cultural Moynihanism of conservative egalitarianism. [62]

Cultural Moynihanism   As in Powellism, Reaganism's rehabilitation of racism involved the appropriation and reaccentuation of the mainstream, ethnicity-based theory of the modern sociology of race. Before 1930, ethnicity theory was an "insurgent approach" to race associated with the "Chicago school" of sociology led by Robert E. Park. Park himself postulated a model of historical development known as the *race-relations cycle* that is still widely acknowledged as an important contribution to the modern analysis of ethnicity. Park considered the four stages of the cycle— *contact, conflict, accommodation,* and *assimilation*—to represent (in Omi and Winant's words) "a way of analyzing group relations and assessing a 'minority' group's progress along a fixed continuum."[63] In contrast to the biologistic paradigm in which differences in intelligence, temperament, and sexuality were deemed to be hereditary in character, this emergent theory—which challenged the eugenicist and social Darwinist thinking of the age—understood race as a social category that was "but one of a number of determinants of ethnic group identity or ethnicity."[64]

Between 1930 and 1965, as the biologistic paradigm suffered from its association with Nazi ideology, the ethnicity paradigm would come to operate as "the progressive/liberal 'common sense' approach to race ... [in which] two recurrent themes—assimilationism and cultural pluralism— were defined."[65] The key works of this period were E. Franklin Frazier's *The Negro Family in the United States*[66] and Gunnar Myrdal's *An American Dilemma: The Negro Problem and Modern Democracy.*[67] According to Rickie Solinger, Frazier (an African American sociologist who studied with Parks) became "a hero and a beacon to those attempting to counter racial biological determinism."[68] Challenging the assumptions of proponents of the "inherent moral degeneracy of the Negro," Frazier developed an environmentalist argument that tied relatively high rates of black illegitimacy to "the social and economic subordination of the Negro" both in the rural South and in the slums of northern cities.[69] However, while Frazier and Myrdal (who drew heavily from Frazier's work) were certainly in step with the most progressive thinking of their age, they championed an assimilationism for blacks that was erroneously based on the European immigrant analogy.

Because of this flaw, ethnicity theory *can* account for rites of inclusion that broaden notions of what it means to be "white" in America beyond the Anglo-Saxon-Protestant formulation, but it *cannot* fully grasp how the controlling visions of rites of exclusion also figure in the same process. In other words, as a dangerous "half-truth," its reliance on the European immigrant analogy underestimates how vital the discourse of discrimination is to perpetuating the dominant social/cultural/racial/gender order. As Patricia Hill Collins demonstrates, this discourse of discrimination is especially salient to the experience of women of color. Often situated in the curious position of the "outsider within," African American women have long played an important symbolic role in the white imagination. In Collins's words: "As the 'Others' of society who can never really belong, strangers threaten the moral and social order. But they are simultaneously essential for its survival because those individuals who stand at the margins of society clarify its boundaries. African-American women, by not belonging, emphasize the significance of belonging."[70] The "significance of belonging," of sharing in the rights and privileges of being "free and white" in America, then, quite literally has a "dark side"—for it is sustained by the ongoing othering, marginalization, and stigmatization of the "nonwhite" category.

Compared to the African American experience, European immigrants' initial contact with entrenched "white" American culture was clearly grounded on a profoundly different *vector of agency* and *angle of identification*—that is, the immigrants themselves were the agency that initiated the contact, often because they identified with the host culture, its racial and economic order, and the promise of opportunity in the New World. For these groups, conditional accommodation by the system was based on acceptance—even endorsement—of a set of color distinctions that figured prominently in a group's interpellation into educational institutions, religious networks, public accommodations, housing arrangements, and labor markets. Indeed, because of their complicity in this system of distinctions, the accommodation and assimilation of non-WASP ethnic groups under the imaginary rubric of "white" have actually, at times, had devastating consequences for African Americans. "More than any other minorities," writes Stephanie Coontz, "blacks encountered periodic increases in discrimination and segregation, first as democratic politicians tried to justify the continuation of slavery, then as blacks were pushed not *up* but *off* the job ladder by successive waves of immigrants":

> No other minority got so few payoffs for sending its children to school, and no other immigrants ran into such a low job ceiling that college graduates had to become Pullman porters. No other minority was saddled with such unfavorable demographics during early migration, inherited such a deteriorating stock of housing, or was so completely excluded from industrial work during the main heyday of its expansion. And no other minority experienced the extreme "hypersegregation" faced by blacks until the present. . . .[71]

Consequently, treating all contact with the dominant order as the same, regardless of whether it takes place in the context of voluntary immigration or territorial conquest or genocide or slavery is a fundamental shortcoming of the ethnicity paradigm that obscures historic disparities in the treatment of "white" and "nonwhite" groups. And yet, despite this fatal flaw, Park, Frazier, and Myrdal's utopian vision still provided the theoretical justification that supported much of the civil rights struggles in the 1950s and 1960s.

After 1965, however, the ethnicity paradigm's view of black culture as "pathological" would increasingly take center stage in "the defense of conservative (or 'neoconservative') egalitarianism against what is perceived as the radical assault of 'group rights.'"[72] Nineteen sixty-five is the key date because it coincides with the publication of a report written by Daniel Patrick Moynihan and issued by the Office of Policy Planning and Research of the U.S. Department of Labor. Titled *The Negro Family: The Case for National Action*, the Moynihan report characterized the black family as a "tangled web of pathology."[73] The significance of the Moynihan report is that it represents an early example of holding the victims of a long history of systematic racial oppression, economic exploitation, and social exclusion accountable for their own misery, for their own failure to "assimilate" as individuals into a supposedly "accommodative" and increasingly "color-blind" society. Since then, a series of studies operating in, or drawing from, the ethnicity tradition (most notably, by Moynihan, Nathan Glazer, Ben Wattenberg, Thomas Sowell, and Charles Murray) have provided the New Right with seemingly "objective" research that justifies eliminating affirmative action programs, cutting welfare funding—and prosecuting a brutal war on black youth under the guise of the war on drugs.[74]

One of the most reactionary and influential of these studies, Murray's *Losing Ground: American Social Policy, 1950–1980,* manages to gloss over the collapse of the Keynesian-Fordist economic order in blaming Great Society initiatives for the rising poverty rate in the late 1970s.[75] Attacking "welfare dependency" for undermining self-reliance, the work ethic, and family values, Murray's book is, in Coontz's estimation, a rewriting of anticharity sentiments popular during the First Gilded Age (mid-1870s to mid-1890s). In fact, for Coontz, Murray's work represents nothing less than a culturalist rewriting of the social Darwinism of the First Gilded Age: "Social Darwinism preached that millionaires exemplified the 'survival of the fittest.' The poor were labeled 'unfit,' a drag on the race. To preserve the unfit in any way was to court disaster. 'Nature's cure for most social and political diseases is better than man's,' argued the president of Columbia University, as did his successors in the 1970s and 1980s, George Gilder and Charles Murray." [76] Advocating the elimination of all social programs aimed at the poor (with the exception of unemployment insurance for the working-age population), Murray replaces the "benign neglect" of Nixon era cultural Moynihanism with a much more malignant and harsher Reaganesque animosity toward the suffering of those in poverty.

Thus, in a strange symbolic inversion, the New Right's appropriation of ethnicity theory converts the "great man" theory's radically individualistic outlook on economic prosperity into a "poor man" theory of personal/familial failure that is then applied to a right-wing revision of the meaning of the Great Society. Murray, Gilder, and other New Right intellectuals redefined the economic turmoil and familial instability accompanying the shift from a manufacturing to a service economy as simply the consequences of cultural pathologies driven by individual immorality. In Reinarman and Levine's phrasing:

> Unemployment, poverty, urban decay, school crises, crime, and all their attendant forms of human troubles were spoken of and acted upon as if they were the result of *individual* deviance, immorality, or weakness. The aperture of attribution for America's ills was constricted. People *in* trouble were reconceptualized as people who make trouble; social control replaced social welfare as the organizing principle of state policy.[77]

Just as this individualist historiography embraces the idea that entrepreneurs must be rewarded, it proposes that the mob of unruly drones must be disciplined.

Impoverished Values and Bitter-Sweet Success   The culturalist vocabulary of ethnicity theory—and its discourse of discrimination—is enunciated in then-Vice President Dan Quayle's assertion that the 1992 Los Angeles uprising was a result of "a poverty of values in the inner city" as well as in the "weed and seed" metaphor of the Bush administration's proposals for rectifying the problems in South Central Los Angeles. But excerpts from the writing of Patrick Buchanan, the darling candidate of the New Right and Religious Right in 1992, provide us with the starkest rendering of ethnicity theory's place in the regressive/conservative "commonsense" approach to explaining racial inequality:

> Why did liberalism fail black America? Because it was built on a myth, the myth of the Kerner Commission, that the last great impediment to equality

in America was "white racism." That myth was rooted in one of the oldest of self-delusions: It is because you are rich that I am poor. My problems are your fault. You owe me!

There was a time when white racism did indeed block black progress in America, but by the time of the Kerner Commission ours was a nation committed to racial justice. . . .

The real root causes of the crisis in the underclass are twofold. First, the old character-forming, consciousness-forming institutions—family, church, and school—have collapsed under relentless secular assault; second, as the internal constraints on behavior were lost among the black poor, the external barriers—police, prosecutors, and courts—were systematically undermined. . . .

What the black poor need more than anything today is a dose of the truth. Slums are the products of the people who live there. Dignity and respect are not handed out like food stamps: they are earned and won. . . .

The first step to progress by any group lies in the admission that its failures are, by and large, its own fault, that success can come only through its own efforts, that, while the well-intentioned outsider *may* help, he or she is no substitute for personal sacrifice.[78]

Blithely dismissing more than two hundred years of slavery and another hundred years of legalized racial discrimination, this blame-the-victim mentality struck a responsive chord of absolution for millions of white, middle- and working-class Americans. In a time of declining expectations, vanishing jobs, and escalating housing and medical costs, the New Right's moral ahistoricism not only released members of the white majority from any sense of responsibility for the many sins of their fathers, but it gave them reason to feel that they were the "truly victimized." In this topsy-turvy view of race relations, Reagan administration officials even cast affirmative action, which began during the Nixon administration, as a new form of "racism." For instance, Clarence Pendleton, Jr., a chairman of the U.S. Civil Rights Commission during the Reagan era, characterized the supporters of affirmative action as "the new racists, many of them black, [who] exhibit the classical behavior system of racism. They treat blacks differently than whites because of their race."[79]

As the antiaffirmative action "Black-lash"[80] by powerful African American conservatives like Pendleton, Sowell, and Supreme Court Justice Clarence Thomas verifies, the coinage of the new culturalist racism has much more exchange value and much wider investment opportunities than that of old-time biologicist racism. The new racism is, in other words, more subtle, more respectable, less inflammatory—and more politically robust than its ancestry. This versatility is evident in Edsall and Edsall's otherwise thorough critique of Reaganism's backlash politics in which the authors, surprisingly, end up endorsing a negative view of the "rights revolution" that condemns the Democratic party for not addressing the cultural pathologies of America's inner cities.[81]

The general political currency of the new racism also circulated in "The Vanishing Family: Crisis in Black America," a CBS News special report that aired in January 1985. Narrated by Bill Moyers, the notoriously "liberal" journalist at the center of the New Right's recent attack on PBS, the report both featured and furthered the cultural Moynihanism of conservative egalitarianism. Moyers, like Moynihan, acknowledges that racism was at one time a decisive force limiting the life chances of African Americans. But, also like Moynihan, Moyers denies that racism goes very far in explaining the contemporary suffering of poor African Americans in the inner city. Describing a black neighborhood in Newark, New Jersey, as "a world turned inside-out," Moyers argues that the alien value system of this world—and the poverty of its inhabitants—are grounded on a moral irresponsibility that is destroying "the" black family.[82] Illustrating his report with men who refuse to support their children, and women who feel they can get along fine without husbands, Moyers concludes that there are very few positive role models in this neighborhood. This negative assessment is perhaps most clearly stated in a passage in which Moyers essentially dehumanizes these transgressors by placing them outside the realm of "people":

> There are successful strong black families in America. Families that affirm parental authority and the values of discipline, work, and achievement. But you won't find many who live around here. Still, not every girl in the inner city ends up a teenage mother, not every young man goes into crime. *There are people who have stayed here.* They're out-numbered by the con artists and pushers. It's not an even match, but they stand for morality and authority and give some of these kids a dose of unsentimental love. [Our emphasis] [83]

Like much of the network news reports of the same period, Moyers's dialogue overtly locates "con artists" and "pushers" as a separate category from "people." Ultimately, for Moyers, the black households he surveils in Newark suffer because they do not measure up to the male breadwinner/female homemaker ideal embraced by Middle America. As we shall explore more fully in chapter 8, this nuclear family ideal is at the heart of a sex-gender system that naturalizes patriarchal authority and converts material advantages into moral superiority: for the two-parent, male breadwinner family structure favored by Moyers is only a real option to households that can afford the luxury of a full-time housekeeper.

There is, of course, something fundamentally venal and smug about "family values" that depict affluence as a virtue and necessity as a sin. Poor families, according to this perverse, self-serving moralism, are not unstable because of material disadvantages; instead, because poor families are unstable, they rightly suffer from material disadvantages. In the top-down moralism of cultural Moynihanism, women of color have been held especially culpable for perpetuating the "cycle of poverty." In Moynihan's own words: "From the wild Irish slums of the 19th-century Eastern seaboard, to the riot-torn suburbs of Los Angeles, there is one unmistakable lesson in American history: a community that allows a large number of men to grow up in broken homes, dominated by women, never acquiring any stable relationship to male authority, never acquiring any set of rational expectations about the future—that community asks for and gets chaos." [84] As the irrational nurturer of chaos, then, the Black Matriarch animates the myth of the all-powerful, domineering mother who robs her lovers and husbands of both their manhood and their proper familial authority.

In the conservative egalitarianism of the New Right, the controlling image of the Welfare Mother represents a subtle, though profound, revision and reversion of Moynihan's Black Matriarch. Where Moynihan's matriarch thesis charged women of color with being too *independent,* the New Right's demonized welfare mothers were too *dependent.* Quoting Collins:

> Like the matriarch, the welfare mother is labeled a bad mother. But unlike the matriarch, she is not too aggressive—on the contrary, she is not aggressive enough. While the matriarch's unavailability contributed to her children's poor socialization, the welfare mother's accessibility is deemed the problem. She is portrayed as being content to sit around and collect welfare, shunning work and passing on her bad values to her offspring. The image of the welfare mother represents another failed mammy, one who is unwilling to become "de mule uh de world." [85]

However, if we believe one of Ronald Reagan's favorite anecdotes, then at least one of these bad mothers shared the enterprising values of her conservative detractors—the notorious "Welfare Queen of Chicago." According to Reagan's often-repeated account, this woman, practicing a kind of "multiple individualism," had "80 names, 30 addresses, 12 Social Security cards," and a "tax-free income" of "over $150,000." [86]

The intersection of "family values" with cultural Moynihanism, then, is one site where the backlash politics of race and gender converge. The discursive power of this convergence explains, in part, why it also provides a common grounding for both Murray's conservative egalitarianism and Moyers's liberal pluralism. In fact, according to Coontz, this common grounding has generated what is often called "the new consensus" among conservatives and liberals about what African Americans need "is not government programs but a good dose of sexual restraint, marital commitment, and parental discipline."[87] In contributing to this consensus Moyers is not, by any means, a renegade liberal: the growing legitimacy of such sentiments among liberals is also evident in the *New Republic*'s Morton Kondracke's declaration that "it is universally accepted that black poverty is heavily the result of family breakdown."[88] But perhaps the most interesting liberal appropriation of conservative egalitarianism's victim-blaming was mouthed by Senator Charles Robb, the philandering Lyndon Johnson's allegedly adulterous son-in-law. Unself-consciously invoking the language of Powellism, Robb claims that in LBJ's time "racism, the traditional enemy from without," was largely responsible for African American economic and social distress. But today, according to Robb, "it's time to shift the primary focus . . . to self-defeating patterns of behavior, the new enemy within."[89]

The contradictions of liberals en masse embracing a neoconservative definition of the situation in the inner city pale beside the ironies of another feature of the 1980s: the right-wing appropriation of the celebrated mainstream media achievements of a handful of prominent African American "individuals"—Bill Cosby, Whoopi Goldberg, Arsenio Hall, Michael Jackson, Michael Jordan, Eddie Murphy, Keenan Ivory Wayans, and Oprah Winfrey. Consider, for instance, how William F. Buckley, Jr., exploits the accomplishments of Bill Cosby. "It is simply not correct," asserts Buckley, "that race prejudice is increasing in America. How does one know this? Simple, by the ratings of Bill Cosby's television show and the sale of his books. A nation simply does not idolize members of a race which that nation despises."[90] This appropriation activates a conservative rite of inclusion that, as Herman Gray argues, confirms "a middle class utopian imagination of racial pluralism"—an imagination based on the comforting deceptions of "open class structure, economic mobility, the sanctity of individualism, and the availability of the American dream for black Americans."[91]

When placed beside the rites of exclusion enacted in the Moyers report (as well as in news coverage of the crack "epidemic"), these images of individual black success tend to give credence to the "recovery" discourse of black progress and middle-class racial pluralism—Entman's third component of the new racism listed earlier (p. 91). As Gray puts it, the success of black stars in popular movies and television programming reinforced a myth that the failure of black people in the urban underclass is "their own since they live in an isolated world where contemporary racism is no longer a significant factor in their lives": "In the [televised treatment of the] world of the urban under class, unemployment, industrial relocation, ineffective social policies, power inequalities, and racism do not explain failure, just as affirmative action policies, political organization, collective social and cultural challenges to specific forms of racial domination, and the civil rights movement do not help explain the growth of the black middle class."[92] Thus, the culturalist framing of familiar images of black failure and the individualist framing of images of black success must be understood as part of the same ideological project—a project that is congruent with Reaganism's great man view of economic history and poor man view of personal/ familial failure. As Entman found in research that resonates with Gray's, the prominence of African American journalists in Chicago's local television news also contributes to the ironies of black success engendering "an impression that racial discrimination is no longer a problem."[93] And we have found, in this study, that the high visibility of African American journalists—especially CBS's Harold Dow—in the reporting of the cocaine narrative has similar paradoxical consequences.

Perhaps the most powerful statement on the bittersweet success of black media stars appears in a memorable scene from Spike Lee's *Do the Right Thing* (1989). In an interchange near the midpoint of the story, Mookie (played by Spike Lee) confronts Pino (played by John Turturro) about his hostile attitude toward black people in the neighborhood:

MOOKIE: Can I talk to you for a second?
PINO: What?
MOOKIE: Pino, who's your favorite basketball player?
PINO: Magic Johnson.
MOOKIE: And who's your favorite movie star?
PINO: Eddie Murphy.
MOOKIE: Who's your favorite rock star? Prince.
PINO: No, it's Bruce.
MOOKIE: Prince.
PINO: Bruce!

MOOKIE: Pino, all you ever talk about is "nigger this" and "nigger that," and all your favorite people are so-called "niggers."
PINO: Magic, Eddie, Prince. . . . They're not niggers. I mean, they're not black. I mean, they're black but they're not really black. They're more than black. It's different.
MOOKIE: Different?
PINO: Yeah. To me, it's different.

In presenting Pino's notions of "difference," Lee succeeds in dramatizing how the discourse of discrimination operates in everyday life. And, after Pino makes disparaging remarks about Minister Louis Farrakhan and the Reverends Al Sharpton and Jesse Jackson, the confrontation degenerates into an exchange of curses:

MOOKIE: Pino, fuck you, fuck your fucking pizza, and fuck Frank Sinatra.
PINO: Yeah? Well, fuck you, too. And fuck Michael Jackson.

This last curse ushers in the controversial montage in which several characters directly address the camera while delivering a string of ethnic slurs and racial epithets. Clearly, Lee's invocation of Michael Jackson to frame this invective is meant to cleverly undermine the crossover superstar's own ongoing denials of "discrimination." Although surgically inscribed on his androgynous and assimilationist features, Jackson's attempts to deny (or, better yet, erase) difference are perhaps most overtly expressed in the lyrics of a recent recording in which he naively affirms that "it don't matter if you're black or white"—an affirmative action that gives credence to Reaganism's conservative egalitarianism and its doctrine of "color-blind" bigotry.

As we suggested at the end of chapter 1, the war on drugs figures prominently in the inclusions and exclusions of Reaganism's economic agenda and racial polarizations. Above all, the crack crisis provided the Reagan administration with an opportunity to tighten controls on the drone population by demonizing a disposable fragment of the peripheral labor market (impoverished, black, urban youth) while it hypocritically pursued a policy of loosening controls on the entrepreneurial class. Recognizing the connections between the seemingly contradictory tightening and loosening of governmental controls during the 1980s provides insight into the Reagan legacy—a legacy that includes not only the highest incarceration rate in the world and the recovery of white racism (as exemplified by the popularity of David Duke), but also insider trading scandals on Wall Street, the savings and loan bailout, the devastation of South Central Los Angeles in a major urban uprising, a worsening health-care crisis, and an overwhelming national debt. As Lawrence Grossberg warns about Reaganism's legacy: "What is at stake here is a practice and principle of redistribution which puts into place a new form of regulation (in the service of both continuous and emergent interests in the surveillance of the population)."[94] Our recontextualization of the cocaine narrative in the first part of this book, then, is meant to demonstrate how the war on drugs is deeply implicated both in the drastic expansion of the modern control culture and in the broader legitimation of the backlash politics of the New Right. In Part II, we interrogate how journalistic coverage of defining moments in the cocaine narrative corroborated the drug control establishment's self-interested promotion of drug hysteria, while, at the same time, it affirmed the New Right's moral framing of economic distress—distress that was further aggravated by Reaganomic's cynical upward redistribution of wealth.

Part II

INTERROGATING THE COCAINE

 NARRATIVE

5

The Trickle-Down Paradigm:

White Pow(d)er and Therapeutic Recovery

> Guided by emotion and empathy, working through ritual and repetition, television's core vocabulary reflects its role as a therapeutic voice ministering to the open wounds of the psyche. As a "close-up" medium whose dramatic and social locus is the home, television addresses the inner life by . . . maximizing the private and personal aspects of existence.—George Lipsitz [1]

Cautionary Tales

Admittedly, Part II is primarily concerned with interrogating defining moments of the journalistic discovery and obsession with crack in 1986. Chapters 6 and 7, in fact, are entirely devoted to analyzing news coverage of the 1986 crisis. Chapter 8, which explores how conventional gender ideology inflects the cocaine narrative, culminates in an examination of a sinister new form of maternity that emerged during this crisis—the crack mother. And chapter 9, which is devoted to "aftermath" coverage, considers how the network news responded to widespread criticism of journalistic performance during the 1986 crack crisis. In this chapter, the briefest of Part II, we propose a reading of precrisis coverage that is designed to foreground disturbing discrepancies in the journalistic treatment of "white offenders" and "black delinquents." For between 1981 and 1985, when cocaine was primarily defined as an elite, white, suburban, "recreational" drug, the dominant news frame conforms to what we term the "trickle-down" paradigm—a frame that, in stark contrast to the "siege" paradigm regulating the crisis coverage of 1986, is organized around discourses of recovery associated with the therapeutic branch of the medical-industrial complex.

Consequently, many news stories in Phase I of the cocaine narrative operate as cautionary tales that articulate the gray area between rites of inclusion and exclusion. Where rites of inclusion tend to celebrate certain individuals, actions, or ways of being as models worthy of emulation by a transformative Us, cautionary tales operate on the border between Us and Them. In giving a narrative form to this contested terrain, the cautionary news story condemns certain problematic individuals, actions, or ways of being as examples of transgression that require purifying solutions (solutions that involve, in some cases, rehabilitative therapy and, in others, punitive separation). The middle-up identification with the economic elite during the 1980s was certainly not without contradictions—and it is in the realm of the cautionary tale that these contradictions were made visible, often verifying that, like Us, the wealthy suffer and sin, too.

In opposition to Horatio Alger's nineteenth-century American Dream sagas, there were cautionary tales in the early twentieth century that began to challenge the ethic of individual success and its promise of happiness. Of these, Edwin Arlington Robinson's "Richard Cory" is perhaps the briefest and the most instructive:

> Whenever Richard Cory went down town,
> We people on the pavement looked at him:
> He was a gentleman from sole to crown,
> Clean favored, and imperially slim.
>
> And he was always quietly arrayed,
> And he was always human when he talked;
> But still he fluttered pulses when he said,
> "Good-morning," and he glittered when he walked.
>
> And he was rich—yes, richer than a king—
> And admirably schooled in every grace:
> In fine, we thought that he was everything
> To make us wish that we were in his place.
>
> And so we worked, and waited for the light,
> And went without the meat, and cursed the bread;
> And Richard Cory, one calm summer night,
> Went home and put a bullet through his head.[2]

In four stanzas Robinson captures all of the basic elements of the modern cautionary tale: the commonsense point of view of the admiring drone; the distinction between the refined taste of the privileged and the coarse tastes of laboring classes; the disparity between public appearances and private realities; the popular beguilement with self-inflicted or reckless death; and, of course, the reassuring conviction that mere worldly success cannot buy happiness.

In the 1980s, alongside the masculinist celebrations of Iacocca, Ueberroth, Jobs, Trump, and Turner were a series of counterexamples that, like "Richard Cory," explored the limits of the success ethic. Ivan Boesky, the Hunt brothers, John Connally, Michael Deaver, Lyn Nofziger, and Charles Keating would all become central figures in cautionary tales that, in another day and age, might have suggested widespread corruption at the top of the world of business. In the realm of drug coverage the most notorious entrepreneurial figure to surface during the 1980s was John DeLorean, the former Detroit automotive executive who failed in his bid to launch a luxury sports car company in Northern Ireland and was busted for trying to smuggle a suitcase laden with fifty-five pounds of cocaine into the United States. DeLorean's arrest would be a kernel event of news coverage in 1982. Later, news reports of his 1984 trial would further expose DeLorean's greed and arrogance by airing undercover audiotape recordings of him bragging that "cocaine is better than gold." But DeLorean would not be the only figure who was "iconized" as a decadent transgressor during this period. Figures in the world of professional sports also would be visualized as public deviants in Phase I coverage. However, we delay our discussion of sports scandals until the next chapter when we interrogate the Len Bias death drama. Here, we focus on another highly significant arena for drug transgression during the early 1980s—the world of show business.

The Hollywood Chronotope

In chapter 3 we introduced the term "chronotope" in a footnote to our discussion of the "newsroom." In chapter 4 we described the 1960s as a chronotope demonized by Reaganism. And later in this chapter we describe the command center of drug hotlines as a postmodern chronotope that is, like the newsroom, everywhere and nowhere. For us, the notion of chronotope is worth rescuing from the realm of literary jargon because it defamiliarizes the taken-for-granted significance that the commonsense term "setting" plays in the narrative experience. As an alternative way of thinking about setting, chronotope (which literally means time-space) emphasizes the *inseparability* of the spatial and the temporal dimensions of the story-world. Put another way, the concept of the chronotope historicizes narrative space. The setting, after all, is what connects the news story to the world of everyday life. It gives the story a concrete existence, a sense of temporal and spatial unity, and sometimes even a sense of tradition, community, culture, and place. But beyond organizing time and space, a news story's chronotope also carries with it an image of human existence that has powerful aesthetic and sociocultural implications. Within this image of human existence, the characters come to life as heroes, seekers of truth, authority figures, normal people, deviants, law-abiding citizens, criminals, transgressors, fools, and lunatics.

Since we do not have the space (or the time) to analyze all of the diverse chronotopes emerging in our study, we want to strategically deploy this concept in an analysis of a crucial news setting that emerged during Phase I coverage: Hollywood—a hotbed of scandal whose very name is synonymous with decadence. In the twentieth-century American imagination Hollywood has in many profound ways displaced the frontier as the supreme liminal space where the discourse of upward mobility at the heart of the American Dream becomes reality. Like the frontier, Hollywood is as much a state of mind as it is a real place. Although it does have a concrete existence in a densely populated area near the coast of Southern California, the idea of Hollywood transcends its spatial and temporal coordinates. In the popular mind, Hollywood is not just a smoggy part of Los Angeles. Instead, it is, again like the frontier, a chronotope that beckons: an imaginary land just beyond the edge of civilization and its constraints—an "out-West" place of magical transformations, of triumphant individuals, of boundless opportunity, of passionate risk-taking, and of exhilarating freedom.

Stardom The liminal figure that gives a body and voice to this state of mind is the star. The star literally embodies the American Dream of "being anything you want to be." For the star, after all, inhabits a world of make-believe. Kevin Costner, for instance, is a big league baseball player in one picture, a Civil War cavalry lieutenant in another, Robin Hood in another, a JFK conspiracy theorist in another, and a bodyguard in still another. In his public appearances as "Kevin Costner, the movie star," he is a booster of Native American causes, a supporter of conservative congressmen like Phil Graham of Texas, and a naive boob who thinks Madonna's stage performance is "neat." [3] In this mix of changing roles and conflicting political positions, the "real" Kevin Costner is at once strange in his familiarity and familiar in his strangeness—something of a public mystery, a phantom that excites both desire and identification.

The familiar/strange dynamic behind star-audience identification is, in fact, central to the ritual function of Hollywood stardom and scandal. Identification results from a fundamental paradox: the star is at once ordinary and extraordinary; in other words, the star is typical enough to be accessible and recognizable, yet individuated enough to be experienced as unique and fascinating. As Richard Dyer suggests, social types available in the culture at large form the symbolic background from which stars emerge as strong figures of identification. In Dyer's words, "What is abundantly clear is that stars are supremely figures of identification . . . and this identification is achieved principally through a star's relation to social type."[4] Star-audience identification, then, involves a ritual interplay between living social relations at work in everyday life and the mediated human representation of those relations.

This typecasting of a star is often a critical factor in the enactment of a scandal ritual in modern news. A scandal that ruptures a carefully crafted star persona is likely to carry more impact than one that only substantiates an existing public image. Consider, for instance, contrasting cases associated with AIDS news: the scandal-coded deaths of Rock Hudson and Liberace. Since the 1950s, Rock Hudson's stardom represented the "maximization" of a masculine, heterosexual type—the "beefcake" or the "hunk." For many of his mainstream fans the notion that Hudson was, in fact, gay was almost unimaginable. Therefore, when his sexual orientation was disclosed in the publicity surrounding the announcement of his infection and approaching death, the news was both tragic and shocking to many of his fans. With Liberace, the situation was certainly just as tragic. But the disclosure of Liberace's homosexuality was not as shocking, not as scandalous, because it merely verified what was apparent to many in the flamboyance of his concert performances. Consequently, of the two cases, Rock Hudson's death scandal was much more traumatic, much more "newsworthy" (because it was coded as more deviant), and, ultimately, much more decisive in the evolving AIDS narrative because it burst Hudson's constructed star image.[5]

Scandal Cycles In more than eighty years as an American institution, Hollywood has been rattled by several major scandal cycles. One of the earliest and most significant cycles occurred in the early 1920s and involved the four "Ds" that helped establish Hollywood's reputation as the New Babylon: Divorce, Drinking, Drugs, and Death. The most sensational scandal of this period centered on the death of Virginia Rappe, an aspiring young actress who died in the wake of a wild drinking party sponsored by the popular slapstick comedian Fatty Arbuckle. Although eventually Arbuckle was acquitted of criminal charges arising from the incident, the notoriety ended his on-screen career. However, a less memorable scandal is more relevant to this study: the overdose drug death of Wallace Reid. Reid's cinematic persona was that of the strong, courageous All-American Boy. Similar to the reaction to Rock Hudson's illness, the American public was, literally, disillusioned when postmortem publicity revealed that, contrary to this image, Reid was a long-time heroin addict.[6]

What is particularly relevant about the outcome of this early scandal cycle is how these disclosures of private "sins" in the movie colony were appropriated by moral and disciplinary forces as evidence of the need to police motion picture content. Across the country these scandals gave new vitality and validity to long-standing state and federal censorship campaigns. In 1922, to mollify the snowballing threats of federal censorship, the motion picture industry formed a protective trade organization known as the Motion Picture Producers and Distributors of America (MPPDA). Under the patriarchal leadership of former Postmaster General Will Hays (an Indiana Republican, a GOP campaign chairman, and an elder in the Presbyterian Church), the MPPDA would become in the 1930s and 1940s the self-regulatory agency that supervised a set of disciplinary mechanisms that effectively enforced tough restrictions on the content of American motion pictures. The centerpiece of this regulatory machinery was the Motion Picture Production Code, a puritanical document cowritten in 1930 by a Jesuit priest (Father Daniel C. Lord) and a Catholic reporter (Martin Quigley). The code, though, would not be strictly enforced until 1934 when, confronted with the threat of boycott by the newly formed Catholic Legion of Decency, the MPPDA authorized the setting up of the Production Code Administration to issue its "Seal of Approval" to all movies that conformed to the code's many prohibitions. And, of course, the whole normalizing system was facilitated by severe sanctions imposed on any company that distributed a film without this bureaucratic approval.[7]

This regulatory action speaks to a normalizing dynamic that might be described as the *political economy of scandal*. Scandals among the famous and the wealthy have a political value that is often used to justify further policing, regulation, disciplining, and normalization of the workaday lives and the popular pleasures of the drone population. Clearly, this economy is alive and well in side effects of a cluster of Hollywood drug scandals in the early 1980s. The central incident in this cycle was the drug overdose death of John Belushi. Belushi's death, however, was not quite as shocking as Reid's or Hudson's because of Belushi's typecasting. Belushi was known as a reckless rogue because of his unusual, multicharacter, comedic performances on "Saturday Night Live"; furthermore, his most famous filmic character, Bluto in the tremendously popular *National Lampoon's Animal House,* was the quintessence of beer-chugging fraternity excess. Given Belushi's image, then, it was not a complete surprise that he would partake of recreational drugs—and overindulge in his quest for thrills. In some ways, actor Stacy Keach's arrest in Great Britain for cocaine possession was more newsworthy than Belushi's death because it contradicted the public image of tough-guy law enforcer projected in Keach's starring role on television's "Mike Hammer." At any rate, the Belushi and Keach scandals together motivated tabloid exposés as well as congressional probes of Hollywood's cocaine consumption. True to the political economy of Hollywood scandal established in the early 1920s, the new cycle also aroused renewed campaigns to "clean up" the movies—the most notable of which was fronted by none other than Nancy and Ronald Reagan.

In keeping with the formulaic nature of television news, almost every report on the Reagans' call for self-regulation in Hollywood's treatment of drug abuse included a clip from the scene in Woody Allen's *Annie Hall* in which Allen's character sneezes and blows a cloud of cocaine all over revelers at an upscale Manhattan party. However, for us, there are major problems with using this clip to illustrate Hollywood's glorification of the drug culture. First, Woody Allen is one of the few American independent filmmakers who has managed to thrive (at least before his public and court battles with Mia Farrow) outside the institutional boundaries of Hollywood proper. To select this clip as typical of Hollywood filmmaking is not only inaccurate, it is deceptive. But more important, the use of this clip completely neglects one meaning of the scene in *Annie Hall.* This scene, in fact, pokes fun at the trendiness of cocaine snorting among New York's pretentious elite; the use of this clip adopts the wrongheaded interpretive position that the very appearance of cocaine in a movie constitutes an endorsement, or glorification, of its use. This kind of absurd literalism speaks to the impoverished outlook of journalism's information-oriented perspective on human communication—a literalism that is often completely blind to such meaningful subtleties as irony, parody, and satiric social commentary.

Modern News and the Therapeutic Ethos

Like the DeLorean story, the Hollywood drug scandal exemplifies the trickle-down paradigm that dominated journalistic depictions of the demand side of the cocaine "epidemic" from 1981 through 1985. According to this news frame, "rot"—in the form of cocaine demand—was seeping from a degenerate element of the economic elite to an envious and ingenuous middle class that studied and admired the life-styles of the rich and famous—and was, consequently, vulnerable to their vices. Operating within this paradigm, journalists generally fabricated cautionary tales about the limits of individualism. But, as Stephanie Coontz observes, like those circulated during the First Gilded Age, such exposés during Reagan's Second Gilded Age generally allowed the "middle class to differentiate itself from the 'amoral' rich without feeling any duty to oppose their actions or construct an alternative political morality":

> [Some] have argued that the popularity of Tom Wolfe's 1987 *Bonfire of the Vanities* reflected a revulsion against the values of the 1980s. Perhaps so, but it was a revulsion that, like its 1880s precursors, promoted a self-righteous, conservative, antipolitical response. In each period, popular social commentary allowed "decent" people to define themselves in opposition to both the dependent or criminal poor and the idle or profligate rich. In comparison, of course, the honest, hard-working "middling sort" who minds his own business and takes care of his own family need engage in no self-criticism. He can only congratulate himself on his freedom from vice—unless, of course, he is so stupid as to give a quarter to a beggar or, in Wolfe's version, allow demagogues from the underclass to make him feel guilty.[8]

Because decadence and its damaging consequences on those who imitate the illicit taste distinctions of the economically privileged did not fit the moral polarities proposed by Reaganism, these stories portrayed "the top" margins of society as the source of cocaine pollution without prompting the widespread demonizing of the opportunistic entrepreneur. In dramatic contrast to crack coverage in late 1980s drug news, the preponderance of cocaine stories on the networks in the early to mid-1980s fit into a genre we characterize as therapeutic in theme and moral tone.

In *Tele-Advising: Therapeutic Discourse in American Television,* Mimi White argues that "confessional and therapeutic discourse centrally figure as narrative . . . strategies in television in the United States" and that "the rise of the therapy or counseling show . . . in the 1980s" represents "a hybrid subgenre of reality programs and talk shows."[9] Where White focuses attention mainly on fictional melodramas, televangelism, and home shopping networks, we contend that even the most conventional news narratives serve a therapeutic function by marking off boundaries between deviancy and normalcy, sickness and wellness. In the choices that conventional reporters make regarding crime and corruption narratives, implicit moral judgments are embedded that help readers and viewers sort out those experiences that either affirm or deviate from their own values. John Cawelti argues that formulaic stories, whether front-page news reports, TV newscasts, or classical detective fiction, allow audiences to survey "the boundary between the permitted and the forbidden and to experience in a carefully controlled way the possibility of stepping across this boundary."[10] So the drug-related murder story on page one of the *Detroit Free Press*, a Ted Koppel interview with an L.A. street gang member, and a "Columbo" detective story share in the maintenance of the thin moral distinction between the normal and the deviant, the permitted and the forbidden—a moral line that marks the commonsense border between Us and Others. The reporter, often functioning as therapist, constructs and maintains the moral border, reveals the deviance, and allows us to make judgments about the experiences under scrutiny. We may then test our own morality in relationship to the experience reconstituted in news reports.

Don Hewitt, creator of "60 Minutes," has long argued for the pastoral power of television's therapeutic role—present at least since John F. Kennedy's assassination:

> Sometimes television is a theater, sometimes it's a cinema, sometimes it's a sports arena, sometimes it's a newspaper, and sometimes it's a chapel. And sometimes when there are national moments of distress, America calls on Peter Jennings, Tom Brokaw and Dan Rather to minister to the needs of the nation. They don't go to Jerry Falwell. They don't go to Pat Robertson. They don't go to Cardinal Cooke. . . . The first time it happened . . . Jack Kennedy was killed in Dallas. They all came to the set and they held hands and the whole country bound up its wounds by holding hands in front of the television set. They went to church in front of television.[11]

The therapeutic dimension in news plays a powerful role in contemporary society and links journalists to other twentieth-century practitioners and professionals.[12] However, unlike Don Hewitt, many modern journalists are uncomfortable overtly discussing the therapeutic dimensions of news, preferring instead to describe their professional work as the dispassionate dissemination of information. Most conventional journalists are more comfortable using the metaphors of "objective" social science than the metaphors of "subjective" psychotherapy. However, the rise of the modern journalist and rise of the modern therapist in the early 1900s are intricately bound together.

Therapy and Modernity   Modernity, of course, is a tricky and contradictory concept since its aesthetic and political meandering has spanned the last three centuries. As David Harvey notes, "the project of modernity came into focus during the eighteenth century" as part of the Enlightenment.[13] The modern condition's associated characteristics include tendencies toward positivism, technocracy, and rationalism, accompanied by, contradictorily, both a belief in and a distrust of progress and organizational planning.[14] One modernist turn in both literature and therapy is marked by a search for the eternal, immutable, universal amid the alienation and chaos of rapid social change and individual "growth."

Psychologist Kenneth Gergen argues that steadily throughout the modern era, psychology—like modern journalism—shifted to a more scientific, rational view of self. Gergen has examined "what the present century has done both to obliterate the romantic preoccupation with deep interior and to replace it with the rational, well-ordered, and accessible self."[15] Freud, who Gergen sees as a transitional figure between the romantic and modern eras, and his attention to "seething and repressed motivational forces, so central to the romantic definition of person, slowly dropped from view." In their place, the ideal self allegedly "functioned exactly like a mature scientist, observing, categorizing, and testing hypotheses."[16] Such rational skills, Gergen contends, constituted the "essence" of the individual for modern psychology, with therapy functioning in modern times "to build or restore essence."[17]

Around 1900 several crises and problems signaled the rise of expert analysis and therapy. These included, according to Jackson Lears, urbanization, which ushered in the "anonymity of the city"; institutionalization, which created "an interdependent national market economy" that cut people off from ties to the land and "primary experience"; technological advance, which introduced "prepackaged artificiality" along with "unprecedented comfort and convenience"; and secularization, which displaced religion and isolated people from the rituals of tradition.[18] The development, then, of therapy's social influence can be traced to turbulent sociocultural shifts from religion to science, from farm to factory, from scarcity to abundance, and from "self-sacrifice" to "self-realization."[19] The modern analyst emerged from these adjustments as a prominent advocate of what Lears calls the "therapeutic ethos," offering renewal of "a sense of selfhood that had grown fragmented, diffuse, and somehow 'unreal.' "[20] The therapist gained legitimacy, in part, by solving problems of self-crises, by riding the prestigious coattails of medical science, and by positioning the individual at the center of modern life. In *Deliver Us From Evil,* Norman Clark discusses "this developing consciousness of individual, rather than communal, dignity" as a "turning inward for new sources of individual direction."[21] As Warren Susman once observed, "One of the things that make the modern world 'modern' is the development of consciousness of self," that is, the development of modern individualism: "It is further a striking part of the turn-of-the-century decade that interest grew in personality, individual idiosyncrasies, personal needs and interests."[22]

In the shift to late capitalism, as Mimi White notes, products were "promoted as having the ability to transform our lives, alter our whole persons for the better, and renew our energies."[23] As part of modern consumer culture, therapists sold respite from the increasing demands of bureaucracy and the pressures on the self cut adrift. Harry Levine's landmark study of the U.S. temperance movement, in fact, traces the construction of modern alcoholism as a social problem requiring maintenance by "a complex and hierarchical web of community relations" in the seventeenth and eighteenth centuries to redefinition as a personal problem of self-discipline that "required shifting social control to the individual level."[24] In the twentieth century, then, both journalism and psychology helped adjust focus from the social and the collective to individual cognition and personality. In modern journalism, for example, the great commercial success of magazines like *People* and the *National Enquirer* and the "newsmaker" or "people" sections of most papers and news magazines exemplify the ultimate tribute to modern journalism's marketing of individual star "characters."

In *Habits of the Heart* Robert Bellah and his colleagues attribute to the modern therapist features that also characterize the modern reporter:

> For all its genuine emotional content, closeness, and honesty of communication, the therapeutic relationship is particularly distanced, circumscribed, and asymmetrical. Most of the time one person talks and the other listens. The client almost always talks about himself and the therapist almost never does. . . . The therapist's authority seems to derive from psychological knowledge and clinical skill, not from moral values. The therapist is there not to judge

but to help clients become able to make their own judgments. The therapist is, nonetheless, even in not judging, a model for the client.[25]

When news functions under therapeutic metaphors, there is a sense of intimacy between reporter and interviewee. There is also closeness and distance. Reporters reveal little about themselves, while interview subjects reveal much. This asymmetrical relationship is not problematic for most reporters, just as it is not problematic for many therapists, since they believe that distance endows them with a surer skill in seeing the "way things really are" in modern life. This stance at least partly explains why none of the nearly forty network reporters responded to questionnaires we sent regarding our study and their central role in constructing the cocaine narrative in the 1980s.

The notion of an "objective," disinterested, nonpartisan press is a thoroughly modern construct, an invention of the nineteenth century and the rise of science.[26] One constructed ritual of modern professionals, including reporters, includes celebrating rationalism and masking their moral take on the world. Professionals do not see such masking as problematic since, through constructed conventions, they erect categories that make it seem as if these two ways of being—living and observing—are separate. Disguising the moral overtones embedded in the standard news report also "protects" the audience and is partly attributed to what W. Phillips Davison has called the "third person effect": the phenomenon of experts, intellectuals, and others believing that only *other* viewers and readers are influenced, affected, manipulated, and oppressed by mass media reports.[27] Michael Schudson adapts Davison to the view that modern experts implicitly believe in the maintenance of elite hierarchies of knowledge and ultimately that they know best.[28]

Therapy and Individualism In contemporary America, therapists, in part, serve as standard bearers for individualism, and all their modern spin-offs—including rational, well-informed reporters—take up the battle against bureaucracy, deviance, and the undermining of self. But increasingly, both professional postures in reporting and therapy maintain the jagged break between individuals and institutions. Rather than provide ways to repair this break, our contemporary narrative formulas intensify and solidify the gap as *natural,* as part of common sense. In drug news, for instance, by isolating interview subjects and focusing attention on individual characters and violations of Middle American norms, journalism does not point to an examination of the individual's standing within institutions in our age of bureaucracy. Nor does journalism usually help to clarify issues in which drug problems are not merely individual, but are attended by structural problems in our culture, such as the growing division between rich and poor exacerbated under Reaganomics.[29] As Susan Mackey-Kalis and Dan F. Hahn suggest, "As the 'just say no' campaign dialogue shifted the responsibility for social problems from the arenas of politics and medicine to morality, solutions to these problems became a matter of individual self-will and restraint generalizable to all areas of private morality."[30]

In a society dominated by the ideology of individualism, news as therapy finally supports the existing social order by facilitating our adaptation to "the way things are." With this prevailing centrist order accepted as a constant, as it is in many of the therapeutic drug stories we tracked on the evening news, TV reports frequently view problems only as individual ones that affirm or deviate from a taken-for-granted center. So the alleged apolitical "neutrality" or "balance" modern journalists often lay claim to merely disguises a covert centrist politics that champions American individualism.

In their critique of therapy and individualism, Bellah and his colleagues warn us about the limits of certain therapeutic contexts: "What is not questioned is the institutional context. One's 'growth' is a purely private matter. It may involve maneuvering within the structure of bureaucratic rules and roles, changing jobs. . . . But what is missing is any collective context in which one might act as a participant to change the institutional structures that frustrate and limit. Therapy . . . lacks any public forum."[31] Since some successful therapies, although focused on individuals or small individual groups, help people reintegrate into larger communities that they previously were cut off from, we do not want to make a blanket indictment that therapy has no relationship to a communal forum. But overall, the primacy of the individual and the secondary status of the public forum remain an important focus in the professional training of modern therapists. In *The Politics and Morality of Deviance,* Nachman Ben-Yehuda argues: "Therapists . . . are engaged in reintegrating patients into the *existing* symbolic-moral universes, not in producing social revolutions. . . . Therapists are not taught to be agents of social change but to facilitate personal change so that deviants would be reintegrated into conformist society."[32]

Our major criticisms, however, apply not to therapy (which is not our area of analysis) but to the ways that news exploits various aspects of the therapeutic ethos. Conventional news reports, operating as therapy tales

and depending heavily on medical expertise, celebrate a problematic feature of the therapeutic: the paramount standing of the individual and the
subordination of the social. In our study, the precrack therapy stories, in
spite of protests that journalists might make about their own balance and
neutrality, are ultimately about reintegrating "deviants" into "conformist
society." When Ben-Yehuda contends that "therapists are not agents of
social change," we can make that same assertion about modern journalists
whose task it is to stand morally apart from (or above) everyday experience, describe it in dispassionate terms, and make it into common sense.[33]
Although journalists pretend not to take moral stands, many clearly value
the morality of individualism.[34] This is what makes the therapy narrative
such an attractive formula for reporters covering the drug story in the
1980s. Reporters could refract the cocaine story—disconnected from most
of its historical, political, and economic contexts—through the lens of expert therapists; this endows a report with the legitimacy of news attributed
to experts and still honors the tenets of a single "reality" celebrated by
modern professionals. Instead of acknowledging multiple and competing
realities, both modern journalists and therapists salute the mythologies of
individualism as more "real" than social, political, collective, and alternative "realities."[35] The therapeutic narrative taps into modern journalism's
belief in the power of the lone reporter to tackle issues in the world and to
make sense for the rest of us. As Gaye Tuchman reminds us, "newswork
emphasizes the primacy of the individual. . . ."[36]

The most troubling aspects of therapeutic news forms involves masking larger economic and political dimensions. Murray Edelman notes in his
critique of therapeutic culture that invoking words such as "community,"
"counseling," "client," "treatment" or the phrase "helping profession," in
fact, disguises the careful hierarchical arrangement of power in the medical
profession.[37] A fairly substantial hierarchy of discourse exists for "clients"
of therapy at the bottom and medicine's top professional classes—those
who control decisions about who is balanced and who is not, who is a
candidate for therapy and who is a pathological misfit. As Ben-Yehuda
suggests, "we may conceptualize therapy as a form of social control and
therapists as control agents."[38]

Conventional journalism, too, with its primary focus on individuals
rather than underlying structures, often accepts the "social control" function of the medical-industrial complex as a given, as natural, as common
sense. And journalists, as "the custodians . . . of common sense," to use
Mary Mander's phrase,[39] lack an analytical, systematic framework for cri-

tiquing taken-for-granted institutional arrangements of power. Common sense, after all, repels self-scrutiny. Status quo values, economic structures, or political arrangements are often taken-for-granted in news stories as "the way things are." By its very opposition to analytic and self-reflexive examination, taken-for-granted common sense contains no abstract strategies for interpreting hierarchical arrangements or competing points of view and therefore certifies class and political (or, in this case, therapist/client and reporter/interviewee) divisions as natural and given. News as common sense, then, inadvertently maintains these divisions through an inability to assess both the ways that news reports are not like therapy and the ways that social hierarchies are *not* natural and given. According to John Fiske, those in power, whether in psychotherapy, journalism, or other modern professions, are able "to naturalize their social interests into 'ordinary' common sense" coded in the taken-for-granted language of the "helping professions" or the therapeutic formula of news narratives.[40]

The democratic potential of network TV news, then, rests in its ability to present therapeutic issues in the public forum of television. But journalism is limited here by its narrow vision of the public forum. Like most therapists, conventional journalists are not trained as agents of social change. In the spirit of modernity, they long ago subordinated their role as provokers of public debate to their function as disseminators of "neutral" information.[41] Reporters and their narratives about drugs typically define the public dimension of the 1980s' cocaine story only in terms of the personal idiosyncrasies of its characters—in terms of drug users and their individual problems. Rather, contemporary journalism ought to lead (or, at least, lead us to) a larger critical forum and public debate over who gets to demarcate the contested borders between balanced and unbalanced, legal and illegal, medication and intoxication, therapy and sin, prescription and addiction, morality and immorality, mainstream and marginal, order and chaos. We agree with historian Christopher Lasch: "What journalism requires is public debate, not information."[42]

The Therapeutic Narrative

Television journalism implicitly invokes the modern authority of therapy to structure and legitimate its drug news narratives. In this section we demonstrate the key narrative themes and patterns of the therapy genre in drug news that appeared regularly in the hundred network drug stories we analyzed from 1981 through 1986. In this textual analysis we examine the narrative strategies first in terms of their narrative characteristics, particularly the roles played by settings, reporters, experts, and interview subjects. Second, we investigate thematic patterns that appear throughout these types of narratives, particularly with regard to class and race.

The Hotline Chronotope   The treatment center, the courtroom, and the emergency room are all important disciplinary settings in the first phase of cocaine coverage. But perhaps the most revealing settings to emerge during this period are the command and control centers of various drug hotlines. One of the earliest of these services, 800-COCAINE, a national hotline headquartered in Summit, New Jersey, was the subject of a report broadcast by CBS on May 10, 1983. In his lead-in, Dan Rather laments, "You can dial a prayer in this country, or dial a joke. You can dial the latest sports news, or dial for a horoscope—but Jane Wallace reports that the latest telephone service reveals a deep sickness in our society." For our study, the hotline represents a postmodern therapeutic chronotope that, like the newsroom, is everywhere and nowhere.

The hotline setting, one example of the most advanced form of surveillance technology, has many disciplinary applications beyond its covert therapeutic intents. At one extreme, it has facilitated the self-incrimination of the cocaine confession; at the other, it has expedited the anonymous tattling of the cocaine informer. In television news discourse these hotlines were, until very recently, always treated as altruistic public services. However, what motivated these services was not "ideologically pure." The 800-COCAINE hotline, for instance, was sponsored by a medical consortium interested in identifying potential candidates with health insurance for expensive treatment programs. However, the ethical problems of a treatment referral business masquerading merely as a public service were exposed in October 1986—not by the networks—but by the trade publication *Advertising Age*. As Linda Cecere observes in her exposé, 800-COCAINE "became a model for direct response healthcare marketers of how to use the phone to mix public service with revenue generation": "Five trained counselors offer callers advice and referrals based on a directory of 4,000 health care facilities. The hotline, which costs $250,000 annually, logs an average 1,200 calls daily and as many as 2,200 when cocaine abuse is discussed on national news programs."[43] Therefore, in 1985 and 1986 something of a symbiotic relationship developed between news organizations and the hotlines (although this relationship would erode soon after the publication of Cecere's article).[44] During this period, apparently blind to the hidden agenda of these enterprises, the network news gave the hotlines uncritical exposure and legitimacy (what amounted to thousands of dollars in free advertising), and the hotlines gave the networks a new and regular source of often flawed "hard data" on the progress of the cocaine epidemic (see evening reports from ABC, 5/17/85; NBC, 5/31/85; CBS, 12/4/85; ABC, 5/27/86; CBS, 11/12/86).[45] Once again, the definitions of the cocaine problem proposed by an entity whose very existence depended on the production and reproduction of drug delinquency were accepted as face-value common sense by most television journalists.

Redeemable Offenders   The "characters" featured in the drug stories during the precrack period span a broad spectrum: from the seriocomic story (NBC, 3/29/83) of "Cocaine Grannies" (see analysis in chapter 8) to therapeutic "horror stories" featuring anecdotes relayed by experts about addicts who sell babies for cocaine or interviews with respectable businessmen who confess to trading wives for coke (ABC, 5/16/83). Another confessional scene features a middle-class nurse who testifies: "I had sex for drugs" (NBC, 8/10/84). More typical are the parade of middle-class folks who fill up these narratives: a white-collar worker from New York who ruined his family because of his addiction (CBS, 5/31/84), a man hooked on cocaine who forgets to shower (CBS, 5/31/84), a computer saleswoman who spent $20,000 a year for cocaine (NBC, 8/10/84), and a mother who snorted coke during her pregnancy (NBC, 8/10/84). Most of these people are white, relatively affluent at one time, now in some kind of formal therapy program where costs range from $1,000 for five-day programs to $6,000 for three-week programs (ABC, 5/16/83).

Although the therapy narrative features a variety and range of characters, one narrative profile of the "typical" cocaine user had developed by 1984, based on reports from the national hotline service: white male, thirty-one years old, some college, salary around $25,000 (CBS, 5/31/84).[46] In addition to this character profile, another trait of treatment stories features the juxtaposition of decadent taboos with violations of middle-class norms and social order. Both news reports and the therapists depicted in these stories join in their affirming efforts to reintegrate offending characters back into "normal" society. Typical is the impassioned confession of one former addict who characterizes cocaine addiction in terms of losing control over middle-class or mediating norms, particularly family and religion: "There's no control whatsoever. There's no control for anybody who does cocaine. I don't care who it is. Despite the fact that I've broken my wife's heart for years, and . . . I've missed a lot of my kids growing up, . . . despite all that and despite all the promises and all the swearing to God and all the praying, it just didn't stop." What makes these confessions particularly potent and useful for journalists committed to factual data is that the confessional narrative has been coded in modern society, according to White, as "an agency of truth."[47] As a result, in the cocaine narratives we analyzed, there was never any doubt expressed by the reporter (who had typically wrung the on-screen confession from a subject) that an interview subject's words were anything other than "the truth."

In modern society, Peter L. Berger and Thomas Luckmann remind us:

> Therapy entails the application of conceptual machinery to ensure that actual or potential deviants stay within the institutionalized definitions of reality. . . . Since therapy must concern itself with deviations from the "official" definitions of reality, it must develop a conceptual machinery to account for such deviations and to maintain the realities thus challenged. This requires a body of knowledge that includes a theory of deviance, a diagnostic apparatus, and a conceptual system for the "cure of the souls."[48]

News, then, is implicated in this therapeutic apparatus to the extent that it implicitly identifies and "theorizes" drug users as deviants from the truth and "reality" of social norms (e.g., family, religion), diagnoses them again in terms of how different they are compared to "normal" middle-class people, elicits from them the truth of a confession, and provides a system for "cure" either by displaying them in a formal group session (CBS, 7/13/81) or by portraying the reporter as therapist drawing pointed confessions from interview subjects (NBC, 3/29/83). "Successful therapy," Berger and Luckmann note, "establishes a symmetry between the conceptual machinery and its subjective appropriation in the individual's consciousness; it resocializes the deviant into the objective reality of the symbolic universe of the society."[49] The therapy narrative suggests that while many of the characters have lost their way, they are still redeemable as "one of us" by virtue of their current and former ties to "objective reality"—the truth of mediating, middle-class values.

While reformed addicts constitute a large group of players in therapeutic drug dramas (as do police who more prominently figure in the siege/gangster/pathology stories), another major performer is the medical expert, generally either a cocaine researcher, phone therapist (hotline counselor), or some other type of treatment mediator who also bears witness to middle-class norms. These experts largely serve to push the strategy of beneficent therapy in restoring moral order and social control through discourses of recovery.

News stories as therapy define normalcy and so represent rituals of inclusion; in turn, these narratives prepare the way for crack stories of exclusion in which certain inner-city populations who often refuse to confess or reform (i.e., gangs, crack-addicted mothers) are portrayed as pathologically unrepentant and therefore outside the reach of the therapeutic ethos. Both cocaine rituals parallel similar temperance movement narratives analyzed by Joseph Gusfield in *Symbolic Crusade*. During the nineteenth century most heavy drinkers were constructed as "troubled people" or "repentant deviants," who did not contest the antialcohol norm of native-born, small-town, and typically middle-class WASPs, and were reintegrated into the society by means of "assimilative reform." But later, during the early twentieth century, the massive immigration of working-class, immigrant, urban Catholics brought a different attitude toward drinking for which many refused to apologize; they were subsequently coded as pathologically unredeemable "people who create trouble" or "enemy deviants" subject to the strict and "coercive reform" of Prohibition.[50]

The Shadows of Race and Class

In terms of thematic import, strong attention to class distinguishes this 1980s' therapy phase of drug stories (as opposed to the role that race plays in the later pathology/siege crack stories). The most telling class metaphor of this particular genre of story is again the Reagan trickle-down analogy. Here we see cocaine use moving top-down from the more elite executive, professional, entertainer class (e.g., John DeLorean, John Belushi, Richard Dreyfuss, Stacy Keach, Peter and Roxanne Pulitzer) to middle- and working-class America. In early February 1983 anchor Dan Rather outlines the terms of the cocaine shift: "Cocaine is widely perceived as the drug of choice for celebrities including professional athletes and show business luminaries. But dangerous living can be contagious." Of course, Rather indicts the culture of excess and locates himself (despite his then $2 million a year salary)[51] in the camp of commonsense Middle America and against "dangerous living." Later in the same story, expert psychologist Philip Herschman applies the Reaganomics metaphor to drug use: "It's trickling down. And more and more people are using it. It's not uncommon now to have the average blue-collar worker in a factory using cocaine on a regular basis" (CBS, 2/7/83). The central theme of these story types maintains that upper-class professional decadence has contaminated the rest of Us (i.e., white Middle America). Reports document cocaine use, once deemed "a private plaything for the privileged" (CBS, 5/16/83), now "soaring among just plain folks" (CBS, 2/7/83). By late 1984 CBS calls cocaine "an underground epidemic attacking America's middle class" (11/27/84). The precrack trickle-down movement seems most complete in this Harold Dow report from Washington, D.C.: "Cocaine used to be a drug of the rich. But it is now called by the experts the drug of choice of middle-class Americans—that includes America's children" (CBS, 11/28/84).

Intriguingly, the racial dimensions of these precrack drug narratives are masked and then superseded by class dimensions. A startling example comes from a March 1984 CBS report describing New York's "Operation Pressure Point," the arrest of 2,700 "dope dealers and addicts" on the Lower East Side, and the overflow conditions at local clinics. Invoking a catastrophe metaphor that dehumanizes addicts as natural (and national) disasters, Chris Kelly concludes his report: "Despite the flood of addicts engulfing the clinics, the city is expanding 'Operation Pressure Point' into Harlem where the problem is even worse." The implication: as long as the problem appears as a class rather than race problem, the news can stop outside the gates of Harlem even though (if we believe CBS in 1984) for black urban America the impact of drugs is "even worse."

Because most of them are white, reporters generally are more comfortable dealing with class rather than racial issues.[52] Presumably, it is easier to achieve the coveted ideal of journalistic "balance" in stories that foreground class. Staking out a middle—or balanced—position offers a double coding for reporters who celebrate sanity and common sense at the same time that they claim to be politically neutral. It is a claim for the middle of the road, which is, after all, a political locus safely between the excess of the decadent upper class and the pathology of the sinister lower class. It also is easier to mask the racial dimensions of a news report when people of color are coded by their "lower" position in the socioeconomic hierarchy rather than explicitly by race.

The claim for the middle also disguises an economic strategy. It is, after all, in the financial interest of modern mainstream news organizations to stake out the middle ground. Such identification does not risk alienating large portions of the population who buy daily newspapers and watch evening newscasts. William Greider, a former *Washington Post* editor, makes the connection between good business and balanced journalism: "If you're going to be a mass circulation journal, that means you're going to be talking simultaneously to lots of groups that have opposing views. So you've got to modulate your voice and pretend to be talking to all of them."[53] Jon Katz, a former CBS News producer turned media critic for *Rolling Stone*, positions this idea historically:

> The idea of respectable detachment wasn't conceived as a moral principle so much as a marketing device. Once newspapers began to mass market themselves in the mid-1880s, after steam- and rotary-powered presses made it possible to print lots of papers and make lots of money, publishers ceased being working, opinionated journalists. They mutated instead into businessmen eager to reach the broadest number of readers and antagonize the fewest. . . .
>
> Objectivity works well for publishers, protecting the status quo and keeping journalism's voice militantly moderate.[54]

Tom Yellin, an original producer of ABC's "Nightline," extends the argument to network TV news professionals: "The political sensibilities of people in network television are mainstream, traditional and conservative; neither far left nor far right. . . . We share the same basic assumptions of bankers, lawyers and the rest of the establishment."[55]

The therapeutic dimension, like the journalistic posture, honors the detached professional who observes experience and makes allegedly balanced, value-free judgments about the nature of drugs in American life. David Eason contends that contemporary journalism supports the modern dualistic distinction between *observed* experience—the reporter as detached and balanced onlooker—from *lived* experience: "To become an observer is to see social reality as composed of active participants, who must take responsibility for their acts, and passive spectators, who bear no responsibility for what they watch."[56] In most of the network therapy narratives, reporters draw a boundary between observing and living. They reconstruct experience—"tell the story"—but it is *other* characters in the narratives who bear the responsibility for *actual* experience. Conventional journalism and the ideal of balance, as Michael Schudson argues, operate as "concrete conventions which persist because they reduce the extent to which reporters themselves can be held responsible for the words they write" (or the pictures they broadcast).[57] The therapeutic narrative, then, remains a powerful strategy for organizing experience since it allows reports and reporters a venue for masking their neither-far-left-nor-far-right centrist politics. Reporters both invoke voices from the therapeutic professions at the same time that they become balanced pseudo-therapists themselves in their stories,[58] comforting us in their closing on-screen stand-ups about what has to be done to solve the "drug problem" and what addicts have to do to become more "like us."

The common thread that stitches together the therapeutic narratives of the early 1980s, the Hollywood scandals of the 1920s, and the nineteenth-century temperance movement is the potential for mending those of Us—tinted white and middle class—who have become part of a deviant and decadent social fabric. But when white powder becomes hard rock, when mainstream journalism discovers the crack in the fiber in late 1985, television news narratives take a dramatic turn. Stories once tinted by class become stained by race. And as the therapeutic turns to the pathologic, news about recovering Us shifts to tales about incriminating sinister Others.

6

The Siege Paradigm:

Rewriting the Cocaine Narrative

At the symbolic level, social control must fulfill the functions of creating scapegoats, clarifying moral boundaries, and reinforcing social solidarity. The primeval form of scapegoating directs aggression towards individuals not responsible for the group's frustration. . . . In the ancient Hebraic ritual, the high priest puts his hand on the goat, confesses injustices and sins, then the animal is driven out into the wilderness. This is to leave people feeling purified and solidified. These functions remain when societies move towards putting blame not on the community or its "arbitrary representative" but on certain individuals who are then properly caused to suffer.—Stanley Cohen [1]

War is, of course, the health of the networks, and of their promotion departments. Scenes from the battlefront play especially well. The war on drugs produces bang-bang and the most vivid pictures—customers cruising the crack houses, trembling crack babies, cops going in for the bust as the hand-held camera shakes along behind. DRUGS, reads the four-and-a-half-inch headline of CBS's full-page ad of September 6 [1989] pushing a week of special reports on the drug story. ONE NATION, UNDER SIEGE. "America is at war," the text declared. "And the enemy has already taken the streets."—Todd Gitlin [2]

Paradigm Shift

According to precrack drug news, the drugs "of choice" for the subordinated groups in America's economically ravaged inner cities were "T's and Blues" (ABC, 4/20/81), PCP (CBS, 8/15/84), "Wack" [3] (CBS, 1/23/86), and "Black Tar" heroin (NBC, 3/28/86). Part of the sense of hysteria that accompanied the introduction of crack was its violation of this system of illicit taste distinctions. A pleasurable substance that was once the province of the wealthy had finally trickled down to the lowest levels of the American socioeconomic racial order. In this violation of taste distinctions, cocaine would not only lose some of its value as a status symbol, but it would become associated with a mode of drug abuse that is more a matter of desperation than recreation. In the symbolic inversion, the preferred purifying solution to the cocaine problem would shift from the therapeutic branch of the medical-industrial complex to its armed disciplinary forces.

Turning Point   On television the journalistic crack frenzy would not begin in earnest until the fifth month of 1986. According to the *New York Times*'s Peter Kerr, a "turning point" in the coverage occurred on May 18. On that momentous Sunday, three of New York City's newspapers (the *Times,* the *Daily News,* and *Newsday*) featured lengthy articles on the crack trend.[4] Within ten days the three national networks followed suit: NBC's Dennis Murphy filed his copycat crack story on Friday, May 23; ABC's John McKenzie and CBS's Harold Dow filed their reports on the following Tuesday, May 27.

Murphy's report features experts from both the law enforcement and therapeutic branches of the medical-industrial complex. Arthur Nehrbass, speaking as a representative of the Dade County, Florida, police, claims that a half-dozen uses will turn a person into a crack addict; and Dr. Arnold Washton, one of the chief definers of the cocaine problem during this phase, speaks as director of a drug hotline in using the metaphor of Russian roulette to describe crack use. Murphy's report ends up as something of a science-fiction horror story with the crack house depicted as a terrifying and alien setting: a place where children are held as collateral; "a modern-day opium den"; "a filthy room with a steel door and armed guards." And the crack user is treated as an alien Other on the order of a space invader. In fact, the report concludes with a scene of a black male taking a hit of crack and exclaiming, "Beam me up to the Enterprise."

On CBS, Harold Dow (*the* chief reporter of the cocaine problem) fashions a report that features the performance of an expert who qualifies as a chief definer in both Phase II and Phase III coverage: Robert Stutman of the DEA. Significantly, Dow opens his story by addressing the audience with the second-person singular—"You." This "You" address is an aspect of the campaigning voice of the crusading journalist—the self-righteous reporter engaged in a clear-cut, good-vs.-evil (and politically safe) mission. In adopting this voice, it is hard to tell where Dow's narration stops and

Stutman's sound bite begins. Consider the ideological fusion of Dow the reporter with Stutman the DEA agent in the first moments of the report:

DOW: This is it. The drug so powerful it will empty the money from your pockets, make you sell the watch off your wrist, the clothes off your back . . .
STUTMAN: Or kill your mother.

Stutman's "atrocity tale" is reminiscent of the horror story circulated by Harry J. Anslinger in the 1930s concerning the dangers of marijuana ("An entire family was murdered by a youthful [marijuana] addict in Florida. . . . With an ax he had killed his father, mother, two brothers, and a sister" [see chapter 3]).[5] But aside from the alarmist sentiments of these words, this fusion of reporter narration and expert testimony stands as an early example of the kind of explicit journalistic identification with law enforcers that exemplifies the pun in the title of our book. In blunt terms, Dow's report is a clear case of what we want to critique as "cracked coverage."

So is John McKenzie's report on ABC—but for different reasons. McKenzie also calls on Stutman and Washton (recurring experts who Jonathan Alter might call the "usual suspects").[6] But what sets McKenzie's report apart as being particularly troubling is its wrongheaded rewriting of the history of cocaine pollution established in Phase I. At one point in the report, McKenzie states that "what was once a ghetto problem has moved into the suburbs." This ahistorical historical observation plays over a dramatic tracking shot that begins on a mean inner-city street and dissolves into a neat suburban setting. For us, this clever dissolve covers a major case of journalistic *amnesia* regarding the class dimensions of cocaine pollution during the early 1980s—amnesia that *replicates* the ahistoricism and victim-blaming of the new racism.

The Crack House Chronotope   One of the actions that facilitated this paradigm shift in the news framing of the cocaine narrative was the journalistic discovery and demonization of a deviant setting that would become depicted as nothing less than a locus of evil in our culture—the crack house. Where Hollywood holds a positively charged space in American culture, the crack house is a negatively charged chronotope—a threatening place of assembly and enterprise where entrepreneurial spirit and the ideology of consumerism (that are so central to Reaganomics) are pushed beyond the limits of decency, of "good taste," of social and moral control.

One way of understanding the symbolic power of the crack house is to contrast it with a vital "site of assembly and space of discourse" in seventeenth- and eighteenth-century Europe—the coffeehouse. According to Peter Stallybrass and Allon White, the coffeehouse was a social setting that helped forge the bourgeois identity that was taking shape during this period.[7] Around 1650 the coffeehouse, which was a cultural feature of the Middle East, became a prominent social institution throughout Europe. Indeed, Stallybrass and White discovered that, less than fifty years after its introduction to England, there were more than two thousand coffee-houses in London alone. In contrast to the tavern, the coffeehouse was a place of sobriety and productivity, of "democratic" accessibility and moral "decency":

> The new coffee-house was a heterodox, contradictory place which provided a mediation between domestic privacy and the grand public institutions of busi-ness and the state. At the same time whilst it stoutly and successfully resisted the interventions and interferences of the State, it was an important instrument in the regulation of the body, manners and morals of its clientele in the public sphere. When it first appeared it was immediately seized upon by protestants as a counter-force to the tavern and the alehouse. The early coffee houses sold no alcoholic beverages and indeed they were initially defined in clear opposition to the tavern. Every coffee house had a list of rules posted and under strong protestant influence the rules (of which copies survive) included no swearing, no profane scripture, no cards, dice or gambling, no wagers over five shillings, no drinking of health. As distinct from the tavern, inn and alehouse, these were *decent* places to go.[8]

The coffeehouse, then, was a *regulated* setting. It was, unlike the tavern and alehouse, a place where people were under control, where pleasures were disciplined. Unlike cocaine in the 1980s, coffee was widely celebrated as a "wakeful and civil drink."[9] In fact, it was treated as a wholesome alternative to ale, beer, or wine, an alternative that, in Stallybrass and White's words, had value "as a new and unexpected agency in the pro-longed struggle of capitalism to discipline its work-force."[10] Ultimately, Stallybrass and White conclude that the coffeehouse "played a central role in the formation" of early modern "bourgeois institutions and manners":

> Its great claim to superiority over the alehouse was that it replaced "idle" and festive consumption with *productive* leisure. . . . They were systematically and subtly retexturing the discourse of the alehouse and other public places of as-sembly to accord with the goals of professionalization, productive economy and "serious knowledge." Jokes, chatter and gossip were flanked by serious, ordered meetings of both an academic and commercial interest. . . .[11]

For us, the crack house represents a kind of postmodern shadow image of the coffeehouse setting. It is first and foremost an *unregulated* place of *undisciplined* pleasure. The people who frequent the crack house are often described, even in their own public confessions, as "out of control." But perhaps the most damning thing about the crack house is its association with an underground economy, a "black" market, that embraces the central principles of capitalism, most notably the profit motive, while also undermining the legitimate economy's long-standing efforts to "discipline its work-force." As Todd Gitlin observes, American culture is "a drug culture":

> Through its normal routines it promotes not only the high-intensity consumption of commodities but also the idea that the self is realized through consumption. It is addicted to acquisition. It cultivates the pursuit of thrills; it elevates the pursuit of pleasure to high standing; and, as part of this ensemble, it promotes the use of licit chemicals for stimulation, intoxication, and fast relief. The widespread use of licit drugs in America can be understood as part of this larger set of values and activities.[12]

In a roundabout way, by giving concrete form to the limits and the contradictions of "this larger set of values and activities," the crack house plays an important role in the maintenance of contemporary bourgeois institutions and manners. In the crack house setting, we see the despair, the exploitation, and the perversity of capitalism writ large: consumerism over the edge; the callous marketing of death; the filth and waste of greed. Or Reaganomics out of control. Journalism's obsession with the site of the crack house as "abnormal" or deviant space, then, masks the actual ways that the crack house mimics the "normal" and routine business workings of capitalist enterprise.

The vilification of the crack house as a sinister place of assembly has, of course, justified the most brutal and excessive of armed responses by the forces of decency and control. In these purifying expeditions into the chaos and filth of the crack house, the network news frequently encourages strong audience identification with the invading police. In such reports the journalist often, literally, adopts the "outlook" of the police—a perspective that, in the context of the inner city, is perhaps best described as the colonizer's point of view on the colonized.

The Policing Apparatus   This angle of journalistic identification often featured an aspect of "the discourse of discrimination" that would become a prominent feature of visualizing deviance during the 1986 crisis—that is, the use of the *stigmatizing clandestine camera*. In news discourse the use of a clandestine camera is almost always associated with rites of exclusion. The clandestine camera marks the transgressor under surveillance as an alien Other who does not have the same rights to "privacy" accorded the ordinary citizen. For this study,[13] we distinguish between two types of clandestine footage—*independent* and *implicated*. The former is footage gathered by news organizations without the cooperation of policing authorities; in most cases, this type of footage uses a hidden camera to record open drug transactions on the streets of New York City, Miami, Oakland, among other large U.S. cities. Implicated footage, on the other hand, is material that can be gathered only with the cooperation of law enforcement organizations; in some cases this footage involves journalists taking cameras along on police sting operations and recording the action; in other cases the footage may actually be generated by government surveillance cameras and then incorporated into the reporter's news package (the most familiar example of this is perhaps the incriminating undercover government footage of Mayor Marion Barry smoking crack in a Washington, D.C., hotel room).

The percentage of cocaine stories featuring clandestine footage rises over the course of the 1980s. In our video sample of Phase I coverage, only about one story in four (eleven out of forty-two) would feature such footage—but in 1987 and 1988 such footage appears in more than half the cocaine stories (twenty-seven of forty-eight) that we analyzed. However, the most striking development in stigmatizing camerawork occurred during the 1986 crisis coverage—that is, the emergence of the *raiding footage* of a hand-held camera accompanying police during the invasion of a crack house. This raiding footage does not appear at all during the first stage of cocaine coverage. But beginning with a report aired by ABC on July 28, 1986, it would be a titillating feature of several crisis updates (see, for example, ABC, 8/5/86; CBS, 8/7/86; NBC, 8/10/86; CBS, 9/10/86; and ABC, 9/18/86). During this period, according to Robert Stutman, it was easy to enlist journalists in the war on drugs because crack was "the hottest combat-reporting story to come along since the end of the Vietnam War." [14] In the postcrisis coverage, as raiding footage became incorporated into the "image bank" available for recycling in any network news story about cocaine, it became something of a visual cliché, appearing in about one of every four stories in our video sample (see ABC, 1/10/87, 3/9/87; CBS, 3/16/87; NBC, 11/4/87; CBS, 4/26/88, 5/2/88, 5/12/88; CBS and NBC, 5/18/88; CBS, 6/3/88, 8/30/88; NBC, 11/27/88).

For us, raiding footage represents nothing less than the convergence of the reportorial outlook with the policing point of view. In this convergence we see what Richard Ericson, Patricia Baranek, and Janet Chan characterize as a shift in press/police relations from a "reactive approach" to a "proactive strategy." Where the traditional reactive approach entailed the police "either defending their actions when questioned or simply enclosing on knowledge," the police now recognize that a proactive strategy "is useful in controlling the version of reality that is transmitted, sustained, and accepted publicly":

> the police have come to appreciate that the news media are *part of* the policing apparatus of society, and can be controlled and put to good use in this respect. The news media are incorporated into the architecture of new police buildings (they are given newsroom facilities there), they are taken into account in police organizational charts, they are subject to the regulations in police operation manuals, and they are part of everyday practice at all levels of the police hierarchy.[15]

The proactive strategy would pay off in the 1980s war on drugs in favorable coverage of high-profile drug campaigns choreographed for the television cameras by police organizations in San Francisco (CBS, 8/17/81), San Diego (ABC, 3/28/84), New York City (CBS, 8/4/86 and 3/18/88), Miami (ABC, 7/28/86; NBC, 8/10/86 and 11/4/87), and New Jersey (CBS, 9/10/86). There was even saturation coverage of police action on Wall Street (ABC, CBS, and NBC, 4/16/87). But all of this proactive police work would pale beside the media coup scored by former L.A. Police Chief Daryl Gates on April 6, 1989—a day that Mike Davis thinks will perhaps "go down in history as the first 'designer drug raid' ":

> As heavily armed and flak-jacketed SWAT commandoes stormed the alleged "rock house" near 51st and Main Street in Southcentral L.A., Nancy Reagan and Los Angeles Police Chief Daryl Gates sat across the street, nibbling fruit salad in a luxury motor home emblazoned "THE ESTABLISHMENT." According to the *Times*, the former first lady "could be seen freshening her make-up" while the SWATs roughly frisked and cuffed the fourteen "narco-terrorists" captured inside the small stucco bungalow. As hundreds of incredulous neighbors ("Hey, Nancy Reagan. She's over here in the ghetto!") gathered behind police barriers, the great Nay-sayer, accompanied by Chief Gates and a small army of nervous Secret Service agents, toured the enemy fortress with its occupants still bound on the floor in flabbergasted submission. After frowning at the tawdry wallpaper and drug-bust debris, Nancy, who looked fetching in her LAPD windbreaker, managed to delve instantly into the dark hearts at her

feet and declare: "These people in here are beyond the point of teaching and rehabilitating." This was music to the ears of the Chief, whose occupation thrives on incorrigibility. Gates fairly beamed as television cameras pressed in: "We thought she ought to see it for herself and she did. . . . She is a very courageous woman."[16]

Our analysis suggests that, long before Nancy Reagan's conversion to Gates's point of view, network news coverage of the 1986 crack crisis had become "part of the policing apparatus," marking off poor, black drug transgressors as "beyond the point of teaching and rehabilitating." After May 1986 the spread of cocaine was no longer primarily constructed as a cautionary tale of rot at the top or of middling individuals making bad choices. Instead, cocaine pollution animated siege narratives in which a color-coded mob of dehumanized inner-city criminals threatened the suburbs, small towns, schools, families, status, and authority of (white) Middle America.

The Len Bias Death Scandal

In chapter 5 we suggested that Hollywood holds a special place in twentieth-century American culture as a kind of utopian setting where dreams become reality. Here, we want to extend that discussion to a consideration of another utopian setting in contemporary America—the "world of sports." Although the world of sports obviously does not exist in a fixed and isolated geographical location that can be plotted at one spot on the map, it can still be analyzed as a ritual space that shares many of Hollywood's symbolic features. Like Hollywood, the world of sports is a place of individuality, opportunity, risk-taking, and freedom: a place where personal achievement is presumed to be a matter of talent and luck, not class or breeding; a place where hard work and excellence are rewarded with glory and fame. And in the 1980s the world of sports also would become a hotbed of scandal.

The sports arena, in fact, has long been a ritual space where the ideals of a people are made manifest in the rules of the game and their application to stylized human activity. As Clifford Geertz suggests in his famous "thick description" of the Balinese cockfight, sport renders "ordinary, everyday experience comprehensible":

What it [the cockfight] does is what, for other people with other tempera-
ments and conventions, *Lear* and *Crime and Punishment* do; it catches up
these themes—death, masculinity, rage, pride, loss, beneficence, chance—and,
ordering them into an encompassing structure, presents them in such a way as
to throw into relief a particular view of their essential nature. . . . An image, a
fiction, a model, a metaphor, the cockfight is a means of expression; its function is neither to assuage social passions nor to heighten them (though, in its
playing-with-fire way it does a bit of both), but in a medium of feathers, blood,
crowd, and money to display them.[17]

Geertz's description of the expressive power of the sporting event provides
an interesting contrast to the ritual function of the sports scandal. Whereas
the sporting event is almost always a conservative display and hearty affirmation of moral beliefs and ideological values that organize the (generally
masculine) common sense of a culture, the sports scandal represents a rupture of those values and beliefs and a challenge to the established order.
When a greedy owner effectively confiscates publicly supported property
by moving a beloved franchise to another city, when a baseball executive
suggests that people of color do not have the intelligence to manage in
the major leagues, when a university must disband its football program
because of institutionalized cheating, the culture at large is forced to reconsider and renegotiate values associated with the ethic of success and the
"free" market economy. In other words, precisely because of the conservative character of sports, a scandal within its domain represents an ideological rupture, a moment when people confront cultural contradictions
that are normally masked by the routines of everyday life.[18]

Therefore, we suggest that the political economy of scandal in the
world of sports has a double edge. In keeping with the cautionary tradition of Richard Cory, it is clear that scandal rituals often give comfort
to the oppressed by encouraging have-nots to seek happiness with their
lot and demonstrating that no one is above the law. As we shall see, like
the deaths of Hollywood stars Wallace Reid (in the 1920s) and John Belushi (in the 1980s), the scandal-coded demise of Len Bias also mobilized
the sins of the famous in the service of the regulatory imperative—in this
case, by the National Collegiate Athletic Association (NCAA) rather than
the MPPDA (see chapter 5). But on the other edge this interpretive focus
on domination and discipline neglects other less hegemonic aspects of the
political economy of scandal. For us, in exploring the triumphs, hazards,
core values, and moral limits of individualism, the cautionary scandal can

at times do much more than simply maintain the status quo and rationalize further normalization and regulation—that is, it also can be a powerful symbolic vehicle for challenging and transforming what it means to be successful, what it means to live the good life, and what it means to share in the American Dream.

During the 1980s a series of scandals associated with cocaine pollution traumatized the world of sports. In football, the Dallas Cowboys, the New Orleans Saints, and superstar Lawrence Taylor would be caught in the center of drug scandals (see ABC, 6/30/82; NBC, 8/25/82; CBS, 7/15/83; NBC, 10/2/85; ABC, 9/02/88). In baseball, the Pittsburgh Pirates, pitchers Steve Howe, Dwight Gooden, and a dozen or so other players—also would become objects of scandal (see CBS and NBC, 6/27/83; ABC, CBS, and NBC, 10/17/83; NBC, 5/7/85; CBS, 5/31/85; CBS, and NBC, 9/5/85; ABC, CBS, and NBC, 9/6/85; NBC, 9/10/85, 9/11/85; CBS, 9/20/85; NBC, 9/24/85; ABC, CBS, and NBC, 4/1/87). Basketball, too, had numerous players suspended, placed in rehabilitation programs for positive drug tests, and charged with drug crimes (see CBS, 4/17/87; NBC, 5/17/87).

However, none of these scandals would approach the uproar that established Len Bias as a posthumous star of the evening news. On June 17, 1986, within a month of the May turning point in the cocaine coverage, the Boston Celtics chose the University of Maryland basketball star as the National Basketball Association's second overall draft pick. Within forty-eight hours of the Celtics' decision, Len Bias died of heart failure believed to be triggered by cocaine intoxication. Tragic and stunning, the timing and circumstances of Bias's death would have dire symbolic consequences—consequences that would make the fatality, without a doubt, the single most important kernel event in the cocaine narrative during the 1980s.

In extreme cases a scandal in the world of sports may develop into what anthropologist Victor Turner calls a "social drama"—a cultural vehicle for, in Turner's words, "manifesting ourselves to ourselves and, of declaring where power and meaning lie and how they are distributed": "As society complexifies, as the division of labor produces more and more specialized and professionalized modalities of sociocultural action, so do the modes of assigning meaning to social dramas multiply—but the drama remains to the last simple and ineradicable, a fact of everyone's social experience, and a significant node in the developmental cycle of all groups that aspire to continuance."[19] Like the rape case of heavyweight boxer Mike Tyson and the AIDS saga of professional basketball's Magic Johnson, the Bias death and its aftermath transcended the scale of the routine sports scandal and displayed all of the characteristic stages of a full-blown social drama: *Breach, Crisis, Redress,* and *Reintegration* or *Separation.*

Breach To organize our analysis of the Bias scandal, we rely heavily on James S. Ettema's application of Turner's work to the study of news as ritual. According to Ettema, the breach stage of a social drama is a moment of rupture "in which some norm, law, or custom is violated in a public setting and in a way that challenges entrenched authority. . . ."[20] Although Bias's actual use of cocaine was not a violation that occurred in a public setting, the Bias drama still meets this description because the death scene was a public spectacle. Many news reports featured cover footage, taken outside an apartment complex, of curious bystanders watching Bias being loaded into the back of an ambulance.

As Foucault suggests, in modern societies where power is exercised by controlling, disciplining, and regulating life, death is the ultimate transgression, the ultimate challenge to entrenched authority. As a living superathlete, Bias was the epitome of the productive, disciplined body, the personification of the competition ethic, and the prototypical triumphant individual. In death, though, Bias would become something of a doppelgänger of his living self—a decaying example of the wasted, undisciplined body, a symbol of how winning at all costs has defiled college sports, and an emblematic representative of "the mob" who perished at the gate on the threshold of becoming one of Us. Bias's status as a "liminal figure" (an "in-between" person who was terminated on the brink of a radical transformation in his cultural status and financial stature) was an absolutely decisive factor in generating the sense of crisis that followed his death/breach/transgression.

Crisis As Ettema observes, in the crisis stage of a social drama "antagonisms become visible and factions form along enduring social fault lines."[21] In news coverage, the Jekyll-Hyde doppelgänger effect—the sense that Len Bias was suspended between two identities, one good and one bad—became the principle that made these "antagonisms" visible. As the crisis developed, one figure would stand in symbolically for the bad Bias and another figure would stand in for the failure of white, male authority. Serving as a narrative surrogate for the bad Bias, Brian Tribble was Bias's "home boy," a black friend from the Washington, D.C., neighborhood where Bias grew up. Accused of supplying the drugs that took Bias's life, Tribble would come to embody the racial, generational, and economic components of the drama's conflict and, therefore, expedite the rewriting of the cocaine narrative from the trickle-down to the siege paradigm. The other transgressing figure was Lefty Driesell, Bias's head basketball coach at the University of Maryland. Driesell was accused of neglecting his athletes' "basic education" and even tolerating drug abuse on the team. As a surrogate for the failure of white patriarchy, Driesell would come to embody the contamination of college athletics by the pursuit of big money and the antiethic of winning at all costs.

The contrasting characters of Tribble and Driesell, then, gave physical forms and human identities to the "enduring social fault line" excavated by the Bias death drama: on one side of the line, the mob of black youth, out of control and in need of discipline; on the other side, white authority debilitated by conflicting interests (commercial viability vs. professional integrity) and competing goals (championships vs. diplomas).

In news reports during the crisis stage the incongruities about the meaning of Bias's life and death often were defined as eruptions of nonsense most foul. Of this coverage, ABC's "Person of the Week" feature, which aired June 27, is a stellar example of journalism trying to make sense of non-sense. In fact, Peter Jennings begins by posing the question: "What does it mean?" The opening narration then goes on to acknowledge, in commonsense language, Bias's status as a liminal, in-between figure and his emergence as a posthumous star of the drug news: "What does it mean? Why does a young man, right on the verge of realizing his dream, risk it all on a date with a killer? Is there a lesson worth telling and retelling? The bare facts are familiar. He had worked so hard to get there. . . . And so in the twinkling of an eye, this person of the week becomes a symbol. A symbol of what?" To explore the person of Bias as a symbol, the report includes sound bites from three experts and one prominent African American newspaper columnist—and all take a slightly different angle on the social drama.

In the ABC report Dr. Steven Feldman speaks as a medical authority who hopes that the "impact of Len Bias's sudden death will be the most strong on young athletes and fans . . . to give them increased cautiousness and conservatism not to try the drug, even for the first time." William Raspberry of the *Washington Post*, though, is not as hopeful. Raspberry's sound bite is positioned as a dialogic response to four brief rhetorical questions expressed in Jennings's voice-over narration.

JENNINGS (v.o.): Unfortunate jock? Or tragic figure? Or both? And will anything really be salvaged from this tragedy?

RASPBERRY: I'm afraid there's not much meaning in this one. It's just brutally sad, desperately tragic, but I don't think there's much meaning there.

Raspberry's denial of meaning, of course, affirms the death as nonsensical.

With this affirmation, Jennings then switches the focus from the personal to the systemic by fashioning an indictment of the excessive emphasis on winning and money in college basketball: "And was he totally to blame? What about the system that appeared to care more about his baskets than about his basic education? He was allowed to be virtually an academic year behind when he was supposed to be in the graduating class. Twelve of the 18 college seniors selected in the first round of the NBA draft did not have diplomas." Jennings's transition is followed by a sound bite from sports sociologist Richard Lapchick who, in fact, treats money as a pollutant in sports: "What we're seeing is that the money has got so tremendous in college sports that too many schools are willing to cut corners. The mythology has become that you can't educate your athletes at the same time." But Lapchick's sentiments are then muddied by Jennings, who switches the focus off economics back to the personal in a confusing lead-in to a sound bite from Harry Edwards, a well-known African American sociologist at UC-Berkeley:

JENNINGS (V.O.): The pressure has led more than one to drugs.
EDWARDS: The problem is exacerbated in sports because you have a situation where individuals are all too often hand-carried through life.

In this collision of seemingly contradictory observations, it is not clear whether athletes need more pressure placed on them to succeed on their own without being "hand-carried through life," or whether such added pressure would encourage even more drug consumption.

The ease of this discursive shift to a general indictment of "the system" is facilitated by several hidden and questionable assumptions about the root cause of the breach. Chief among the assumptions is the unstated assertion that more "basic education" would have somehow prevented the tragedy. In this assertion we can begin to see the operation of the political economy of scandal. Just as in the Hollywood scandals of the 1920s and 1980s, Bias's dramatic death became the vehicle that helped revitalize and validate calls for the implementation of new regulations and new normalizing procedures. In concert with other scandals involving African American players like Chris Washburn, Kevin Ross, and Hart Lee Dykes— and white coaches at the time like North Carolina State's Jim Valvano,

Kentucky's Eddie Sutton, and the University of Nevada at Las Vegas's
Jerry Tarkanian—the Bias/Driesell transgression would be taken by the
NCAA as justification for introducing Proposition 42, which was designed
to establish severe new restrictions that were primarily meant to police the
recruitment and eligibility of minority athletes.[22] Therefore, in this shift
to the systemic we can perceive both edges of the political economy of
sport scandal: in Lapchick's incrimination of money as a pollutant in the
world of sports, we see the expression of a utopian impulse that challenges
the status quo—an impulse that runs counter to the entrepreneurial ethic
of Reaganism; but Lapchick's diagnosis is quickly undercut by Edwards's
indictment of permissiveness—a charge that not only rationalizes further
normalization and regulation by the control culture, but that reinforces the
moralism of the New Right.

In fact, Edwards (who is a former minister of propaganda for the Black
Panther party and achieved national prominence as the organizer of the
famous Black Power protest of the 1968 Olympics) has become a central
figure in what Mike Davis calls "Black-lash"—a "qualitatively new and
disturbing dimension of the war on the underclass."[23] Consider, for instance, an interview with Edwards published in *San Francisco Focus*. After
Edwards, who is now a highly paid consultant to professional sports, suggested that, "the reality is, you can't" save a "13-year-old kid selling crack
in the streets," the interviewer asks, "So then what?"

EDWARDS: You gotta realize that they're not gonna make it. The cities, the culture,
and Black people in particular have to begin to move to get that garbage off
the streets.
S.F. FOCUS: How?
EDWARDS: It means we have to realize that there are criminals among us and we have
to take a very hard line against them, if we're to preserve our next generation
and future generations. Even if they are our children.
S.F. FOCUS: So what do you do if you're a parent and you discover your 13-year-old
kid is dealing crack?
EDWARDS: Turn him in, lock him up. Get rid of him. Lock him up for a long time.
As long as the law will allow, and try to make it as long as possible. I'm for
locking 'em up, gettin 'em off the street, put 'em behind bars.[24]

Therefore, Edwards exemplifies the swelling support of African American
leadership for the stigmatization and even the dehumanization of black
urban youth caught up in the crack economy.

After Edwards's sound bite in the "Person of the Week" report,

Jennings moves to his concluding statement, which, in many profound ways, reverberates with the cautionary tale of Richard Cory (see chapter 5):

> The trouble is we do not know the answer to so many questions. The same questions people asked when John Belushi died, and Jimi Hendrix, and Janis Joplin, and the other idols of a generation. We wonder if this young man and others have ever read a poem by A. E. Housman called, "To an Athlete Dying Young"—

> The time you won your town the race
> We chaired you through the market-place.
> Man and boy stood cheering by,
> And home we brought you shoulder-high.

> To-day, the road all runners come,
> Shoulder-high we [brought] you home,
> And set you at your threshold down,
> Townsman of a stiller town.

> Smart lad, to slip betimes away
> From fields where glory does not stay
> And early though the laurel grows
> It withers quicker than the rose. . . .

Like the Robinson poem, Housman's is told from the point of view of the admiring drone, and it explores the ironies of an unexpected death. The differences, though, are even more relevant for what they say about the news and its rewriting of experience. Where "Richard Cory" ends on the shock of nonsense ("And Richard Cory, one calm summer night, / Went home and put a bullet through his head."), this tale ends by trying to translate the nonsense of an athlete dying young into, quite literally, garden-variety common sense ("And early though the laurel grows / It withers quicker than the rose."). The point is that here the poet Housman and news anchor Jennings share the same mission—making common sense out of nonsense.

On July 28, just the day after ABC News named the late Len Bias as its person of the week, breaking news seemed to verify William Raspberry's worst fears that Bias's death was without any redeeming qualities. It was reported that Don Rogers, a defensive back for the NFL's Cleveland Browns, had died of cocaine intoxication. Although Rogers and Bias played different sports and were at different stages in their careers, their deaths would be linked forever in the news. Rogers, too, was an African American and from a humble background—which means that his death scandal conformed to the social and racial delineations that were already visible in the playing out of the Bias scandal. In fact, Don Rogers's death scandal is best thought of as a subplot in the social drama initiated by the Bias breach—a subplot that would extend the scope of the breach and its aftermath to include all of professional sports. At least in the world of mainstream journalism, the timing of his death, coming so soon after Bias's, precipitated a sense of escalating chaos, a sense that the nonsense had, indeed, gotten out of hand and measures needed to be taken to restore order and normalcy—and control by the power bloc.

·*Redress*   According to Ettema, in the redress stage, "adjustive mechanisms ranging from personal advice to formal juridical procedures to rituals of sacrifice are invoked." [25] Our analysis of this stage of the social drama is complicated by the fact that the Bias scandal takes on its power and meaning, not in isolation, but in dialogue with other related conflicts and struggles. One way of thinking about this dialogue is to treat the Bias affair as a social drama within the larger social drama of the crack crisis within the larger continuities and disruptions of the cocaine narrative within the larger antagonisms of racial and gender conflicts within the larger oppositions of the politics of rich and poor in America. In these progressively broader frames, relevant redressive actions licensed by the Bias scandal might include everything from President Reagan's new emphasis on demand in his war on drugs to Peter Ueberroth's championing of new, more invasive, random drug testing in professional baseball.

In fact, Turner's work on the operating rules of the social drama directly addresses the symbolic importance of such chief definers as Reagan and Ueberroth in constructing the cocaine narrative. In Turner's terminology, these chief definers are "star groupers"—"the ones who develop to an art the rhetoric of persuasion and influence, who know how and when to apply pressure and force, and who are most sensitive to the factors of legitimacy." And, for Turner, the star grouper's ritual performance is especially crucial during the redress stage. For star groupers are the ones "who manipulate the machinery of redress, the law courts, the procedures of divination and ritual, and impose sanctions on those adjudged to have precipitated the crisis, just as it may well be disgruntled or dissident star groupers who lead rebellions and provoke the initial breach." [26]

Ueberroth's performance in an ABC report airing on June 30 (three days after the Rogers death) serves as an example of the star grouper in action. Jennings's lead-in makes direct connections between the trauma of the recent deaths and the frustration about how to proceed with redressive action: "Well, the death of Don Rogers following so soon after the death of Len Bias has given many people a new sense of urgency about the use of drugs among athletes. But as ABC's Jim Lampley reports, what has been most urgent so far is only the debate on what to do." In the context of a debate about drug testing and the civil liberties of athletes, baseball commissioner Ueberroth is called on to practice the star grouper's art of the "rhetoric of persuasion and influence." In this noteworthy instance, Ueberroth's We vs. They rhetoric labels drug use as an "un-American" activity: "We've got to educate them so that they know what they are doing can terminate their life, mostly rearrange and disarrange their families, their children. It causes most of the crime. It's un-American. We've got to stop it." For us, Ueberroth's invocation of the language of McCarthyism gives a voice to the hysteria of the moment and spotlights the similarities between the excesses of the infamous McCarthy witch-hunts in the 1950s and the "Crack Scare" in the 1980s.

Separation/Reintegration According to Ettema, in the final stage of the social drama "the attempts at redress are seen as either to succeed or fail and ceremonies may be enacted to mark reconciliation or else permanent cleavage."[27] Within the narrow frame of news that dealt specifically with the Bias case, the redressive stage took shape as an open-ended melodrama with two primary lines of action. One series of reports followed the prosecution of Brian Tribble; the other documented the tribulations of Lefty Driesell. These developing story lines would introduce their own minor set of star groupers responsible for legitimating, mobilizing, and executing redressive action: Dr. John Smialek, the coroner; Arthur Marshall, the state's attorney; Allan Goldstein, defense attorney for two of Bias's roommates; and John Slaughter, chancellor of the University of Maryland.

One of these story lines would end in separation. In late October Lefty Driesell was officially separated from the University of Maryland when he announced his forced resignation. But, surprisingly, the Tribble story ended much like DeLorean's. In June 1987 Tribble, after being tried for possession of cocaine, possession with intent to distribute, distribution of cocaine, and conspiracy to distribute, was acquitted of all charges.[28]

Although it would be hard to completely characterize Tribble's controversial acquittal as a healing instance of ritual reconciliation, there was at least one poignant moment of reintegration in television news coverage of this social drama. Almost two years after his death, the memory of the "good" Len Bias would be somewhat redeemed in a news segment on the victims of the war on drugs. Broadcast May 19, 1988, on NBC, the report documented the antidrug work of Lonise Bias, Len Bias's mother. Ms. Bias is, in fact, convinced that she is now "on a mission from God to help save the youth and the families of this nation." In this motherly ministry, her son is transformed into something of a Christ figure, a sacrificial offering for our collective transgressions. In her words, "Len Bias was born to die the death he did to save so many other young people. One man's life to save so many millions." In that brief moment of reintegration, Len Bias is once again a triumphant individual—and a transcendent, purified figure— at least in the eyes of his mother.

Self-Fulfilling Prophecy

While we certainly sympathize with Lonise Bias's struggle to make sense of her son's short life, our analysis takes a considerably different perspective on the meaning of the Len Bias's death drama—a perspective informed by a contrasting example of a sports death in 1984 that did not trigger a full-blown social drama. That year, early one June morning in Winona, Texas, Miami Dolphins' running back David Overstreet perished in a violent explosion after he fell asleep at the wheel of his sports car and crashed at high speed into a row of gasoline pumps. The autopsy established that Overstreet was "legally intoxicated" at the time of his death with a blood-alcohol count of .12. But, despite the spectacular circumstances of his fiery death, Overstreet's passing would merit only forty seconds of airtime on the networks' evening newscasts—two scanty twenty-second reports delivered on CBS and NBC on the day of the accident. This is barely a trace of coverage compared to the more than thirty-five minutes of precious newscast time that the networks devoted to the Bias scandal in 1986 (see NBC, 6/19/86; ABC and NBC, 6/20/86; CBS and NBC, 6/23/86; ABC, CBS, and NBC, 6/24/86; ABC, 6/27/86; CBS, 7/6/86; ABC, 7/18/86, 7/21/86; NBC, 7/25/86; ABC, CBS, and NBC, 10/29/86). Indeed, the newscast time accorded the Bias coverage outweighed the Overstreet coverage by a factor of 50.

Of course, several temporal, spatial, and economic factors contribute to this disparity. Overstreet died in the off-season, while Bias had recently been in the news because of the NBA draft. There were also important "proximity" factors—Maryland is right next door to the nation's capital and within a day's drive of the East Coast headquarters of the networks;

Winona, Texas, is out in "nowheresville." There were economic considerations, too: the cocaine business does not provide advertising revenues for the newspaper, magazine, and broadcasting industries; nor does it bankroll a powerful lobbying operation in Washington, D.C.; nor does it, at least overtly, make campaign contributions to influential politicians; nor does it hypocritically provide funding for Partnership for a Drug Free America (the "This is your brain on drugs" ad group). The alcohol business, though, is actively engaged in all of these sanctioned forms of profit-sharing-with-strings-attached. Because of these economic distinctions, Overstreet was "*legally* intoxicated," and his death, though just as permanent as Bias's, was more or less "licit" and therefore less scandalous.

But the most decisive factor in generating the extraordinary coverage of the Bias affair was the pivotal role it played in authenticating news coverage of the crack crime wave. As Mark Fishman notes, crime waves are first and foremost a state of mind: "we are talking about a kind of social awareness of crime, crime brought to public consciousness. . . . One cannot be mugged by a crime wave, but one can be scared. And one can put more police on the streets and enact new laws on the basis of fear. Crime waves may be 'things of the mind,' but they have real consequences." [29] In studying the drastic increase in 1976 of coverage of crimes against the elderly in New York City, Fishman discovered what he calls "the crime wave dynamic"— a dynamic that is essentially fueled by what we would term "dialogic" or "reciprocal" relations within the community of newsworkers and between this community and its official sources.

In this dynamic a crime theme (mugging, prostitution, terrorism, etc.) is discovered by some news organization or promoted by some official source, and this theme begins to act as an organizing principle in newswork itself. In many ways the "discovery" of crack by mainstream journalism in 1985 was a lot like Columbus's "discovery" of the New World in 1492. In both cases, discovery is a matter of point of view. For the people already living in the New World, Columbus's errant voyage was not so much an expedition as an intrusive probe by technologically superior forces who, ultimately, were bent on conquest. Similarly, for the people living in the underworld economy of the drug trade, journalism's detection of crack was not so much about the discovery of a new epidemic as it was about their coming under the unwelcome scrutiny of the optics of power, the surveillance of economically superior forces who also were bent on conquest. In fact, according to Terry Williams's ethnography of a group of young drug dealers—*The Cocaine Kids*—crack had been marketed in New York

City and Los Angeles as early as 1983. So what was a breaking story for networks in 1985 and 1986 was old news to the people on the street.[30]

There are, of course, many possible explanations for this two-year lag between the introduction of crack on the street and its discovery as a news theme by the journalistic establishment. It may have taken several months for the consequences of crack abuse to register on the official drug surveillance system. It also may have taken some time for dealers to build demand for their new product. However, we suggest that two other factors were even more decisive in delaying the discovery of the crack theme. First, crack and its target market did not fit the trickle-down paradigm governing the cocaine narrative during 1983 and 1984. And, second, the crack theme was not heavily promoted—yet—as a drug epidemic by expert and official news sources. In fact, one reference to the crystalline form of cocaine appeared as early as October 26, 1984, in an ABC report (by Judd Rose) about Los Angeles street gangs and their "rock houses" (rock was the West Coast term for crack). But this early report did not succeed in rewriting the cocaine narrative; nor did Jennifer McLogan's first crack report (ABC, 12/1/1985) discussed at the start of chapter 7. However, after the launching of NIDA's "Cocaine, the Big Lie" campaign, when such moral entrepreneurs as Arnold Washton and Mitchell Rosenthal joined Robert Stutman on the crack bandwagon, newspapers and newsmagazines found the necessary collusion to completely rewrite the race and class contours of the cocaine narrative. In the late May 1986 turning point the network news organizations followed suit.

After the initial discovery, definition, and promotion of the news theme by some combination of opportunistic journalists and enterprising sources, incidents that fit the theme are more likely to be reported. In other words, at a certain point in its evolution the crime theme becomes a kind of self-fulfilling prophecy. As other news organizations pick up on the prophetic theme, the original news judgment that broke the story is ratified. According to Fishman, it is this combination of reportorial initiative, cooperating sources, preordained confirmation, and organizational conformity that transforms the crime theme into a crime wave:

> Journalists who have not seen a particular crime theme learn to see it simply by watching their competition. They are able, using the consistency rule, to report the crime theme their competition taught them to see. At this point, when a crime theme is beginning to spread through more and more media organizations, the "reality" of the theme is confirmed for the media organizations who first reported it. They now see others using their theme.

Moreover, as the theme persists, new organizations already using the theme will not hesitate to report new instances, because they confirm a past news judgment that "this thing really is a type of crime happening now." Thus, each use of a theme confirms and justifies its prior uses.[31]

The Bias death proclaimed that the crack theme "really is a type of crime happening now." It corroborated the reigning news judgment regarding the seriousness of the cocaine threat, and that, more than anything else, substantiated the newsworthiness of the death breach.

David Overstreet's death, on the other hand, despite the efforts of MADD, did not occur at a moment when the drunk driving news theme was as trendy (or economically viable)—and therefore, it was not as newsworthy (or financially safe). It was not as economically sensible and financially safe because, as we suggested, Overstreet was "legally intoxicated"—his death involved a type of transgression that was "sanctioned" by a major source of advertising revenues. Ironically, at a time of such hysteria about inner-city drug abuse, the tobacco and alcohol industries were cynically engaged in contributing to the Partnership for a Drug Free America while aggressively test-marketing products like Uptown cigarettes (R.J. Reynolds Tobacco Company) and PowerMaster malt liquor (G. Heileman) intended primarily for the "black consumer market."[32] Although these notorious products were eventually pulled off the market after widespread protests, this "legal" drug targeting of people of color continues, especially in billboard advertising. In 1987, for instance, black neighborhoods in St. Louis had three times as many billboards as those found in white neighborhoods; of these, 62 percent of billboards in black neighborhoods advertised tobacco and alcohol compared to only 36 percent in white neighborhoods. A similar pattern was discovered in Detroit: 43 percent of billboards in the predominantly black city advertised alcohol and tobacco against only 24.7 percent in surrounding white suburbs.[33]

In presenting the Overstreet/Bias contrast, then, we mean to criticize the news treatment of sports scandals without in any way diminishing the suffering or seriousness of either death. In the words of William Raspberry, the untimely demise of both of these talented young men was "just brutally sad" and "desperately tragic." Yet, as the Overstreet case suggests, the brute fact that Len Bias died young and died recklessly does not sufficiently explain the extraordinary news interest. The point is, in and of itself the news of an athlete dying young does not a social drama make. It is, instead, the interpretation of that death within the current regime of journalistic "truth" and its interpellation within an existing scheme of newsworthiness that constructs a death event as a relevant breach—a breach that, in turn, foments crisis, demands redress, and culminates in either separation or reintegration.

Orthodoxy of Nostalgia

The influence of Reaganism on journalistic coverage of the domestic drug problem is especially evident in the "crisis updates" in 1986. During a time of waning East-West tensions, the war on drugs compensated for the death of the cold war by providing journalism (and the New Right) with a new host of enemies, both foreign (the Medellin cartel, Manuel Noriega, etc.) and domestic. For many journalists the crack crisis would become a highly addictive stimulant that, as Todd Gitlin observes, induced the euphoria of "battlefield" reporting: "Sometimes, watching repetitive footage of murders, drug busts, and 'record seizures' one gets the feeling that the so-called war on drugs is itself a drug, keeping the population high and the promoters in business, serving to keep terribly painful truths at bay." [34] Clearly, in the amphetaminelike journalistic frenzy of the summer of 1986, many reporters became gung-ho recruits in the war on drugs who, consciously or unconsciously, also championed both the nostalgia and the backlash politics of the New Right.

The Contaminated Generation   Consider, for instance, a crisis update reported by Nadine Burget (CBS, 7/3/86): a story about pollution of middle-class family life by a generation of transgressors. Burget treats the 1960s generation as a contaminated mob that continues to perpetuate America's drug problem. Showing archival footage of hippies smoking marijuana, Burget characterizes the 1960s as a time when an *entire* generation of young people was "dropping out and getting high." The report suggests that this sent a message to America: "Illegal drugs are OK." The Reaganesque dimensions of Burget's view of the 1960s is especially clear when she fast-forwards to the 1980s. In narration that resonates with memorable lines from 1980s movies (*The Terminator*'s cybernetic threat, "I'll be back"; *Poltergeist*'s innocent warning, "They're here"), Burget reports that the contaminated generation is still an undead force in contemporary society: "*They* drop back in, but some of them are still getting high . . . . The experts say the millions of parents who use drugs aren't just abusing themselves. *They* are teaching the next generation to self-destruct—one line, one drink, one toke at a time" (our emphasis). In using such We-vs.-They rhetoric, Burget marks off the 1960s generation as a collective Other (not unlike a pack of terminators or a tribe of poltergeists) that has traveled through time, returning from the midnight decade to threaten the American family.

This negative view of the 1960s generation would continue after the 1986 crisis coverage. For instance, a prime-time special in 1988 would take this negative view of the 1960s to an absurd extreme. Titled "Drugs: Why This Plague?" (ABC, 7/11/88), the special features a segment reported by Joe Bergantino that anticipates the future time travel of the contaminated. Over archival footage of Woodstock, Bergantino asserts that "having grown up in the 1960s and early '70s at the height of the drug culture, today's generation of adults, more than any before it, seems more willing to take from drugs what drugs will offer." After presenting a confession from an anonymous white male transgressor and a sound bite from "drug expert" and hotline guru Mark Gold, Bergantino concludes by reporting that "doctors fear" the illusion that "illicit drugs are really not that bad" will "follow this generation of baby boomers into old age, eventually transforming the adult drug plague of today into *a geriatric drug epidemic* of a not too distant future" (our emphasis). Back in the studio a grim Peter Jennings then ushers in the commercial break with grand narration that comments on the prospect of old boomers getting high: "And there can be no thinking person in this nation who wants that to happen. . . ." In "thinking" about this prophecy, we wonder how Jennings's advertisers might exploit this future epidemic. We imagine that the famous plug line of First Alert commercials would certainly have to be rewritten—from "I've fallen and I can't get up" to "I'm tripping and I can't come down." We also wonder, somewhat more seriously, how transgressing boomers would manage to live long enough to spawn a "geriatric drug epidemic" if illicit drugs are so dangerous.

The Out-of-Control Campus   Mark Nykanen's satellite story (NBC, 7/7/86) takes essentially the same conservative view of history and applies it to the drug pollution of a setting that is closely associated with the political unrest of the 1960s: the college campus. Nykanen animates long-standing campus conflicts by using students at the University of Michigan to illustrate the results of a new study released by Michigan's Institute for Social Research (ISR). According to the study, "Cocaine use among America's brightest and most privileged young people is greater than ever before. More than 17 percent of the students questioned, more than one in six, report using the drug in the past year." Lloyd Johnston, the author of the report and a survey researcher at ISR, speaks as a medical authority on addiction: "I don't think there's any question among scientists in this field that cocaine is one of the most addictive substances known to man." The story then reports on animal research at Concordia University in Montreal in which rats, given unlimited access to cocaine, take it in preference to food, lose half their body weight, and finally die. Quoting Nykanen's narration: "Scientists say the only explanation for this behavior is that the rats like the way cocaine makes them feel. So do college students." After this painfully obvious observation, Nykanen interviews three anonymous college students who essentially say they like the way cocaine makes them feel.

The "objective researcher/transgressing student" contrast in Nykanen's report speaks directly to the contradictions that continue to make the college campus a conflicted space in American culture. On the one hand, the college campus is a location devoted to the transmission of tradition and the mobilization of administrative research. In the transmission mode, the educational mission involves indoctrination in the legitimate taste distinctions associated with "high culture" and the more practical acquisition of exclusionary professional credentials. Using Pierre Bourdieu's terminology, education, in this sense, is geared toward the accumulation of "cultural capital."[35] In the mobilization mode, the bulk of the research mission is geared toward the generation of so-called value-free knowledge that can be usefully applied to developing natural resources, streamlining production processes, stimulating economic growth, rationalizing taste hierarchies, establishing artistic standards, enhancing national security, normalizing international relationships, improving weapons systems, managing flexible labor markets, and, of course, producing and disciplining all manner of deviancy. Clearly, Johnston's surveillance research is a prime example of the kind of "objective," "value-free" scientific enterprise devoted to what Foucault terms "the normalization of the power of normalization."[36]

On the other hand, the hedonistic students who transgress in their feel-good use of cocaine speak of the college campus as a liminal setting that is situated "in between" the institutions of the bourgeois family and the bourgeois profession. Before the 1950s and 1960s, the college campus was primarily devoted to reproducing the economic elite. A college education was the mark of privilege, an important status definer that separated the professional class from the middle and working classes. However, with the progressive democratization of higher education under Fordism in the 1950s and 1960s (first with the GI Bill and later with affirmative action), the college diploma lost some of its aura. Believing in education as a means of social mobility and economic salvation, mobs of working- and middle-class overachievers, both women and men, invaded the quasisacred space of the college campus—and after they made it through the gate, a funny thing happened: the Eurocentric, patriarchal and antidemocratic values of the ruling elite came increasingly under scrutiny and attack in campus reform movements of the 1960s and 1970s.

In Nykanen's report, the question of how to control the transgressions of youth who live in that contested, in-between space is explored in sound bites from two experts who hold differing opinions on the subject: Dr. Robert DuPont and University of Michigan vice president Henry Johnson. For us, DuPont (even more so than Lloyd Johnston) embodies the modern expert devoted to the expansion of disciplinary power. Following Nykanen's introduction, "Dr. Robert DuPont says colleges should crack down on cocaine by searching dormitory rooms with drug dogs and subjecting students to urine tests," DuPont's sound bite asserts, "What they do is exactly the opposite. They establish a climate of a kind of safe house in the dormitory where a person can use drugs with immunity." To achieve journalistic balance (and construct narrative tension), Nykanen then calls on university administrator Johnson. And Nykanen's dialogic framing of Johnson's sound bite is relevant to our analysis:

NYKANEN: University of Michigan Vice President Henry Johnson says that while the college should try to discourage student drug use, it should not act as a policeman.
JOHNSON: I don't think we have to tell students how they should behave as individuals.
NYKANEN: But most experts agree that the campus cocaine problem is getting worse. The survey shows a significant increase in cocaine use as young adults grow older. By the time they reach age 27, 40 percent have tried the drug. And scientists say many of them are likely to become addicts.

For the record, it should be noted that Nykanen and DuPont are white—and Johnson is black. In following Johnson's bite with a "But," Nykanen indicates reportorial disagreement with Johnson's *permissive* sentiments, and, at least implicitly, lends support to DuPont's hard-line approach to purifying the campus.

Furthermore, in concluding with statistics that seem to suggest that the situation on college campuses is out of control, Nykanen also supports the general moral paranoia of the New Right. In the moral universe of the New Right, the college campus is an out-of-control space where generation after generation of young people challenge authority, reject traditional and "family values," embrace hedonism, and court radicalism. Academia's diminishing value as a place for indoctrinating, disciplining, and tempering the youthful bourgeois soul has been the subject of much breast-beating and teeth-gnashing. For some, it has signaled nothing less than the precipitous decline of Western civilization.[37] In the 1980s and 1990s the controversy over the control of the college campus and the purpose of a college education exploded toward the end of the Bush era in raging debates about the imposition of a "core curriculum" that instills Western values and the censuring of "politically correct" "tenured radicals" who pose a threat to "Reagan's Children."[38]

The Reactionary Privatized Community "Reagan's Children" also figure prominently in a story reported by Jennifer McLogan that echoes Reaganism's "Take Back America" theme (NBC, 6/21/86). Covering with open approval "community action" to "take back the streets" from crack dealers, McLogan fashions a report that displays images of antidrug demonstrations in the Bronx, Miami, Baltimore, and Corpus Christi, Texas. At one point, a young Texan appears who carries a poster with the motto, "Say Nope to Dope! Drink Coke." This fleeting image is "naturalized" in the taken-for-granted news package and McLogan never comments on the sign in voice-over. Illegal drug entrepreneurs modeling their businesses on American corporations like Coca-Cola, with illicit versions of entry-level positions (lookouts), middle managers (marketers), high-level managers (distributors), and CEOs (overlords and kingpins) is never at issue. Furthermore, in this particular image of the Coke poster, Coca-Cola's historic connection to cocaine is lost (for nearly twenty years the beverage contained traces of cocaine; in 1903 during the first cocaine "epidemic," caffeine replaced cocaine).[39]

McLogan's report also approves of the religious consecration of expanded police action—a consecration that seems to transform the anti-crack crusade into something of a holy war:

> The "Take Back the Streets" demonstrations in the Bronx parallel the efforts of many other protest groups that have cropped up across the country. In Miami, at the request of various church and civic groups, police added special anti-crack motorcycle patrols. . . . In Baltimore, it was a religious vigil, a plea

for peace among street gangs involved in drug-related violence. In New York,
Father Eric Mattson, caught unawares in his rectory by burglars high on crack
and was shot at, believes he has an answer.

Young Father Mattson's "answer" to the crack crisis, perhaps unintentionally, invokes the paranoia of McCarthyism in the 1950s. Calling for
what amounts to the establishment of a police state, Mattson argues that
the crack crisis "should become a national security issue headed by the
President of this country." McLogan's closing narration provides a reportorial "Amen" for the priest's invocation: "Many police officials across the
country said that they're amazed at recent citizen response to the crack epidemic. The public's support will back up their efforts to combat the drug
problem." As we discuss in more detail in chapter 7, McLogan's crusading
narration is perhaps best thought of in terms of a kind of ventriloquism in
which the reporter becomes a mere mouthpiece for policing forces.

In this particular instance of ventriloquism, notice how McLogan reports as fact what could only be a suspicion. For how could the priest know
for certain that the burglars were "high on crack"? As we discussed in
chapter 1, this tendency to attribute all criminality to the current chemical
scapegoat was also a feature of the newspaper coverage of the turn-of-the-
century cocaine "epidemic." It is worth noting that, in speaking for the
police, McLogan fails to make a distinction between the vigil in an African
American church in Baltimore and Father Mattson's vindictive comments
in New York. And yet these are two very different religious responses to
the drug problem: the former oriented toward the loving hope of reconciliation and redemption; the latter directed toward the hateful desire for
retribution and damnation. McLogan's report conflates both responses
into a general call for more police intervention in the inner city—a call that
naturalizes the "Vengeance is Mine, Saith the Lord" response of Father
Mattson.

Naturalized in such "community action" stories are images of obedient children under the watchful eye of adult control, marching military-like
in antidrug parades—or rallying in high school auditoriums, in the Rose
Bowl, and on the lawn of the White House under the apparently benign and
approving gaze of Nancy Reagan (see, for instance, ABC's coverage of "Say
No to Drugs Week," 5/22/86). McLogan's report and others like it verge on
what Susan Sontag terms a "fascist aesthetic." Celebrating images that at
times are reminiscent of the mass demonstrations in Leni Riefenstahl's *Triumph of the Will,* news coverage of antidrug marches and "Just-Say-No"

rallies often, borrowing Sontag's words, "flow from (and justify) a preoccupation with situations of control, submissive behavior, extravagant effort, and the endurance of pain; they endorse two seemingly opposite states, egomania and servitude."[40] Indeed, the "Just-Say-No" motto is almost a pure expression of the fusion of egomania with servitude. The quick behavioral fix to the drug problem, according to this motto, is reducible to strong egos simply obeying the arbitrary rules imposed from above. Just saying "no," then, is not only a matter of "self-control" but one of strict compliance with existing social controls—of, literally, internalizing the modern police state into a "normal" and "normalized" personality structure.

Ultimately, McLogan's story is underwritten by a profound sense of nostalgia for a time when "the community" was in control of deviance. As Stanley Cohen observes, such nostalgia for a bygone era of organic community control "is a rather more complicated phenomenon than it first appears; nor is it quite the same as sentimentality or romanticism." According to Cohen, the form that nostalgia takes in crime-control ideology "is a look back to a real or imagined past community as providing the ideal and desirable form of social control":

> This impulse is reactionary and conservative, not in the literal political sense, but in always locating the desired state of affairs in the past which has now (usually *just* now) been eclipsed by something undesirable. As in all forms of nostalgia, the past might not really have existed. But its mythic qualities are profound. The iconography is that of the small rural village in pre-industrial society in contrast to the abstract, bureaucratic, impersonal city of the contemporary technological state.[41]

Cohen goes on to argue that nostalgia for preindustrial forms of community control is flawed on several counts:

> In the first place, the content of the visions themselves is often historical and anthropological nonsense: neither the pre-industrial rural village nor the tribal or folk society were exactly communities in the ways that are idealized. In the informal justice literature, for example, radicals often underplay the paternalism, the fixed lines of authority and the arbitrary nature of justice. Conservatives forget the high degree of conflict and disorder that were tolerated. Both sides tend to ignore the implicit threat of violence (natural or supernatural) which often lay behind the submission to community or informal justice. And despite the evidence from revisionist historiography, about the unequal, arbitrary and random nature of eighteenth-century policing and punishment, there

is still a tendency to look for a "Golden Age" of pre-capitalist community control.[42]

In other words, what is masked by the orthodoxy of nostalgia in social control discourse is the not-so-innocent past of "community action" in America—the forgotten history of lynching, tar-and-feathering, and witch-hunting that made many small towns and villages into places ruled by racial violence, religious intolerance, and vigilante "justice."[43]

So, despite the fact that our study is generally aligned more with communitarianism than with individualism, we are still alarmed at the way "community action" is framed in the drug news. McLogan's report, for instance, in failing to make any distinctions between the Baltimore prayer vigil and "take back the streets" demonstrations in the Bronx, Miami, and Texas, publicizes a narrow view of community that is contaminated by Reaganism. Unlike President Franklin Roosevelt's vision of America as a national community, Reagan proposed that America was a nation of neighborhoods—a utopian vision that, as Reich argues, "offered a way of enjoying the sentiments of benevolence without the burden of acting on it":

> Since responsibility ended at the borders of one's neighborhood, and most Americans could rest assured that their neighbors were not in dire straits, the apparent requirements of charity could be exhausted at small cost. If the inhabitants of another neighborhood needed help, they should look to one another; let them solve their problems, and we'll solve our own. The poor, meanwhile, clustered in their own, isolated neighborhoods. By the 1980s many of America's older cities—forced to take even more financial responsibility for the health, education, and welfare of poor inhabitants—were becoming small islands of destitution within larger seas of suburban well-being.[44]

Reaganism's idealized vision of the neighborhood as a kind of "privatized" community, then, facilitated the guiltless upward redistribution of wealth during the 1980s by releasing people living in the suburbs from any sense of responsibility for—or identification with—the economic distress experienced in the inner city. Deployment of the "neighborhood as community" creed, like its implementation of the doctrine of "family values," was crucial to the divide-and-conquer/unite-and-mobilize strategies that helped the New Right realign the electorate along racial, rather than class, lines.

In practice, then, this narrow vision of the privatized community was anything but benevolent—or for that matter Christian, at least to those believers who take the parable of Good Samaritan seriously. The famous parable features an interchange with a hostile lawyer who questions Jesus' knowledge of the Old Law. After Jesus agrees with the lawyer's "reading" of the Old Law—that eternal salvation involves loving "the Lord thy God with all thy heart, and with all thy soul, and with all thy strength, and with all thy mind; and thy neighbor as thyself"—the lawyer continues to test Jesus by asking, "And who is my neighbor?" Jesus' response to this question takes form as a familiar parable that we hope to make strange once again by putting it in dialogue with Reaganism:

> A certain man went down from Jerusalem to Jericho, and fell among thieves, which stripped him of his raiment, and wounded him, and departed, leaving him half dead.
>
> And by chance there came down a certain priest that way: and when he saw him, he passed by on the other side.
>
> And likewise a Levite, when he was at the place, came and looked on him, and passed by on the other side.
>
> But a certain Samaritan, as he journeyed, came where he was: and when he saw him, he had compassion on him,
>
> And he went to him, and bound up his wounds, pouring in oil and wine, and set him on his own beast, and brought him to an inn, and took care of him.
>
> And on the morrow when he departed, he took out two pence, and gave them to the host, and said unto him, Take care of him; and whatsoever thou spendest more, when I come again, I will repay thee.

Jesus concludes the story by asking the lawyer, "Which now of these three, thinkest thou, was neighbour unto him that fell among the thieves?" [45]

If the lawyer had been indoctrinated in the Gospel according to Reagan, he might very well have answered, "the priest and the Levite." For, under Reaganism, the privatized community became a place that was devoted much more to "exclusion" and callous self-interest than to "inclusion" and righteous benevolence. Contrary to the welcome-wagon image of neighborhood voluntarism promoted by Reaganism, the neighborhood association became, in the 1970s and 1980s, a vehicle for mobilizing "not-in-my-backyard" activism and "neighborhood watch" vigilantism. The war on drugs has played a prominent role in transforming the neighborhood into an arena for panoptic surveillance, where neighbors watch out not only for threatening strangers but for transgressing neighbors. "One of the most disturbing aspects of the War on Drugs," writes Christina Jacqueline

Johns, "is the extent to which it is creating a society of informers": "The fact that the state is encouraging citizens to become informants is less disturbing than the relish with which the population is doing so. In a poll conducted by the Washington Post/ABC News in 1989, 83 percent of the respondents favored encouraging people to phone the police to report drug users even if it meant turning in 'a family member who uses drugs.' "[46] The "informant society" that has flourished in this atmosphere of fear, intolerance, and suspicion has blanketed postmodern America with police hotlines. Heavily promoted by local news organizations (Crimestoppers) and even network programming ("America's Most Wanted"), these hotlines provide a flow of anonymous tips from neighbors watching television watching neighbors watching neighbors. Perhaps even more disturbingly, in the local informant networks set up by the drug control establishment, children are encouraged to police their parents, and vice versa. For instance, one popular program known as Drug Awareness Resistance Education (DARE), pioneered by Daryl Gates and the LAPD, puts uniformed police in the elementary classroom and encourages students to not only just say "no," but to snitch on those who dare to say "yes."[47] Although, in the face of evidence to the contrary, DARE officials deny accusations that they teach children to inform on parents and friends, the program still speaks to the dramatic expansion of the police state into America's schools and neighborhoods during the Reagan era. Beginning as a modest pilot program in Los Angeles in 1983, the DARE curriculum is now taught to an estimated 5 million students nationally every year.[48]

Urban Corruption vs. Provincial Virtue   While the contrast between idyllic village and impersonal city is not explicitly a part of the McLogan report, it is incorporated into the cocaine narrative in a series of contrasting crisis updates that demonize the threatening inner city while sanctifying threatened Middle America. As Reider argues, Middle America and its Silent Majority are products of political backlash born in the 1960s:

> In the 1960s and its aftermath, conservatives and Republicans found a responsive audience among once-Democratic constituencies: southerners, ethnic Catholics in the Northeast and Midwest, blue-collar workers, union members, even a sprinkling of lower-middle-class Jews. Out of this maelstrom of defection there emerged a new social formation, Middle America.
>
> Middle America did not really exist as a popular term before the 1960s. In part, it emerged out of the center's own efforts to name itself. But it also emerged from the efforts of others to capture and beguile it, most notably

from the oratorical flourishes of Republicans, reactionaries, and conservatives who had their own ideological projects in mind.[49]

Middle America was the target of much of the divide-and-conquer politics of the New Right (what Spiro Agnew called a "positive polarization" of the electorate)[50], and, increasingly, Middle America became the target of network drug news in 1986.

On the threatening hand, the adverse vision of the inner city appears in a classic crisis update reported by John Quinones (ABC, 7/28/86). It opens with a police raid, shows the horrors of crack, depicts crack as primarily a "minority" problem, and celebrates reactionary community action. The story is set in Miami and New York City. The Miami scenes show a seventy-year-old Latina grandmother being arrested at a crack house—and then Miami Police Chief Clarence Dickson compares packets of crack to bombs in World War II that wiped out whole neighborhoods. Such "semantic escalation" of the drug war has even reminded some commentators of another "police action" that left America at war with itself—the Vietnam war. For instance, the *Nation*'s Ralph Brauer argues that all the "bellicose rhetoric" of authorities like Chief Dickson, William Bennett, and Daryl Gates has resulted in the dehumanization of drug offenders. "They become a cypher, a concept, not human beings," writes Brauer. "Drug dealers are referred to as 'scum,' 'creatures,' 'creeps,' and are painted with the evil countenances of subhuman life forms, in the same way that World War II propaganda stereotyped the Japanese and television caricatured the Vietnamese."[51] In fact, the New York scenes of Quinones's story feature a sound bite from police commissioner Benjamin Ward that is noteworthy for its animalization of the buyers and sellers of crack: "It's like a bunch of tigers, and wild cats, and lions have moved into your neighborhood and you're the prey."

On the threatened hand, many crisis update stories appear as infiltration warnings charging that cocaine was not just an inner-city "plague" isolated in metropolitan Los Angeles, Oakland, New York City, Detroit, Washington, D.C., and Miami. In a chain of stories directly addressed to a complacent Middle America, the networks drafted alarmist reports of spreading cocaine contamination in small cities and towns across the United States: Paw Paw, Michigan (ABC, 8/5/86), Evansville, Indiana (NBC, 8/27/86), Fort Pierce, Florida (CBS, 8/7/86), even pastoral Vermont (ABC, 10/6/86). These news stories support a powerful conservative American myth of Middle America being an idealized place of "community" where human nature is innocent and simple, uncontaminated by the harsh realities of dehumanized, urban America.

From its sanctification of Middle America as a place of endangered purity, to its dehumanization of urban drug transgressors, to its approval of the reactionary privatized community, to its indictment of educational permissiveness, to its unfavorable portrayal of the 1960s, the routine news coverage of the crack crisis, in many instances, represented mainstream journalism's ideological convergence on—and moral conversion to—the once-extremist views of the New Right and Religious Right. In repackaging the backlash politics of Reaganism, such crisis updates were nothing less than ringing endorsements of Reaganism's call to "take back America." Naturalizing this orthodoxy of nostalgia as common sense, mainstream journalists lent a veneer of "truth and objectivity" to the New Right's paranoia and prejudices. Ultimately, such cracked coverage made it easier to dismiss the disadvantages experienced by the black and brown urban poor as self-inflicted—matters of individual choice and self-indulgence, not matters of economic history, social structure, and racial inequality.

7

Captivating Public Opinion:

The Ventriloquist Turn

The indiscriminate hysteria over drugs reflects that old anxiety at the heart of the middle class: the fear of falling, of losing control, of growing soft. "Drugs," as an undifferentiated category, symbolize the larger and thoroughly legal consumer culture, with its addictive appeal and harsh consequences for those who cannot keep up or default on their debts. It has become a cliche to say that this is an "addictive society," but the addiction most of us have most to fear is not promoted by a street-corner dealer. The entire market, the expanding spectacle of consumer possibilities, has us in its grip, and because that is too large and nameless, we turn our outrage toward something that is both less powerful and more concrete.—Barbara Ehrenreich [1]

Seismic Hysteria

It was November 29, 1985, when the *New York Times* printed its first front-page story on crack. By December 1, NBC's Jennifer McLogan had crafted the first crack report to appear on a major network television newscast.[2] In that prophetic report McLogan characterized crack as the "frightening wave of the future." Unfortunately, McLogan's words were even more portentous when applied—not to crack—but to the ensuing news coverage of the so-called crack epidemic; for, in 1986, the American public would witness what quantitative researchers term an "intermedia convergence"[3] on the crack issue.

The "seismic jump"[4] in drug stories during the spring and summer of 1986 was documented in a study of major newspapers, newsmagazines, and network newscasts conducted by Lucig Danielian and Stephen Reese. In a forty-week period beginning March 30 and ending December 31, 1986, the researchers found 406 news stories dealing with cocaine:

Of the major newspapers, the *New York Times* printed the most with 139 stories, followed by the *Washington Post* (60), the *Los Angeles Times* (38), and the *Wall Street Journal* (13);

In the news magazine category, *Time*'s nine stories barely edged out *Newsweek*'s eight;[5]

And of the 129 stories aired on the nightly network newscasts, ABC ran 34 stories, CBS 49, and NBC 46.[6]

Among the news magazines, *Newsweek* would set a dubious standard of crass exploitation by hyping the crack scare with three drug abuse covers in just five months. Boosted in June by the drug-related deaths of athletes Len Bias and Don Rogers, newspaper coverage of cocaine issues would reach a feverish peak in July, only to quickly decline in August after Reagan announced a shift in the strategy of his drug war from an almost complete preoccupation with eliminating supply to an emphasis on demand reduction.[7] On the television networks the fervent coverage would be sustained longer, staying at a high level until the early September airing of prime-time documentaries on CBS ("48 Hours on Crack Street") and NBC ("Cocaine Country").

Not surprisingly, the seismic jump in drug stories was accompanied by a parallel jolt in measurements of public opinion. In April 1986, only 2 percent of the respondents to a New York Times/CBS News Poll picked drugs as the nation's most important problem. By early September 1986 another survey by the same authority found that drugs now topped the list of perceived ills with 13 percent of 1,210 adults interviewed choosing drugs as the nation's most important problem.[8] These results were matched by an ABC News Poll, also released in September, that found that 80 percent of those responding believed that America was caught up in a national drug crisis. In the same poll, however, 62 percent also reported that drugs were not a major problem in their own hometowns.

The drastic shift in measurements of public opinion in 1986 provides a textbook case of the normalizing power of a widely promoted crime wave. As Stuart Hall and his colleagues found in their study of the social construction of "mugging," crime is one of the areas where the news media are most successful at mobilizing public opinion because, after all, crime is less open than most other public issues to competing or alternative definitions of the situation:

a police statement on crime is rarely "balanced" by one from the professional criminal, though the latter probably possesses more expertise on crime. . . .

By and large, the criminal, by his actions, is assumed to have forfeited, along with other citizenship rights, his "right of reply" until he has repaid his debt to society. Such organized opposition as does exist—in the form usually of specific reforming groups and experts—often shares the same basic definitions of the "problem" as the primary definers, and is concerned merely to propound alternative means to the same objective: the returning of the criminal to the fold.[9]

The point is that, although Jesse Helms favored more death sentences for pushers and the Reverend Jesse Jackson favored more educational programs for youngsters, both still essentially shared the basic definitions of the problem and the same ultimate objective. And in a situation where both of these Jesses become soldiers in the same cause, the conditions were obviously ripe for the spawning of a powerfully normalizing consensus.

Indeed, according to Adam Clymer's front-page report of this transformation in poll results: "The only comparable spurt in concern over an issue recently is the record of terrorism, cited as most important by 1 percent in a December 1985 Times Poll, and then by 15 percent in the April 1986 Times/CBS News Poll."[10] Terrorism, by the way, fell back to 1 percent in the September 1986 survey, but that precipitous decline did not generate much journalistic interest. The ascending drug numbers, though, were big news, attracting the attention not only of journalists, but of politicians of every stripe.

How do we interpret these numbers? Or, perhaps more important, what meaning do we attribute to these polls? If we apply the "poll as mirror of reality" metaphor, as many journalists do, then the sharp increase in public concern about drugs should be somehow related to a corresponding increase in drug abuse. But even the government's own statisticians are quick to acknowledge that illicit drug use overall has been generally in decline since peaking in 1979–80.[11] If this extraordinary public concern does not reflect rising gross rates of abuse, then what does it reflect? In this chapter we try to make sense of this question by considering how a government information campaign helped set the stage for the crime wave dynamic that prompted discriminatory hysteria in both mainstream journalism and Middle America in the wake of the Bias death drama. In analyzing this dynamic, we also address the "politics and poetics" of opinion polls in the cocaine narrative.

The Big Lie

> Objectivity is authority in disguise. . . .
> —John Fiske [12]

Before the June death of Len Bias the rising number of cocaine stories during the spring of 1986 can be attributed largely to the spectacular success of a government campaign orchestrated to "increase public awareness" of the drug problem. After all, it was in April 1986 that NIDA launched the first phases of a multimedia Cocaine Prevention Campaign called "Cocaine, the Big Lie." [13] NIDA is, in fact, part of the legacy of an earlier war on drugs declared by President Nixon in the 1970s. Established in 1974, NIDA is a part of the U.S. Department of Health and Human Services. According to Susan B. Lachter and Avraham Forman (who are researchers associated with this agency), NIDA, as "the primary federal agency responsible for reducing the demand for illicit drugs," actively sponsors and conducts research "on the consequences of drug abuse and on the development of effective treatment and prevention approaches." [14] This research includes three surveillance projects that are common and recurring statistical rituals featured prominently in the networks' framing of cocaine stories: the National Household Survey on Drug Use; the National High School Senior Survey; and the Drug Abuse Warning Network (see, for example, the following newscasts: NBC, 1/18/82; CBS, 6/29/82; ABC, 10/25/82; NBC, 8/9/84; CBS, 11/27/84, 5/31/85; ABC, 7/16/85, 3/20/86, 8/5/86; ABC and NBC, 7/7/86; ABC, 7/10/86; CBS, 9/2/86; NBC, 2/27/87, 1/13/88, 5/16/88, and 5/17/88). Designed both to monitor drug use in the population and assist in facilitating federal efforts to discipline that use, all of these projects illustrate the way in which survey research is used as a normalizing technology.

Monitoring the Future Consider the survey of high school seniors (also known as the "Monitoring the Future Survey"), conducted by the University of Michigan's ISR, which routinely receives million dollar-plus grants from NIDA (see chapter 5). Beyond measuring rates of drug abuse, this survey also reports attitudes and beliefs: how harmful the seniors think drugs are; how much they personally approve of drug use; how apt they are to use various drugs in different circumstances; among other questions. And according to Susan Lachter and Avraham Forman, "NIDA uses this information about student attitudes when designing public education programs, because the reported attitudes often provide clues about the kind of messages needed to encourage attitude or behavior change." [15]

One such message was the "Just-Say-No" campaign. "After reviewing the national trends in attitudes and behaviors of the late 1970s and early 1980s," write Lachter and Forman, "it became clear to NIDA that public awareness of the dangers of drug abuse was very low":

> The Institute was concerned that the broad spectrum of the public was involved in drug use, particularly teenagers, and that parents and other adults had to take action to prevent the onset of drug use among young children. In 1982, NIDA used all of its drug use and attitude data to design the "Just Say No" information campaign that was aimed primarily at junior high school students and their parents. Its goal was to present a drug-free life as a healthy norm for teenagers; drug use would no longer be a rite of passage for America's youth. [16]

This provided the groundwork for the "Cocaine, The Big Lie" campaign. Work began on this anticocaine propaganda in July 1985 when NIDA signed a contract with the National Advertising Council, which, in turn, acquired the voluntary creative services of the same agency that helped execute the "Just-Say-No" effort—Needham Harper Worldwide. But in contrast to "Just Say No," this effort would target the young adult audience. The first series of public service announcements featured repentant ex-users like former Miami Dolphins running back Mercury Morris, but by the end of 1986, other spots also featured celebrities with clean records (presumably nonusers of cocaine): First Lady Nancy Reagan and baseball stars Mike Schmidt and Reggie Jackson.

The surveillance research sponsored by NIDA and its public awareness campaigns, then, are intimately linked. We would further suggest that the public opinion polls conducted by media organizations are also a part of this system of regulation, exhortation, and persuasion. Like NIDA research, polls commissioned by news organizations are common and recurring statistical rituals in the framing of cocaine stories (see ABC, 5/17/85, 8/4/86; CBS, 9/1/86; ABC, 9/15/86, 9/16/86, 9/17/86, 9/18/86, 9/19/86; CBS, 4/25/88, 5/18/88, 8/30/88). For us, the polls showing that the American public considered drugs to be the country's number one problem in 1986 elaborate and replicate the kind of surveillance and control procedures enacted by ISR for NIDA. Like ISR's Monitoring the Future Survey, the New York Times/CBS News Poll is an example of what Foucault terms a "meticulous ritual of power"—a ritual that activates normalizing mechanisms that are not so much committed to expressing the collective "will of the people" as they are devoted to *monitoring* and *manipulating* that will and channeling it in directions that, ultimately, promote the expansion of the narco-carceral complex.[17]

Governing Opinion This interpretation of the meaning of the polls points to what Benjamin Ginsberg describes as "the domestication of mass belief." "While contemporary western governments do listen and defer to their citizens' views," writes Ginsberg, "the public opinion to which the regimes bow so assiduously is not the natural and spontaneous force that confronted their predecessors. Instead, the opinion that contemporary rulers heed is in many respects an artificial phenomenon that national governments themselves helped to create and that their efforts continue to sustain."[18] The dramatic rise in public concern about drugs in 1986 is certainly a conspicuous example of a national government taking an active role in generating public opinion. Clearly, self-interested agents of the government sector of the medical-industrial complex played a crucial role in authorizing and promoting the 1986 crisis. For example, although he does not explicitly take any credit for coauthoring the crisis, Lloyd Johnston does readily admit to participating in some two hundred to three hundred press interviews in 1986.[19] And it is no accident that the newspaper and magazine coverage of cocaine really takes off only after NIDA launched the "Big Lie" campaign.

As Benjamin Ginsberg also observes, polling allows officials "a better opportunity to anticipate, regulate, and manipulate popular attitudes": "Ironically, some of its early students believed that polling would open the way for 'government by opinion.' Instead, polling has helped promote the governance of opinion."[20] For us, the connection between news polling and NIDA surveillance is evident in the companion questions asked by the same poll that found drug control to be at the top of the public agenda. As if they were surrogates for NIDA, the pollsters also asked if people were willing to spend more taxes to jail drug sellers and if they were willing to take drug tests on the job. Two-thirds of the respondents answered in the affirmative to the first question. And, three-fourths of those working full-time said they were willing to submit to drug tests.[21] But what is not at all clear about these results is whether the percentages represent a genuine response on the part of the respondents or whether they speak more accurately of the views and the ulterior motives of the polling authority. For "polls do more than simply measure and record the natural or spontaneous manifestation of popular belief," writes Ginsberg:

data reported by opinion polls are actually the product of an interplay between opinion and the survey instrument. As they measure, the polls interact with opinion, producing changes in the character and identity of the views receiving public expression. The changes induced by polling, in turn, have the most profound implications for the relationship between public opinion and government. In essence, polling has contributed to the domestication of opinion by helping to transform it from a politically potent, often disruptive force into a more docile, plebiscitary phenomenon.[22]

This interaction between opinion and the survey instrument is even apparent—and acknowledged—in the aforementioned New York Times/CBS News Poll. Fifty-one percent said drug testing in the workplace would reduce drug use a great deal, but support was less when the question included the issues of privacy. Only 44 percent favored drug testing when asked, "Would you favor a policy that would require workers in general to be tested to determine whether they have used illegal drugs recently, or would that be an unfair invasion of privacy?" The inclusion of the privacy question, then, resulted in a significantly lower approval response to the notion of drug testing.[23]

Convenient Self-Deception In taming public opinion the polls have essentially substituted a simulation of the will of the people for active political involvement—and, in this substitution, measurements of public opinion become mechanisms of normalization and control. After all, "the public" does not design the poll; instead, it is cast in the docile role of passively responding to a survey instrument that mainly measures matters of pressing interest to government, business, academics, mainstream media, or other controlling forces. In other words, as Ginsberg observes, poll questions "have as their ultimate purpose some form of exhortation":

> Businesses poll to help persuade customers to purchase wares. Candidates poll as part of the process of convincing voters to support them. Governments poll as part of the process of inducing citizens' to obey. . . .
>
> In essence, rather than offer governments the opinions that citizens want them to learn, polls tell governments—or other sponsors—what they would like to learn about citizens' opinions. The end result is to change the public expression of opinion from an assertion of demand to a step in the process of persuasion.[24]

And why do journalists poll? Because, knowingly or unconsciously, conventional news reporting is also caught up in this process of persuasion. Journalists poll to persuade both their publics and themselves that they really are telling the "Truth," that the prevailing news judgment is invested with some measure of "objectivity," that they are not, in the words of Mark Hertsgaard, "mere stenographers of power."[25]

The journalistic reliance on polls, then, can be seen as part of what Shanto Iyengar and Donald R. Kinder identify as the "American media's deep commitment to 'objectivity' ":

> As a professional ideology, objectivity includes three commitments: to independence (journalism should be free from political pressure); to balance (journalism should present without favor the positions of all contending parties); and to objectivity (journalism should simply present the facts, without passing judgment on them). These ideological commitments have become intertwined over the last several decades with a set of working routines that have promoted the development of intimate relationships between the press and members of the government. Journalists have increasingly come to rely on government officials and agencies as their primary sources of information, and to focus on their activities as the basic subject matter of news. Government officials, for their part, need journalists to communicate both with the public and with other elites.[26]

As a statistical ritual, the opinion poll provides both *legitimation* for journalism (as a scrupulously "objective" profession) and apparently "independent" *validation* of journalism's reliance on (and compliance with) official definitions of the situation. As we shall demonstrate below in our analysis of the politics and poetics of polls, journalists tend to ignore both their own role in "priming" public opinion as well as the government's part in setting the "news agenda."[27] In indulging in this common *ignore*-ance, mainstream journalism is engaged in a kind of continuing and convenient self-deception—a "Big Lie" in its own right.

To summarize, NIDA's "Big Lie" campaign is significant for several interlocking reasons. First, the timing of the "Big Lie" media blitz provides at least a partial explanation for the drastic increase in cocaine stories in May 1986 as news organizations initially responded to the campaign with magazine covers and feature stories. Second, in the "Big Lie" campaign we can see how the discourse of discrimination of a government "public awareness" crusade links and mediates two intimately related but temporally distinct statistical rituals: NIDA-sponsored surveillance of the population (in the realm of the past); and journalistic measurements of public drug hysteria (in the realm of the future). But most important, the campaign demonstrates how apparently "neutral" government-funded administrative research, like NIDA's National High School Senior Survey, is not simply an instrument for monitoring the population, but it is a mechanism of social control; and, at least during the 1980s, this control was used to manipulate public opinion in ways that contributed to the New Right's framing of social predicaments, economic failures, and racial inequities in individualist moral terms. Finally, the "Big Lie" campaign was an important "pre-text" for the Len Bias death drama.

The Politics and Poetics of Polls

The crime wave dynamic that sustained crack as a major news theme in July 1986 would come to a head on August 4 when all three network news organizations opened their Monday evening telecasts with news of a White House press conference. Two of these stories (Sam Donaldson's on ABC and Bill Plante's on CBS) are of particular interest because together they speak to:

> 1. The place of public opinion polls in the reciprocal relations that produce and reproduce social problems, crime waves, and political spectacles;
> 2. Salient differences between the networks in their treatment of Reagan and their orientation to partisan politics;
> 3. The activation of humanitarian rhetoric in justifying more surveillance and discipline of the population;
> 4. Two variants of ideological closure.

Tom Jarriel, substituting for Peter Jennings, introduces Donaldson's report with a direct reference to polls: "President Reagan proposed a broad new initiative to combat what polls have found to be the number one problem in the country—drug abuse. Mr. Reagan wants mandatory drug testing for some government officials and suggested cabinet members volunteer to take drug tests to set an example. At the White House with the full story, here is Sam Donaldson." This matter-of-fact lead-in hints at the ideological utility of treating polls as the spontaneous will of the people. Notice that Jarriel's language makes polls into a self-determining, objective expression of social reality. In other words, polls are "naturalized." Polls are not treated as human products that are subject to the distortion, tampering, and manipulation of the survey instrument. Instead, like the confession in the therapeutic narrative, polls take on the aura of the direct and unmitigated "truth."

In this sanctification of polls Jarriel confuses the measurement of per-ceptions of reality with reality itself. The phrase, "what polls have found to be the number one problem in the country—drug abuse," essentially invests public opinion with the status of objective truth. In other words, Jarriel's lead-in asserts that if most people believe that drug abuse is the "country's number one problem," then that belief, in itself, constitutes an authentic reflection of what is *really* wrong with the country. Truth, then, becomes a matter of measured popularity—and the most popular responses to a series of survey questions are taken to be a direct reflection of the world and its dysfunctions. In taking the polls to be the authentic expression of reality (instead of a possibility among competing, multiple realities), Jarriel depicts Reagan as someone who is in touch with his public and responsive to its demands.

This view of the "in-touch" Reagan is not shared by Jarriel's counter-parts at CBS. In fact, Dan Rather's lead-in to Plante's story is saturated with skepticism: "Another ratcheting up, today, of President Reagan's high publicity tough talk against illegal narcotics in America, including the endorsement of the idea of mandatory drug tests for millions of Ameri-can citizens. CBS News White House correspondent Bill Plante begins our coverage of the election-year war on drugs—the next phase." While Jar-riel implies that Reagan is responding to the will of the people as chan-neled through polls, Rather portrays the president's crusade as a seemingly tough, but empty, gesture that is more a matter of "high publicity" political talk than a genuine good faith commitment. Rather also puts a different spin on the proposed mandatory drug testing. In Jarriel's lead-in the new testing appears to be very limited in its application, confined to "some government officials"; in Rather's narration the same program appears to have much broader implications, affecting "millions of American citizens." The striking contrast between Jarriel's paltry "some" and Rather's expan-sive "millions" coupled with Jarriel's *They-like* "government officials" and Rather's *We-like* "American citizens," expresses important differences in the social and political allegiances informing the two newscasts.

Although these differences are maintained in the discourse of the two competing news packages, Donaldson and Plante use many of the same sound bites. Both reports, for instance, display Reagan's use of the humani-tarian rhetoric of helping people "get well" to justify further disciplining of the drone population by way of drug-testing programs. Consider this transcribed passage of the opening of Donaldson's report:

DONALDSON: The President came to the White House Press Room to kick off his campaign. . . .

REAGAN (bite): . . . the national crusade against drugs . . .

DONALDSON: A crusade, he made plain, aimed primarily at reducing the drug demand, instead of mainly trying, rather un-successfully, to reduce drug supply.

REAGAN (bite): You're not going to succeed until you take the customer away from the drugs.

DONALDSON: The President said law enforcement should be stepped up against the peddlers, but not, he suggested, against the users.

REAGAN (bite): To those people who are found to be using drugs, we don't threaten them with losing their jobs or kicking them out of school. What we say to them is, "We want to help you get well."

DONALDSON: Mr. Reagan cited protecting society from drug users and producing drug-free schools and work places as goals. But the only specific measure he offered today was mandatory drug testing for some, not all, government employees.

REAGAN (bite): Mandatory testing is justified where employees have the health of others and the safety of others in their hands.

DONALDSON: Aides say the President has in mind people like air traffic controllers and anyone who has a security clearance. A White House drug expert wasn't certain of the legal authority for mandatory drug testing, but in any event most federal workers will be asked to take a test voluntarily. The President said he and his cabinet have agreed to take the test themselves.

REAGAN (bite): To show the way to others, yes. We've all agreed to do it.

The invocation of humanitarian rhetoric couched in the medical/therapeutic language of healing and purification is, of course, a standard ploy in justifying the exercise of modern power and sanctioning the expansion of normalizing procedures and disciplinary technologies like drug testing.

The first part of Plante's report is similar to Donaldson's. But, in keeping with the pattern established in Rather's framing narration, Plante places more emphasis on the number of people involved in this proposed drug-testing program. According to Plante, testing everyone with a national security clearance would have included "over two and a half million people in government and the military alone."

The next part of both reports explores the war on drugs as a partisan political issue. But Donaldson and Plante have very different accounts of how the crusade is playing out on Capitol Hill. Interestingly, both of these accounts feature the same sound bite from the once-powerful Texas Democrat, Jim Wright. First, the Donaldson version:

DONALDSON: On Capitol Hill, members of both parties were enthusiastic about the new Reagan Drug Crusade.

REPUBLICAN REPRESENTATIVE JERRY LEWIS (bite): You've got to have the public with you. And the President is very good about sensing this sort of timing.

JIM WRIGHT (bite): We welcome what the President has said. We want his help.

Like Jarriel's lead-in, Lewis's sound bite naturalizes the poll results as a genuine articulation of public sentiment and portrays Reagan as a responsive and responsible leader with a legendary sense of "timing." In dialogue with Lewis's enthusiasm, Wright's words seem to give unqualified support to the crusade.

In Plante's account, though, this image of an "in-touch," decisive president is almost completely inverted. Instead of framing Wright's comments with the words of a cheerleading Republican, Plante seeks out the critical comments of Peter Bensinger, a former DEA administrator:

PLANTE: With federal statistics showing as many as 5 million regular cocaine users, 20 million regular marijuana users, and 10 million alcoholics, officials acknowledge it's hard to drive down demand. The President points to movies and music videos which portray drug use as normal among the young.

BENSINGER (bite): What was lacking for me was a . . . [garbled word] . . . of penalty against the user.

PLANTE: Democrats in Congress, also sensitive to the public mood in an election year are also demanding action.

WRIGHT (bite): We welcome what the President has said. We want his help. We are encouraged that he is awakened to the realities of the problems.

Therefore, while Donaldson presents Wright's comments as an enthusiastic endorsement, Plante frames the same comments as the forthright demand of a worthy political rival. Furthermore, in Plante's report, Wright's final words (strategically and suspiciously absent from the ABC package) fetch up an alternate image of Reagan—the image of an inattentive, nap-taking dunce who is out of touch with reality and must be rudely "awakened to the realities of the problems."

Plante's conclusion does nothing to counter Wright's words: "In fact, the President's anti-drug campaign is the result of polls that show the public is fed up. But the White House is more interested in providing moral leadership than money. And without money, there's not likely to be any massive mandatory drug testing, but there'll still be political benefit from calling attention to the problem." Notice that this skeptical view of Reagan's political gestures is anything but skeptical about the significance of the polls. For Plante, the polls are a credible measure of public frustration that the White House is trying to appease with the hypocrisy of hollow words. Plante's assertion that polls mean "the public is fed up" provides precisely the kind of closure that makes possible the further "normalization of the power of normalization," including the expansion of police powers.

In stark contrast to Plante's skepticism, Donaldson closes the ABC report with a humorous reference to Nancy Reagan's own well-known anti-drug campaign and a final endorsement of the poll as an agent of truth. In voice-over narration, Donaldson poses the question, "And what happens to the old Reagan Drug Crusade—the First Lady's?" To answer this question, the report includes a brief interchange at the end of the press conference between an anonymous reporter and Reagan:

ANONYMOUS REPORTER: Now that your staff is working on this issue, you're not going to take this away from Mrs. Reagan are you?
REAGAN: Do I look like an idiot? [The assembled reporters break out in appreciative laughter.]

Jocular and seemingly extemporaneous moments like this, of course, helped establish and sustain Reagan's popularity, not so much with the general public, as with the Washington press corps. In fact, Michael Schudson has argued that the mythical popularity that helped make Reagan the Teflon president was an impression that was initially generated by the way Washington insiders embraced the Reagans' cosmopolitan good tastes after four years of Georgia peanuts and Billy Beer. Reagan, in other words, charmed the press establishment long before he won the hearts and minds of Middle America.[28] The systematic display of down-home populist humor, in fact, would become a major political tool in the Reagan administration that the spaniel-loving, broccoli-hating George Bush struggled mightily but failed to master.

As this report suggests, Sam Donaldson, although he cultivated an image of tough antagonist, was at times as vulnerable to and appreciative of Reagan's contrived whimsy as any of the other hacks on bended knee at the White House. In fact, Donaldson's genuine admiration of and mock antagonism toward Reagan are both evident in his conclusion to the report: "The President looks like a man who has sensed the public mood. The polls have told him of the concern and started a push that almost everyone agrees with. But so far, it is mainly generalities. The specifics and the money are still to come." Again, Donaldson gives the polls a life of their own. There is absolutely no consciousness—or at least admission—of the combined roles of the government and news media in generating and maintaining this "public mood." Ultimately, in Donaldson's conclusion, Reagan comes across again as a decisive leader for merely holding a press conference and volunteering to urinate in a bottle.

The Campaigning Voice

If we put aside, for the moment, important differences among the networks, then we can see in simplified form how reciprocal relations among state agencies (NIDA and DEA, in particular), the nation's leading newspapers and news magazines, the major television news organizations, and, finally, public opinion polls helped achieve what Stuart Hall and his colleagues term "ideological closure" on the topic of cocaine pollution. Quoting Hall et al.:

> Once in play the primary definition [of a crime problem] commands the field: there is now in existence *an issue of public concern,* whose dimensions have been clearly delineated, which now serves as a continuing point of reference for subsequent news reporting, action and campaigns. For instance, it now becomes possible for the police, who are somewhat circumspect about appearing to involve themselves in matters which are not yet settled, to *demand wider powers* to act on an issue of crime control which has now been installed as an urgent public matter.[29]

In the case of the 1986 crisis, though, many network reporters came to identify so thoroughly with the police that the police were never actually in the position of having to appear to make demands. Like a ventriloquist's dummy, the network news voiced police demands as dire objective necessities or public mandates.

Rewriting a concept discussed by Hall and his collaborators, we call this demanding voice of alarm the *campaigning voice,* a voice more often associated with the newspaper editorial. However, here the concept is applied to television news reports in which the reporter drops any pretense of being a disinterested observer. In these stories the reportorial performance is marked, instead, by moral disgust, righteous exhortation—or some combination of the two.

Reportorial Disgust   The display of reportorial disgust in drug news is an important aspect of how certain people and groups are marked off and excluded as Others, as Not Us. But the symbolic potency of disgust is certainly not limited to the contemporary. As Peter Stallybrass and Allon White found, the mobilization of bourgeois disgust was a prominent feature of nineteenth-century campaigns to clean up the city and control its inhabitants. Drawing from the work of Cynthia Chase and Julia Kristeva, Stallybrass and White argue that disgust activates "the 'demarcating imperative' [that] divides up human and non-human, society and nature, 'on the basis of the simple logic of excluding filth' ":

> Differentiation, in other words, is dependent upon disgust. The division of the social into high and low, the polite and the vulgar, simultaneously maps out divisions between the civilized and grotesque body, between author and hack, between social purity and social hybridization. These divisions . . . cut across the social formation, topography and the body, in such a way that subject identity cannot be considered independently of these domains. The bourgeois subject continually defined and redefined itself through the exclusion of what it marked out as "low"—as dirty, repulsive, noisy, contaminating. Yet that very act of exclusion was constitutive of its identity. The low was internalized under the sign of negation and disgust.[30]

Perhaps the most outstanding example of reportorial disgust that speaks to the "simple logic of excluding filth" appears in "48 Hours on Crack Street" when Dan Rather takes a tour of a crack house. Looking nauseated, Rather tells his police guide, "I'd be scared to death to come in a place like this."

But the crack house was not the only locus of moral disgust in news coverage of the 1986 crisis. For instance, disgust was the common denominator in back-to-back reports on the August 8, 1986, installment of the "CBS Evening News" that documented the coast-to-coast nonsense of the drug culture. The first report, narrated by John Blackstone, is set in California and covers the disturbing response of many of Oakland's African American residents to the funeral of a notorious drug kingpin named Felix Mitchell (see chapter 9's discussion of "Boyz N the East Oakland Hood"). Many of the people observing the funeral are treated as representatives of non-sense. One black woman says, defiantly, "he was a hero in my eyes." A black man, impressed with what Blackstone describes as the "line of Rolls Royce limousines," "the fancy tuxedos," "the horse-drawn antique hearse," and "the solid bronze casket," says wistfully, "he was popular, it look like to me." According to Blackstone's narration, which often lapsed into self-righteous disgust, the legitimate authorities were powerless in the face of such popular non-sense: "In a city that has been battling to stop the sale of drugs in its parks and school yards, the glorification of a pusher became the center of a controversy. There were many, including city officials, who thought the funeral should not be allowed. But the City Council and the police were powerless to stop it." Blackstone then quotes an older black woman on the street who provides what is meant to represent common sense instead of non-sense. Referring to the funeral, the woman declares (with righteous disgust), "I think it's a disgrace to society. They didn't do this for Martin Luther King." Blackstone's conclusion clearly situates the people who mourned Mitchell's passing as being outside the bounds of the appropriate: "Among those who disapproved of Mitchell's showy funeral, there were some who hoped that at least his death would provide proof that crime doesn't pay. But on the streets, they fear the message will be different." A final non-sense sound bite is interjected here from a jubilant black man in his early thirties: "This is great. I mean I hope I have a funeral like this." Then Blackstone delivers his final campaigning words: "A small-time criminal proved he could get very rich pushing drugs in one poor neighborhood. In the campaign against drugs, Mitchell's funeral was an advertisement for the other side."

The "other side" is also explored in the very next story in the newscast by reporter Bruce Hall. Set in Dade County, Florida (on the "other side" of the country from Oakland), the report focuses on a news character who is surpassed only by the drug dealer as a figure of dis-identification and disgust: the crack mother (see chapter 8 for a more detailed analysis of this sinister image of maternity). According to the opening narration of Bruce Hall's report, "33-year-old Vickie Kay Morgan was told today she may lose custody of her 3-month-old baby forever because she used drugs while she was pregnant." Morgan is portrayed as an unsympathetic, witchlike figure who is unreasonably angry at the court's decision to place the child into temporary shelter care. Confronting the distraught woman as she angrily fled the courthouse, the news team gathered a brief sound bite that reads as a sensible though emotional response from a mother: "We all make mistakes. OK? I made one. That's it. But there ain't nobody taking that baby from me. There ain't no one taking my baby." However, when these words are filtered by the hoarse, barfly voice of a harsh-looking, emaciated, poor woman who gives concrete existence to the stereotype of "white trash," the motherly anguish seems somehow evil and threatening and full of non-sense. Although Hall's performance is, admittedly, more neutral and less overtly judgmental than Blackstone's, there is a kind of residual disgust that bleeds over into the second news report. And the structure of the narrative treatment of the "bad mother," especially in the dramatic confrontation outside the courthouse, is clearly calculated to inspire revulsion in the audience.

The Exhortative "You"   In the exhortation mode of the campaigning voice, anchor and reporter narration often feature the preaching second person "You" in ways that are terrifically patronizing and condescending. Here, the journalist and his friends, the "authorities," are situated as well-informed or knowledgeable adults, and the implied audience (the "You") is invested with juvenile naïveté. It is the same childish "You" that is inscribed in a famous American Christmas tune, "Santa Claus Is Coming to Town": "You better watch out; You better not cry. . . ." It is also the same "You" address used in certain "public awareness" campaigns commissioned by NIDA. In the "[Hey, You!] Just-Say-No" campaign the "You" is, admittedly, not stated but understood. But the exhortative "You" is stated in the famous "public service announcement" that features a tough, white, macho, drill-instructor type cracking an egg and pouring it into a hot skillet while directly addressing the camera and saying, belligerently: "One last time. This is your brain. This is your brain on drugs. Got it?"

Perhaps the most ideologically complicated report to use the campaigning voice of exhortative "You" was prepared by CBS's Bernard Goldberg. Airing on August 7, 1986, just three days after Reagan's press conference on drugs, the report warns that "you better watch out because crack is coming to town." What makes it complicated is that this basic message is integrated into a package that calls into question the "reality" of the crackdown on crack. In keeping with CBS's partisan political orientation, Goldberg acknowledges the crack crisis while criticizing the current efforts to contain the spread of crack as "flashy" but ineffective. Thus, Goldberg adopts a "campaigning voice" in a double sense: in one sense, it is an example of the broader antidrug campaign conducted by NIDA as well as all three networks; but, in another sense, it is linked to the political campaigns of the approaching congressional elections in which the war on drugs commanded center stage.

The semantic structuring of the report is also highly complex compared to the typical news package. The report actually opens with Goldberg using the first-person plural, "We," to identify himself with his implied audience.

In the sound bites that Goldberg integrates into his exhortation it is clear that the police also are firmly located in the realm of "We":

GOLDBERG: We see and hear a lot about the so-called crack-down on crack. In Houston, police use a wrecker to rip down a steel door protecting alleged crack dealers. . . . We see and hear a lot about the so-called crack-down on crack, but listen to the reality behind the flashy arrests.

ANONYMOUS POLICE OFFICER: We can only bust them so many times. They re-open almost immediately after we, uh, as soon as they put another steel door on there, they'll open back up.

However, about midway through the report Goldberg shifts voices into the exhortative "You." In this particular case, though, the shift is camouflaged by a transitional sound bite from a pusher who also uses the second-person singular. Goldberg introduces the transgressor by calling attention to his concealed identity:

GOLDBERG: The pushers won't show their faces, but they can't hide their arrogance.

ANONYMOUS PUSHER: If you were a client and you came to my house, and you started smoking and, for example, you had five hundred dollars, I would make sure you wouldn't leave with a penny. And it wouldn't be hard to do it. Just that one hit and I'll take every penny you got.

GOLDBERG: And if you thought crack was a big city problem, think again. This is Fort Pierce, Florida. Population, 37,000. Officials here say crack is out of control.

In this shift, the "You" rhetoric is quickly hooked up to the authority of the "official" definition of the situation in Fort Pierce.

After interviewing two transgressing teenagers, the report then features three sound bites from confused and distraught mothers at a Fort Pierce town meeting. In the context of this report the mothers come to stand in for the implied naive "You" in need of education and protection:

ANONYMOUS MOTHER #1: What I'd like to know is, how did it get such a big foothold in Fort Pierce?

GOLDBERG: Mothers and fathers hold town meetings in Fort Pierce trying to come to grips with crack.

ANONYMOUS MOTHER #2: My kid never did drugs. He's not an angel, but he didn't do drugs. All of a sudden, he's on this "Rock." He didn't even know what it was.

ANONYMOUS MOTHER #3: And it's gonna cover this country if it's not stopped. This entire country. I don't care whether it's small town, big town, what—it's going to be there.

In keeping with the gender trends in the news, hysteria is generally expressed by women who are speaking as mothers. These mothers do not simply represent common sense—they represent common sense in crisis, common sense in need of guidance and security. It is, therefore, a vulnerable version of common sense (see our discussion of Reaganism and Thatcherism's "composite she" in chapter 8).

In its concluding moments, the report demonstrates how the campaigning voice permits an ideological fusion of the discourse of Goldberg, the reporter, with that of an anonymous undercover officer wearing a mask to conceal his identity:

GOLDBERG: You can be sure that we will continue to see flashy arrests in this so-called crack-down on crack. You can be sure of something else, too.

UNDERCOVER OFFICER: As soon as we put them in, they're gonna be right back out again.

GOLDBERG: Back out again, crack-down or no crack-down.

This fused discourse is perhaps best thought of in terms of a complicated double ventriloquism that begs the question: Who's the real dummy? On the surface it appears that the reporter is a kind of ventriloquist giving a voice to both public and police frustrations. In claiming to present the "reality behind the flashy arrests," the reporter seems to be the agent who articulates and frames the words of the public and the words of the police. But if the reporter's narration is understood to be demanding the expansion of state powers to really crack down on crack, then that puts the state in the role of the ventriloquist and the reporter in the position of, well, the dummy.

Jar Wars

Ironically, then, the anti-Reagan spin to CBS's reporting of the war on drugs helped sustain the sense of crisis beyond the August 4 moment of ideological closure. As figure 6 suggests, the coverage of cocaine as a social problem on NBC and ABC peaked in June and July, respectively, and then rapidly dissipated. CBS's coverage, in contrast, would not peak until after Reagan announced his urinary crusade, and it trailed off only gradually at a rate of two stories per month in September, October, and November.

For us, figure 6 provides a graphic representation of the differing social and ideological allegiances of the three news organizations. NBC, the most

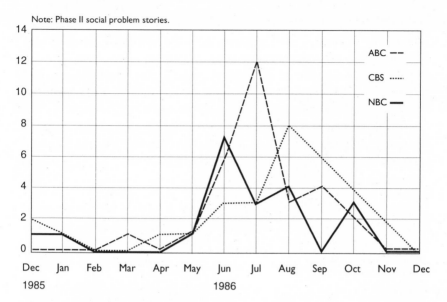

Note: Phase II social problem stories.

FIGURE 6   Network Crisis Coverage

politically conservative of the network news outfits, was more active in the discovery phase of the crisis; and ABC, which is perhaps the most "professionally neutral" of the networks, corroborated the crisis with extraordinarily heavy coverage, featuring twelve stories in July. Both networks treated the August 4 White House press conference as a climax in the crisis when the president supposedly recognized the explosion in public concern about the seriousness of the drug problem and expressed a willingness to undertake and sponsor corrective actions.

The CBS coverage, though, represents a different response to the same kernel event. As a crumbling bastion of old-time liberalism, CBS treated Reagan's press conference not as the end of a social crisis, but as the beginning of a political spectacle. In fact, the New York Times/CBS News Poll cited earlier included questions about the "political implications" of rising antidrug sentiment. Given that the war on drugs was a pet project of the New Right, it should come as no surprise that the poll found that the Republicans benefited the most from this trend: 29 percent of the respondents said that Republicans were better at handling the drug problem to just 17 percent for the Democrats. But, since more than 50 percent of the respondents cited neither party as better, many Democrats saw the drug issue as an opportunity to undermine the Reagan coalition.[31]

A study by the Analysis Group supported by the National Educational Association and the United Automobile Workers arrived at the same conclusions. The study warned that unless Democrats were careful, concern about illegal drugs could help the Republicans in the long run. As quoted in the *New York Times,* the study counseled: "Republican and Democratic candidates can use the issue as a vehicle to establish their character and to show their understanding of some of the voters' deepest feelings about life today. It would be a grave error if, in particular, we allowed older and women voters to slip away on the issue they care about most." [32] The *New York Times* also quotes Stanley B. Greenberg, the president of the Analysis Group, as observing that perhaps the most important political side effect of the antidrug hysteria was that it seemed to be displacing issues "where Democrats elicit greater confidence—reducing the threat of nuclear war, improving education and ensuring a secure retirement." [33]

Republican pollsters most assuredly were dispensing the same advice. And in the desperate attempts by both parties to colonize and exploit the antidrug hysteria, many political races deteriorated into "Jar Wars" in which rival candidates challenged each other to follow the moral leadership of the president and pass a urine test. Although it would be reborn in 1988, the partisan political one-upmanship culminated in 1986 with the preelection passage of the $2 billion Drug Free America Act, which "expert" Lloyd Johnston admitted "can at best be described as hastily drafted Federal legislation." [34]

Although we do want to hold CBS accountable for playing a major role in sustaining the crisis as a political spectacle after August 4, we do not want to bash CBS at the expense of letting the other networks off the hook. In fact, figure 6 not only illustrates important differences in the distribution of coverage among the three networks, but also points to a weakness in our research design. By focusing solely on the news stories indexed as appearing on the daily evening newscasts of the three major networks, we neglect to take into consideration coverage that appears on morning shows, in prime-time news specials, or throughout the day on CNN. That is why our graph of television coverage looks different from one presented by Danielian and Reese in a study that does at least account for prime-time news specials aired in September (CBS's "48 Hours on Crack Street" and NBC's "Cocaine Country"). Whereas our graph peaks in July, the Danielian and Reese graph peaks in September.[35]

In fact, others have identified NBC as the chief network culprit in hyping the 1986 drug scare. According to some estimates, in the seven months leading up to election day, NBC devoted fifteen hours of airtime to cocaine and crack.[36] Clearly, NBC's "Cocaine Country," in declaring that cocaine and crack had become "pandemic," was at least as excessive in exaggerating the threat posed by drugs as CBS's "48 Hours on Crack Street." And Tom Brokaw's disgust reaction artfully displayed when introducing drug stories is also at least as convincing as Dan Rather's. In assigning demerits for the cracked coverage of 1986, there is, indeed, more than enough blame to spread around.

8

Family Matters:

Nurturing Normalcy/Reproducing Delinquency

> Overall, it may be that, in a time of economic and political insecurity, the
> desire to control one's children (since other aspects of life seem out of control)
> becomes a kind of conduit for other fears—of racial "mix," of economic loss,
> of sexuality itself. In this context, invocations of "the family" communicate a
> complex set of meanings and moral sanctions. "The family" provides a new
> moral thrust, a new legitimation, for older right-wing values such as racism
> or prayer in the public schools, which, since the 1960s and the civil rights
> movement, are no longer so easily justified in their own terms. Thus, to argue
> that "moralism" is used by the New Right merely as a gloss to obscure its
> real political aims is misleading, for economic pressures, class divisions, and
> racism are intimately tied to people's concerns about "family autonomy" and
> to sexual politics.—Rosalind Pollack Petchesky[1]

The Functional Family

Despite the many disruptions in the journalistic treatment of the war on
drugs during the 1986 crisis coverage, one major recurring theme would
provide a degree of continuity for the cocaine narrative throughout the
Reagan era. As the news frame shifted from the therapeutic detachment of
the trickle-down paradigm to the proactive policing of the siege paradigm,
issues related to unsettling developments in American family life continued
to play a central role in the journalistic construction of the cocaine problem.
In many cases the language of functionalism is invoked in these reports—
"dysfunctional" families reproduce delinquency; and "functional" ones are
the first line of defense in the war on drugs, the means by which children
should first learn to just say no.[2] However, for us, the symbolic power of
"the Family" as a moral ideal and political project is much more persis-

tent, much more "functional," than its actual existence and operation in the world of everyday life. As Rayna Rapp argues, the concept of family "carries a heavy load of ideology":

> "Family" (as a normative concept in our culture) . . . is a socially necessary illusion which simultaneously expresses and masks recruitment to relations of production, reproduction, and consumption—relations that condition different kinds of household resource bases in different class sectors. Our notions of family absorb the conflicts, contradictions, and tensions actually generated by those material, class-structured relations that households hold to resources in advanced capitalism. "Family," as we understand (and misunderstand) the term, is conditioned by the exigencies of household formation, and serves as a shock absorber to keep households functioning. People are recruited and kept in households by families in all classes, yet the families they have (or don't have) are not the same.[3]

Because of differences in resources associated with class stratification, all families are not created equal. Nor are all household structures ideologically equal in the contemporary politics of gender, sexuality, and moral control. Clearly, the antifeminist and antigay backlash of the Reagan-Bush pro-family agenda treats the nuclear family as a healthy norm and marks off alternative household structures headed by single parents, same-sex couples, or two working heterosexual parents as somehow "dysfunctional," "unnatural"—even un-American.

We have noted in previous chapters how concern about "the family" contributed to the journalistic stigmatization of the contaminated generation that "dropped out" and "turned on" in the 1960s, how the loss of family is a prominent feature of confessions voiced by white, male, middle-class transgressors in therapy stories, and how the African American family is treated as a "tangled web of pathology" in 1986 crisis reports influenced by cultural Moynihanism. In this chapter we want to expand on these observations in concrete analyses of the contrasting images of maternity projected in news coverage featuring First Lady Nancy Reagan and crack mothers.

To begin to grasp how these controlling visions of motherhood are deployed in the drug stories, we first return to the history of Modernity and the politics of Reaganism that have provided the two grand contexts informing our interrogation of the cocaine narrative. Connecting the nuclear household ideal to the turbulent history of Modernity allows us to see what Judith Stacey calls the "peculiar, ephemeral, and internally contradictory" character of this "once-revolutionary gender and kinship order."[4] Treating "the Family" as a key "normative concept" in Reaganism's orthodoxy of nostalgia allows us to appreciate how the New Right exploited anxiety about the insecurities of more flexible, postmodern living arrangements in promoting its so-called pro-family agenda.

Domestic Revolutions

> The experiences of Americans during the depression of the 1930s had paved the way for two different family forms: an emergent form based on two egalitarian breadwinners; and another, more traditional form based on a married heterosexual couple with polarized gender roles—the male breadwinner and the female housewife. But enormous numbers of young adults chose the latter form, determined to make it work for them, even as increasing numbers of women became part of the paid labor force. The potential for developing new ways of structuring families and for radically altered gender roles withered in the face of obstacles to women's employment—poor working conditions, low wages, prevailing social attitudes, and a government policy that encouraged a "family wage" for men and, as a result, female dependence on men.— Jackie Byars[5]

Although the New Right tends to treat it as a fact of nature, or an instrument of God, the privatized nuclear household with its male breadwinner, female homemaker, and dependent children is a relatively new household formation that, like journalism and therapy, is both a product and a producer of the modern condition. In other words, as an ideological project and a practical accomplishment, the Modern Family works as a major consciousness-producing instrument and a powerful vision of social control in the political economy of contemporary America.

The strictly delineated sex roles of this domestic model first emerged in the United States among white, middle-class people in the late eighteenth century (the decades following the American Revolution when Enlightenment thinking was ushering in the Modern Age throughout the Western world). During the course of the nineteenth century its mode of organizing the life-world would come to be hailed as an insurgent ideal. However, until the midpoint of the twentieth century, it was by and large an impractical and luxurious way of running a household that was well beyond the reach of most Americans; most poor white, ethnic households in industrialized urban centers could ill afford to subsist on one laborer's paltry wages; most farming households were still organized according to the residual forms of older patriarchal and extended family patterns; and most African American households were either subordinate to that "peculiar institution" or, after emancipation, too poor to realistically have mothers devote full time to caring for their own young (although a good many African American women cared for other people's children as full-time domestic servants).[6]

Modern Living   For several complicated reasons associated with the yearnings for stability born during the hard living of the Great Depression, the bottled-up sexual energy released with the return of servicemen after World War II, the "containment" thinking of atomic-age paranoia, and the general prosperity of the postwar expansion (as well as its general "rootlessness"), the nuclear family would peak as something of a norm during the "cold war culture" of the 1950s.[7] In that aberrant decade, about three-fifths of the households in the United States conformed to its prescription for living.[8] As many scholars have observed, the triumph of the nuclear family in the 1950s also represented the triumph of Fordism's ethic of consumption in which an older producer ethic based on work, self-sacrifice, and saving was displaced by a new consumer morality based on leisure, self-fulfillment, and immediate gratification.[9] In this transformation the nuclear family's "egoism of possession" was meant to compensate for conditions under which, in modern corporate America, work had lost much of its intrinsic value as a source of pride, accomplishment, and identity.[10] As many social commentators of the day fully recognized, for the organizational men in their gray flannel suits caught in the impersonal white-collar world of corporate hierarchies and for the relatively affluent laborers in their assembly-line jobs caught in the routinized blue-collar world of mass production, work was often an alienating experience.[11]

Richard Nixon, one of the leading political figures of this cold war culture, was also a leading spokesman for its ethic of consumption. In his famous "Kitchen Debate" with Nikita Khrushchev, Nixon conflated the American Way of Life with its standard of living, declaring that the U.S. system worked because "44 million families in America own 56 million cars, 50 million television sets, 143 million radio sets, and . . . 31 million of these families own their own homes."[12] As Elaine Tyler May observes, Nixon's remarks "struck a responsive chord among Americans at the time":

> He returned from Moscow a national hero. . . . Clearly, Americans did not find the kitchen debate trivial. The appliance-laden ranch-style home epitomized the expansive, secure life-style postwar Americans wanted. Within the protective walls of the modern home, worrisome developments like sexual lib-

eralism, women's emancipation, and affluence would lead not to decadence but to wholesome family life. Sex would enhance marriage; emancipated women would professionalize homemaking; affluence would put an end to material deprivation. Suburbia would serve as a bulwark against communism and class conflict, for, according to the widely shared belief articulated by Nixon, it offered a piece of the American dream for everyone. . . . [13]

Nixon's consumerist version of the American Dream, it should be noted, also was being hawked by Ronald Reagan who worked for General Electric at that time and appeared with his second wife in magazine ads as GE's model couple relaxing in their model "GE all-electric home." [14] According to Garry Wills, GE was a pioneer "at creating a demand before supplying the demanded items" because it realized, early on, that "its task was not simply to sell appliances but to create a desire for the kind of life that craves such appliances." [15] The 1950s family ideal represented by the Reagans was precisely the kind of all-consuming home life that GE was devoted to electrifying. [16]

Along with the frontier family of the American West, the 1950s family ideal also has operated, in the popular countersubversive imagination, as a model of the kind of "we-always-stood-on-our-own-two-feet" self-reliance that scoffs at people who receive public assistance at "taxpayers' expense." But this myth of family self-reliance is based more on popular delusion than on historical fact. Indeed, according to Stephanie Coontz, the nineteenth-century frontier family and the twentieth-century suburban family "probably tie for the honor of being the most heavily subsidized [family types] in American history, as well as for the privilege of having had more of their advantages paid for by minorities and the lower classes." [17] Contrary to the anticommunitarian disinformation [18] circulated in the *Little House on the Prairie* books, pioneer families who established homes in the American West in the 1800s "owed their existence to massive federal land grants, government-funded military mobilizations that dispossessed hundreds of native American societies and confiscated half of Mexico, and state-sponsored economic investments in the new lands." [19] The 1950s white suburban family enjoyed similar favoritism from the federal government. The GI bill, the National Defense Education Act, the Federal Housing Authority, the Interstate Highway Act—all benefited the suburban family by expanding educational opportunities, creating new, high-paying jobs, underwriting home ownership, and facilitating suburban development by making commuting to work a way of life—and a way of exclusion. "Such federal patronage," writes Coontz, "might be unobjectionable, even laudable—though hardly a demonstration of self-reliance—if it had been available to all Americans equally. But the other aspect of federal subsidization of suburbia is that it worsened the plight of public transportation, the inner cities, poor families in general, and minority ones in particular." [20] The shift of resources from the maintenance of urban infrastructures to the stimulation of suburban expansion, the informal racism of federal red-lining policies that declared many urban areas ineligible for loans, and the steady deterioration of public transportation and educational services would sow the seeds of the "hyperghettoization" whose ugly bloom currently casts a devastating shadow over most large American cities. [21]

In the three decades following the 1950s, with the collapse of Fordism and the triumph of flexible accumulation, the Modern Family would go into steep decline. Divorce rates soared. Second-wave feminism challenged the institutionalization of male dominance in the workplace. [22] The family wage of Fordism's affluent worker was ravaged by deindustrialization, inflation, and the defeat of the labor movement. And more and more women entered the paid labor market, a privileged few by an affirmative choice to pursue the personal autonomy of a career outside the home; many others by a more reluctant decision to achieve the higher standard of living enjoyed by dual income households; and still others by necessity brought on by the responsibilities of single parenthood. By 1985, only 7 percent of U.S. households conformed to the model of the fully employed breadwinning father and full-time homemaking mother. [23] Although this figure admittedly does not take into account the number of families that pass through a life-cycle stage with this pattern, it still underscores how the "Father Knows Best" norm of the 1950s has become, once again, a widely shared ideal that does not express the facts of life encountered by most American families.

Therefore, the social history of the nuclear family follows roughly the same arc as that of the rise and fall of modern industrial society. Over the course of that trajectory, the family was transformed from its premodern corporate form as the primary production unit into its contemporary privatized form as the primary consumption unit. According to Stacey, this transformation was marked by "four radical innovations":

> (1) Family work and productive work became separated, rendering women's work invisible as they and their children became economically dependent on the earnings of men. (2) Love and companionship became the ideal purposes of marriage that were to be freely contracted by individuals. (3) A doctrine of privacy emerged that attempted to withdraw middle-class family relationships from public scrutiny. (4) Women devoted increased attention to nurturing

fewer and fewer babies as mothering became exalted as both a natural and a demanding vocation.[24]

In the public-private split that accompanied this transformation, the public sphere was defined as a masculine world of instrumental production governed by rational self-interest and the ethic of competition; and the home, a feminine domain of "expressive" consumption devoted to gratifying self-fulfillment and the ethic of nurturance. As Margaret L. Anderson suggests, this represents a sex-gender system that "organizes not just family relations but all social processes and institutions."[25] As we reported in chapter 3, this system is evident in the gendering of power and knowledge of expert sound bites appearing in our study: men are clearly authorized to "know" more in the world of network drug news.

Modern Motherhood In the modern transformation of family ties, the weakening of patriarchal authority in the home would become a mixed blessing for the modern mother. On the one hand, the cult of modern motherhood championed male-female relations that were more egalitarian than the more rigid sex-gender hierarchies and subjugations of the older patriarchal corporate units and family economies of Puritan, colonial, and agrarian America. In fact, many first-wave feminists of the Progressive era embraced the ideal of the nuclear family in crusading against prostitution, supporting campaigns to "clean up" the movies, lobbying for child labor laws, and pushing for a number of educational reforms. As Eli Zaretsky put it: with few exceptions, Progressive era feminists "sought to combine greater individual opportunities for women with protection of the family. Both goals involved governmental action, and in both cases, whether for individual women or families, they took for granted that the basis for independence would be a wage or a salary."[26]

On the other hand, the psychic costs of being confined to the home were considerable. In this sex-gender system, women were essentially reduced to their reproductive identities. During the 1950s this identity structure often caused white middle-class women to suffer in suburban isolation from what Betty Friedan termed the "problem with no name"—the double bind of being both valued and devalued as mothers, of being at the same time idealized as the source of all nurturance and condemned as the root of all pathology.[27] "It was suddenly discovered," writes Friedan, "that the mother could be blamed for everything": "In every case history of the troubled child; alcoholic, suicidal, schizophrenic psychopathic, neurotic adult; impotent, homosexual male; frigid, promiscuous female; ulcerous, asthmatic, and otherwise disturbed American, could be found a mother. A frustrated, repressed, disturbed, martyred, never satisfied, unhappy woman. A demanding, nagging, shrewish wife. A rejecting, overprotecting, dominating mother."[28] Modern motherhood, then, involved a maddening disparity (literally) in parental duties that charged women not only with being responsible for preserving the life of children and fostering their growth, but also for shaping acceptable—or normal—subjectivities. Consequently, although the modern mother may be physically removed from the world of wage labor, she is still implicated in a system of normalization devoted to reproducing a docile work force and stabilizing a dependable marketplace for mass-produced durable goods like television sets, station wagons (now minivans), refrigerators, and microwave ovens.[29] In this system it is her "function" to shape children who are obedient, skilled in the discipline of "being good," and ready to conform to their responsibilities as private citizens and confident consumers by assuming their sexually determined roles as male providers or female nurturers. Almost by definition, families with mothers who fail to normalize their young into this sex-gender system are "dysfunctional."

Engendering Journalism Unfortunately, the taken-for-granted assumptions of this concept of motherhood not only informed Dan Quayle's 1992 election campaign condemnation of TV's "Murphy Brown," but they also pervade much psychiatric and social scientific research on mental illness, work, and family. In the case of psychiatry, as David Spiegel argues, "*family* is a code word for *mother*" in most family explanations of mental illness:

> The mother is seen as responsible for the family environment, for the psychological well-being of her children, and for the emotional atmosphere in the home. To whatever extent we have believed that mothers do something special for their children, we have been more willing to believe that failures in mothering do something devastating to children. In line with this conventional—and distorted—view, perceived or real defects in the emotional environment of the family or the child's psychological development have usually been attributed to some failure in the mother rather than in the father.[30]

In the case of social science, as Rosabeth Moss Kanter discovered, official reports in the 1970s tended to interpret labor statistics according to a double standard: male *unemployment* was regarded as a social problem, while female *employment* was treated as a family problem.[31]

These assumptions, not surprisingly, also underpin journalistic accounts of family problems in contemporary America. Take, for example, the first installment of Bill Moyers's four-part series titled "America's Kids: Growing Up Too Fast" (CBS, 10/22/84)—one of the "noncocaine" reports that we analyzed in our study. Early in the package Moyers visits Covenant House (a Catholic youth shelter in New York City), where Father Bruce Ritter (who a few years later would be forced to give up his position because of charges that he molested some of the children under his care) proposes that "something is different about the new legion of children crying out for help":

RITTER: The kids are much harder, more hopeless, more despairing.
MOYERS: What has changed in American society that has increased this trend?
RITTER: Without a doubt, we are witnessing the increasingly rapid acceleration of the break-down of the American family.

To lend support to Ritter's "expert" diagnosis, Moyers implicitly equates the "American family" with the nuclear household in reporting that "nearly half of all children grow up in single-parent or divorced homes." Then, Moyers visits two boys "who spend part of each week with their father and the other part living with their mother." "They are normal kids," observes Moyers, "and they see their divided lives as just that"—the implication being that the kids are deluded into thinking that the lives they lead as part of a "broken family" are "O.K."

But, for ponderous Moyers, the prevalence of divorce is only part of the problem. "Life is different, too," he reports "even for all those children whose parents have stayed together": "Well over half of the grade school children have mothers who work. And that means when the kids come home from school, they're often home alone." This inference is the opening argument for an indictment of the phenomenon of "latchkey kids" that manages to put more blame on working mothers than on working fathers. According to the demonology of the rest of Moyers's series, working mothers take their place beside drug abuse and television viewing as a clear and present danger to the childhood, the mental health, and the physical well-being of "America's kids."

The sexual disparities of Moyers's report also are evident in the treatment of women who appear as drug transgressors in our study. Later, we consider how the journalistic visualization of deviancy during the crack crisis often reduced women to their reproductive identities in the controlling image of the "crack mother." However, this treatment of female

transgression was also apparent before the 1986 crisis. For instance, consider a seriocomic story on NBC titled "Cocaine Grannies" that appeared
on March 29, 1983. In keeping with the trickle-down paradigm, the story is
about the decadent upper-middle class in our society who entered the drug
business for "adventure," in the word of anchor Tom Brokaw. Featuring
"four middle-aged, divorced women—grandmothers," apparently bored
with shopping and "looking for easy cash and a little excitement," the package investigates a cocaine-trafficking operation that was headquartered in
a sprawling house "in the canyons above Beverly Hills."

In the major therapeutic scene, "Grandma Joy," fifty-eight, confesses
to reporter Brian Ross over coffee in the safety of her kitchen. She had been
arrested for, in her terms, "babysitting" thirty-seven pounds of cocaine
stored in her closet: "A lot of us felt that, 'Oh, well we can do the same
thing men can do. . . . Men do it, why can't we.' Uh-uh. It doesn't work
that way. And I made a very big mistake in my life. And it was a very
big price that I paid." We found, of course, no equivalent example of a
"cocaine grandpa" story, in which older male characters were defined only
in terms of their gender and familial identity. The moral viewpoint here
associates the notion of grandmothering with specific "family values" and
the gender roles that these women had violated in their drug deviancy.
"Grandma Joy," at least in the NBC story, now seeks a return to "shopping for her grandson," apparently a sensible consumerist "cure" for her
deviant, drug-dealing past.

The Profamily Agenda

Although Reagan invokes traditional family values, the actual family of our
first divorced president comprises divorce in three generations (Nancy Davis's
parents, Reagan himself, and Maureen Reagan); one daughter who has been
married three times and another who lived out of wedlock with a member of
a rock group; one son who starred in high school football and baseball while
his father failed to see a single game and another who married a woman his
parents had met but once (and who informed them of the wedding night only
after it was over); and a grandchild whose grandfather did not see him for the
first two years of his life. The split between what *Ronald Reagan* represents
and who he is, I believe, reassures Americans who are also confronted with
difficulties in living out traditional family values but who preserve them in the
realm of the signifier.—Michael Rogin[32]

In keeping with the moral ahistoricism of its racial politics, the sexual politics of the New Right strips the nuclear family of its modern credentials
and erroneously treats it as the "traditional family" or, even more erroneously, the "biblical family." From this myopic perspective, the New Right
portrays the nuclear family as a shelter under siege by enemies both without and within, a haven of self-reliance terrorized by liberated women,
effeminate men, and the intrusive forces of the welfare state. With the fabric of its "profamily agenda," the New Right fashioned a big tent able
to accommodate many of its single-issue constituencies. For this broadly
antiwelfare, antiabortion, anti-ERA, antiaffirmative action coalition, the
forces arrayed against the family are actively involved in: eroding the family
wage by raising taxes and redistributing scarce resources to undeserving
welfare chiselers; taking good jobs away from benevolent breadwinners
and giving them to selfish career women or, even worse, to "less qualified
minorities"; ignoring parental educational preferences by busing children
to achieve unwelcome racial integration; undermining parental control of
family morality by making sex education, contraceptives, and abortion services freely available to teenagers. In concert with the divide-and-conquer
strategies of single-issue politics, then, Reaganism deployed a unite-and-
mobilize strategy in forming a powerful coalition around the idea of "protecting" the family from these forces.

The Doctrine of Corporate Privacy   Following Petchesky, we suggest
that the organizing principle coordinating both its mobilizing and conquer-
ing strategies was Reaganism's peculiar doctrine of privacy. The New Right
found a common grounding for its pro-family and pro-business agendas
by promoting an innovative angle on privacy that represents a drastic departure from classical conservatism. Where classical conservatism valued
individual privacy, the New Right, although it often calls for personal freedom from state interference in economic matters, is more than willing to
sponsor such interference when it came to matters of personal morality. The
justification for this seemingly contradictory position is a corporate privatism that serves business, church, private schools, and privatized nuclear
families—a privatism that, in Petchesky's estimation, "is much closer to
fascism than to classical libertarian doctrine and is thus perfectly compatible, in theory as well as in practice, with a program of massive state
control over individuals' private lives." [33] It is this spin on privacy, then,
that animates Reaganism's loosening of regulatory constraints on entrepre-

neurial interests while simultaneously tightening the margins of illegality on post-Fordism's peripheral, reserve, and demonized labor markets.

The way in which the notion of "choice" played out in the rhetoric of the Bush administration further illustrates the inclusions and exclusions of this doctrine of corporate privatism. In the realm of individual privacy the Bush administration was staunchly against "choice" when it comes to abortion rights—which means that it did not support a woman's freedom to choose whether or not to terminate an unwanted pregnancy. Following Reagan's lead, Bush did his best to fulfill his commitment to stack the Supreme Court with appointees who might vote to overturn the privacy provisions of *Griswold* v. *Connecticut* and *Roe* v. *Wade*. However, in the realm of corporate privacy, the Bush administration supported a policy of "choice" in education—a policy that involved dispensing tax credits to parents who send their children to private schools with promises to give other less fortunate parents more say about which school their children attend in public school systems. It was, in other words, a policy meant to federally subsidize the further privatization of the American education system and morally legitimize the further de facto segregation of public schools without any pretense of being separate but equal.[34]

Ironically, given its demonization of the Fordist economic order, the New Right's pro-family agenda embraced an ideal that was only a reality for most American families during the heyday of Fordism and its affluent worker—and is not actually a viable option for most of the working poor caught up in the peripheral labor markets of Reaganomics.[35] As Petchesky observes, this profamily agenda "has achieved a mass following and a measure of national political power because it is in fact a response to real material conditions and deepening fears, a response that is utterly *reactionary*, but nevertheless attune."[36] The greatest travesty of the Reagan era may be that the administration elected, at least in part, by celebrating the virtues of the nuclear family, would then pursue a doctrine of economic privatization that was further destructive of the very securities required to maintain the "settled living" of such a household.[37]

New Right Womanhood   One of the bitterest ironies of the Reagan-Bush age may be that many of the New Right's most outspoken proponents of an agenda that naturalizes the second-class status of women were, in fact, women. The most prominent women of the New Right—Phyllis Schlafly of the Eagle Forum, Connaugh "Connie" Marshner of the Heritage

Foundation, and Beverly LaHaye of the Concerned Women of America (CWA)—have mastered the hypocrisies of making a career out of bashing career women and of guiltlessly working to make mothers feel guilty for working. Schlafly, the earliest, most famous, and most extreme case, is a Harvard-educated lawyer, author of numerous books, and a two-time congressional candidate who opposed the ERA because it "would take away the marvelous legal rights of a woman to be a full-time wife and mother in the house supported by her husband." Marshner, a graduate of the University of South Carolina and author of *The New Traditional Woman,* advises in her lectures that "women need to know that somebody will have the authority and make the decision" and that their "job" is "to be happy with it." Author of *The Spirit-Controlled Woman* and married to Tim LaHaye (a cofounder of the Moral Majority), Beverly LaHaye is herself in firm control of CWA, a kind of "one-woman fiefdom" with an estimated 150,000 to a half-million members.[38] In Susan Faludi's words, these reactionary activists "could report to their offices in their suits, issue press releases demanding that women return to the home, and never see a contradiction": "By divorcing their personal liberation from their public stands on sexual politics, they could privately take advantage of feminism while publicly deploring its influence. They could indeed 'have it all'—by working to prevent other women from having that same opportunity."[39]

But beyond expressing the hypocrisies of the New Right, Schlafly, Marshner, and LaHaye give a concrete form to a reactionary maternal discourse that supported the patriarchal excesses of the Reagan counterrevolution. Stuart Hall's discussion of how this discourse played out in Thatcherism is relevant to our critique of the New Right's cult of motherhood. In the British context, according to Hall, women are often represented as a *composite "she."* The "emblematic mother of conventional sexual ideology," this Thatcherite "she" is the generalized guardian of the family, keeper of common sense, and custodian of traditional values:

> "She" is . . . concerned about the educational chances of her children: the woman alone in the streets at night, who cannot go about her ordinary business unmolested: the housewife whom the state and the permissive educators would seek to detach from her traditional role and "force" to "abandon" her children and her hearth and go out to work: and, properly addressed, she is the wife of the militant trade unionist on strike, who brings home to him the harsh realities and consequences of living without the weekly wage, and urges a "speedy return to work"—for the sake of the children, of course.[40]

In Reaganism, as in Thatcherism, this composite "she" has played a crucial ideological role in a series of moral panics organized around family, crime, drugs, sex, rock 'n' roll, and rap. The Reaganite "she" is above all in need of protection—even from her own equality. Schlafly, for instance, used the idea of protection to discredit the ERA, which she claimed would "strike at the heart of women's family support rights."[41]

This "emblematic mother," then, is one who accepts the "patriarchal bargain" of trading autonomy and equality for the protection and love of someone to watch over her.[42] In the nuclear family this bargain is binding in a double sense: for women, it ties them to housework and child care; and for most men who enter into this contract in good faith, it carries with it a life sentence at alienating labor with parole only at death or retirement. In other words, this sex-gender system is mostly a bargain for privileged men who are not vulnerable to the daily oppression, desperation, humiliation, and hazards of working for someone else for a living. For the vast majority of men, "family obligations" are a major disciplinary force that capitalism can, and does, take full advantage of in its ongoing struggle to control labor markets. This is, precisely, the double bind that connects the New Right's profamily and probusiness agendas.

As Faludi argues, Schlafly, Marshner, and LaHaye enjoy their prominence only because they accept the terms of this patriarchal bargain:

> As long as the [New Right] women raised their voices only to parrot the Moral Majority line, as long as they split chores [with their husbands] only so they could have more time to fight equal rights legislation, the New Right male leaders (and their New Right husbands) were happy to applaud and encourage the women's mock "independence." The women always played by their men's rules, and for that they enjoyed the esteem and blessing of their subculture. On the other hand, working and single women in the mainstream, who were more authentically independent, had no such cheering squad to buoy their spirits; they were undermined daily by a popular culture that parodied their lifestyle, heaped pity and ridicule on their choices, and berated their feminist mistakes.[43]

The experience of one discarded and discredited New Right woman, though, bears witness to the limits of this arrangement. For three years in the 1970s, she was a national symbol and proud leader of the antigay movement. However, after divorcing her husband, she was denounced by the same male-dominated religious organizations that raised millions by exploiting her familiar name. As former Miss America Anita Bryant discovered, the "love" in the New Right's patriarchal bargain is anything but unconditional.[44]

The Velcro First Lady

> Ronald Reagan was an undeserving beneficiary of the profamily reaction, as humorist Delia Ephron observes in a review of Maureen Reagan's dutiful memoir: "It is funny and a bit pathetic that Ronald and Nancy Reagan keep finding out their family secrets by reading their children's books. It is also ironic that this couple who symbolized a return to hearth, home and the 1950s innocence should, in reality, be candidates for a very 1980s study on the troubled family." The former president's less dutiful daughter, Patti Davis, agrees: "Anyone who hasn't been living in a coma for the past eight years knows that we're not a close-knit family." It seems an astonishing testimony to Reagan's acclaimed media magic, therefore, that despite his own divorce and his own far-from-happy blended family, he and his *second* lady managed to serve so effectively as the symbolic figureheads of a profamily agenda, which his economic and social policies helped to further undermine.—Judith Stacey[45]

In the drug news, the women who appear in commonsense sound bites often give voice to the composite "she" in need of patriarchal protection. By far the most important figure to give a voice to reactionary maternal discourse in the 1980s cocaine coverage was a very specific "she"—First Lady Nancy Reagan. One of the most controversial public figures of the era, Nancy Reagan was, to her credit, much despised by the New Right. Unlike Schafly, Marshner, and LaHaye, the former First Lady was both "in control" and "out of control." She exercised a tremendous amount of personal influence over the scheduling and scripting of the "acting" president's daily performances. She was also a ruthless foe between and behind "the scenes" (as Lyn Nofziger, John Sears, and most notoriously, Donald Regan, found out on being dismissed or "forced out" of campaign or staff positions after incurring her wrath).[46]

Mommy Dearest   But the New Right was not alone in its displeasure. Mrs. Reagan was also almost universally despised by the news media. Although we *do not* come to praise her, we *do* want to distance ourselves from, and even condemn, the generally hostile journalistic treatment of the former First Lady during the 1980s. We suspect that the press response to

Nancy Reagan was not all that different from the New Right's. In both responses, we detect the distinct odor of the misogynist fear of an omnipotent, controlling mother who has turned the patriarchal bargain on its head.

If her husband was the Teflon president, then Mrs. Reagan should go down in history as the Velcro First Lady. Where Jacqueline Kennedy made decorating the White House into a "cause" and was loved for it, Nancy Reagan was roundly bashed as the Porcelain Princess for simply raising money to buy new china for the executive mansion; and where Mrs. Kennedy was greatly admired for her fashionable taste in clothing that could only be quenched by the occasional Paris shopping spree, Mrs. Reagan was greatly condemned for taking advantage of her position to fill her wardrobe with donated designer gowns. The negative press would climax with inflated accounts of her disputes with Raisa Gorbachev that were reported as "cat fights" with global ramifications. Over the course of the 1980s she came to occupy roughly the same symbolic news space as that of hotel "queen" and tax dodger Leona Helmsley. Both were often depicted as ruthless, controlling, shrewish wives with regal pretensions. Both also activated myths of the all-powerful, overprotective, emasculating mother—the demonized mommies of Philip Wylie's "momism"; the mothers who Christopher Lasch once vilified for imposing their "madness on everyone else." [47]

The contrasting public images of Ronald and Nancy Reagan have something of the classic Mutt and Jeff dynamic to it. In the running of his campaigns and his White House staff, he wore the white hat, she the black; he patted people on the back, she sent them packing. In a strange inversion of conventional gender relations, he came off as the warm, affable, emotional, dependent, expressive one (he is reportedly known to address her as "Mommy," even though he did not call his own mother by that name),[48] and she as cold, calculating, cerebral, independent, rational. However, it was this dynamic, perhaps more than anything else, that gave their duet performances such power. For in their joint performances the shrew is not only tamed, but locked in an adoring spell. "Though Nancy Reagan never became a Hollywood star," writes Garry Wills, "she did not waste the eight years she spent trying":

> She learned a thing or two about the Look. She practices a severe economy of expression that makes three or four compositions of the features cover all the necessary social tasks—the smile of delighted surprise at the top of the air-

plane stairs, the concerned gaze at redeemable druggies, and—most of all—
the upward stare at her husband. In photograph after photograph, one finds
the same expression, not varied by a centimeter. The mask has been perfected.
It takes unsleeping vigilance to remain a somnambule, a political chastity symbol. She is the conduit of feeling for those who jump and demonstrate when
Reagan appears or finishes his speech. She must not break the spell; but the
voltage of political loyalty travels through her fixed expression, just as surely
as sex flowed through the cool white face of Garbo.[49]

But, as Wills further observes, it is difficult to say whether Mrs. Reagan's
entranced Look is "that of the hypnotized or the hypnotizer."[50] Her reputation for being a "difficult woman," for being a woman "in control" and
"out of control," gives the Look even more power and gives her husband
even more credibility and stature—for it is no mean accomplishment to
command the undivided "attention" and "loyalty" of such a seemingly
strong-willed and powerful person.

Laser Eyes   As three network reports airing on April 24, 1985, illustrate, Mrs. Reagan's "Mommy Dearest" persona certainly cast a shadow
over her activism in the war on drugs. Covering her hosting of an international drug conference for the wives of leaders of other governments,
all three packages were reported by women journalists (ABC's by Carole
Simpson who is African American; CBS's by Leslie Stahl; and NBC's by
Andrea Mitchell; Stahl and Mitchell are white). Of these, the CBS report is
especially noteworthy. Devoting scant attention to the actual conference,
Stahl presents a surprisingly sympathetic retrospective on the First Lady's
"image problems" in which the reporter sides with Mrs. Reagan against
her critics. The competing packages on ABC and NBC also provide affirmative accounts of Nancy Reagan's hosting of the antidrug conference. But
compared to Stahl's existential quest for the "real" Nancy Reagan, ABC and
NBC offered rather conventional treatments of a newsworthy event. And
yet both are worth considering because of the way Nancy Reagan was able
to use the conference to define the drug problem in terms consistent with
Reaganism.

Mrs. Reagan's influence is evident in the key oppositions that structure both reports—oppositions that confirm the instrumental-masculine/
expressive-feminine organization of modern knowledge. On NBC this
opposition is explicitly developed in Tom Brokaw's introduction: "Nancy
Reagan is a champion of the need to fight drug abuse and today she brought
attention to this world-wide problem by holding a summit of First Ladies

from around the world. Andrea Mitchell reports that they focused on the human rather than the political side of drug abuse." The human/political split, of course, conforms to the private/public split of the modern sex-gender system. According to this system, a summit of First Ladies would, "naturally," be devoted to the subjective human emotions—rather than the hard political facts—of drug abuse. ABC, although also observing these oppositions, is much more subtle and self-conscious. For instance, Peter Jennings's lead-in includes a shrewd tip of the hat to the feminist critique of these distinctions: "Here in Washington, today, a group of what we still call First Ladies from a number of countries sat down with Mrs. Reagan to see what they could do about children and drugs all over the world." In the phrase, "what we still call First Ladies," Jennings plays a crafty game of journalistic irony in which he identifies "First Lady" as a problematic term, without betraying his own position on whether recent attempts to eliminate sexist language in the newsroom are "silly" (the opinion of many conventional male journalists) or necessary. With this slippery phrasing, Jennings is also able to communicate Mrs. Reagan's framing of the conference ("to see what they could do about children and drugs all over the world") without appearing to necessarily approve of her conventional view of caring for children as women's work. Such is the genius of journalistic irony.

Interestingly, both reports featured the striking confession of a young African American teenager who was identified as a graduate of STRAIGHT, a private, tough-love drug treatment program greatly admired by Mrs. Reagan. On NBC, Andrea Mitchell's framing of the confession explores the sex-gender oppositions of Brokaw's grand narration: "Usually it is their husbands who are guests of honor at the diplomatic entrance. This time, an extraordinary gathering of 17 First Ladies, including some from countries that export huge amounts of cocaine, heroin, and marijuana to the United States. Some of what they heard was dry. Lots of statistics. But 16 year-old Robin Page was emotional, talking about guilt, about being afraid to face her mother after taking a drug overdose." Here, Mitchell depicts the "dry" material with "lots of statistics" as a type of masculine knowledge that did not engage the prominent group of women as much as the "emotional" testimony—testimony that directly addresses mothering. ABC's Carole Simpson also acknowledges the emotional impact of Page's confession. For Simpson, Page's words "proved" Nancy Reagan's "point" that "young people on drugs care for nothing—not country, not family, not even themselves." Simpson continues, "The First Ladies listened with tears in their eyes as she told how she turned to drugs to fit in with her classmates, and then nearly died from an overdose." The reporter's emphasis on teary eyes is justified by cover footage of several conferees visibly moved by Page's story, highlighted by a close-up of Nancy Reagan emoting what Wills described as the "concerned gaze at redeemable druggies."

The Page confession, for us, is especially provocative because it also invokes the maternal gaze. ABC's package provides the longest excerpt of Page's moving performance:

> I was laying in bed and I heard my mom's voice. And I kept thinking, "You know . . ." [Page sobs] I don't . . . [she sobs again] I kept thinking, "Dear God, if you just don't let me wake up, I can't see her face again." Because *I was so sick of the disappointment every time she looked at me*. And I remember she came in and she held my hand, and, you know, I was just, I felt alone. Even though she was there, I felt afraid because it had been so long since we were close, or since we held hands, you know? [our emphasis]

The "disappointment every time she looked at me," what might be called the laser eyes of the disciplinary gaze, provides a striking contrast to the wholehearted approval of Mrs. Reagan's adoring Look. The former is a controlling vision—an extension of the optics of modern power into the private sphere of the family. In the approval mode, the maternal gaze aggrandizes; in the disappointment mode, it stigmatizes; in both modes, it is focused on normalization. As the "guardian" of the family, then, the modern mother is responsible for "shaping an acceptable child" with her disciplinary gaze—a gaze that implicates her in the surveillance vectors of the panoptic machine. Expressing the New Right's "traditional family values," this view of mothering, it turns out, was the central message of the summit, as Marie Saega, First Lady of Jamaica confirmed in ABC's package: "We have a message to take home to the mothers in our countries. It's vital that we get close to our children."

But Nancy Reagan also voiced a crucial secondary message grounded in the New Right's doctrine of corporate privatization and consistent with its "profamily agenda"—that fighting drug abuse is the family's responsibility, not the government's. As Mrs. Reagan put it in a sound bite appearing in the NBC package: "Government can't supply the things that parents' groups can supply. It can't supply love, affection, attention." This position is at once naive and savvy. It is naive, especially considering the conference included representatives from Bolivia, Pakistan, Jamaica, and Mexico, because it assumes universal middle-class family controls are in place and only need to be engaged in order to eliminate drug abuse. Even in the U.S. context, this is a naive assumption. It also assumes that the family itself is not part of the problem. For instance, it is not at all clear in Page's confession whether or not her mother's look of disappointment actually contributed to her drug overdose. But because these assumptions are shared by the New Right, it is also a politically savvy position—one that adroitly avoids confronting the paradox of Ronald Reagan's slashing funding for federal drug treatment programs at the same time that his wife holds an international summit on drug abuse.

Crusading Public Relations   On CBS, Dan Rather's delivery of the lead-in to Leslie Stahl's story is strangely stilted. After briefly mentioning the conference, Rather introduces the idea of two Nancy Reagans that works as the structuring principle of the package: "Tonight, White House correspondent Leslie Stahl looks beyond the photo-opportunity Nancy Reagan, to a Nancy Reagan of changing style and image and a lot of clout where it counts." Stahl begins with images of Mrs. Reagan, "alone, front and center on the world stage" as she ceremoniously greets her guests at the First Ladies' summit on drug abuse. "Drugs rip right through the moral fiber of our countries," Mrs. Reagan says in a sound bite extracted from her address to the conference. "They twist values and prevent others from developing." And with that brief opening narration and sound bite, Stahl leaves the conference behind and embarks on a *Citizen Kane*-like quest for the "real" woman lurking behind the multiple negative images of Mrs. Reagan circulated by her many critics.

In keeping with the structuring of *Kane*, Stahl begins her search for the true Citizen Nancy with archival footage of her previous public performances in the news. As we see a glamorous First Lady attending formal functions, boarding airplanes, hugging Frank Sinatra on stage after a presidential performance, and "Look-ing up" to her husband, Stahl explains that the First Lady now hosting the international drug conference "is a far cry from the early images": "Lavish glittering gowns, hobnobbing with the rich and famous. And always what the columnists label 'the adoring gaze.' Has she changed?" In the sound bite, Mrs. Reagan answers that she is "still the same person," but that she has grown: "I don't know how anybody could be in this position and not grow." Stahl reports that bringing two Korean children back to the United States for heart surgery in 1983 was a "significant turning point" for Mrs. Reagan, a "revelation" in which she realized that she "could save a life." A sound bite from former press secretary Sheila Tate (who is hardly a "disinterested" observer) confirms that this was something of a religious experience for the First Lady: "I know she knew what she had done. You could just feel it. It was a nice thing to witness." And thus we have set in motion a rite of inclusion, complete with the discourse of recovery.

Through the rhetoric of rebirth, Stahl goes about rehabilitating the tarnished image of the First Lady. Over a strange image of Stahl and Mrs. Reagan kneeling together on the floor of the White House inspecting what appears to be tile samples laid out on a rug, Stahl's narration addresses the most damaging image of all: "Another major step forward. She began to show she's not the Porcelain Princess her critics called her. And she began to laugh at herself." Mrs. Reagan shares an intimate "ha-ha" with Stahl as this friendly encounter is replaced with a close-up of a novelty postcard showing the First Lady decked out in royal garb complete with crown. In voice-over, Mrs. Reagan delivers the opening line of the kind of Reaganesque self-deprecating joke that served her husband so well: "There's now a picture postcard of me as a queen. . . ." The postcard is replaced by a shot of her addressing a formal gathering. "Now that's silly," Mrs. Reagan continues, "because I'd never wear a crown."

Although the punch line[51] is too obvious to recount here, Stahl's advocacy isn't. Not only is it unusual to see the First Lady, in a dress and hose, kneeling on the floor like a sorority pledge at a slumber party, but it also is strange to see a reporter sharing such an intimate moment with an interview subject. Although this report is clearly in the therapeutic genre, Stahl does not observe the detachment of the conventional reporter-analyst. Instead, Stahl presents a view of the First Lady that is much more intimate than the standard news package. It is, instead, much more in keeping with the kind of human interest stories told on "60 Minutes," where reporters perform overtly as story characters and where Stahl now works.[52]

Perhaps most important, by this point in the report Stahl has positioned herself as an advocate defending Mrs. Reagan from her critics—a role she performs throughout the rest of the package. Including the joke in the package helps Stahl humanize and soften the woman with a reputation for being "ruthless." However, in backing the human Nancy Reagan, Stahl also ends up supporting key elements of Reaganism, most notably its demonization of the 1960s and its advocacy of "traditional family values."

If we believe Stahl's charitable account of Mrs. Reagan's motivation, then the lost generation of the 1960s has cast a shadow over Nancy Reagan's life in much the same way that the lost sled ("Rosebud") haunted Charles Foster Kane. "Even while Mrs. Reagan's critics were accusing her of insensitivity," Stahl charges, "she was quietly devoting herself to fighting school-age drug abuse, a cause she says she embraced after raising two children in the 1960s." This narration is followed with a clip from an antidrug film, *Chemical People,* in which Mrs. Reagan states, "I don't think we can afford to lose a generation of our young people to chemicals." In the final half of the report Stahl also privileges Mrs. Reagan's conventional maternalism by defending her from detractors on both the left and the right. In this defense, Stahl does a brilliant job of centering the much-maligned First Lady on a moral middle ground—a centering that justifies her feminine wiles in terms of old-fashioned Middle American values.[53]

In repositioning Mrs. Reagan in the center, Stahl acknowledges that critics on the right are concerned about the "influence she seems to have over her husband." This influence is illustrated by a famous clip from an episode at the Reagan ranch that occurred during the 1984 campaign. According to Wills's account of the incident, after being asked what America could do to bring Russians to the negotiating table, the president "stood in apparent thought for a full five seconds": "At last, Mrs. Reagan, after looking up at him, whispered, 'Doing everything we can.' He lifted his head then and grinned. 'Doing everything we can.' " Although, later, Nancy Reagan denied prompting her husband (she claimed to be thinking out loud), and Reagan asserted that, contrary to appearances, he was not pausing to think of something to say, but was "refusing" to answer the question because it was a "photo opportunity" and not a press conference, the incident seemed to confirm the impression that Nancy Reagan was the real power behind the throne.[54]

In retrieving and recycling this clip from the network's image bank, Stahl does nothing to disconfirm this impression. But, in contrast to the more common journalistic view that Nancy Reagan is both "in control" and "out of control," Stahl portrays Mrs. Reagan's power as a "moderating influence" that encourages the president to develop "his role as peacemaker" and leads him "away from the narrow social agenda of the Religious Right." Although Stahl concedes that "members of the President's staff admit privately that they are afraid of her," the First Lady's clout is cast in terms of maternal protection with the inclusion of a sound bite provided by Maureen Reagan. "She has an excellent sense of when somebody is not serving the President well," beams the admiring stepdaughter. "She also has a very good sense of letting people know that she knows that, and that they better straighten up."

So Stahl does not deny Nancy Reagan's power. But she does frame it in terms that depart from the myth of the all-powerful, domineering mother. For Stahl, it is, instead, a moderating power—a power that, presumably, any loyal wife would exercise to encourage her husband's better angels and protect him from the incompetence of fools or the manipulation of fanatics. In other words, it is a power that is consistent with "traditional family values"—values that Stahl also invokes in setting Mrs. Reagan apart from her feminist detractors on the left.

After claiming that Mrs. Reagan stuck with the drug issue "against the advice of her media experts who said it was too depressing," Stahl sets up dueling sound bites between Betty Friedan, icon of liberal feminism, and James Rosebush, a deputy assistant to the president. According to Stahl's framing narration, "some say pictures of a warm and caring First Lady are done with a purpose":

BETTY FRIEDAN: There's a theory among some women that this may be even a conscious PR campaign to change her image.
ROSEBUSH: This has got to come from the heart. You can't manufacture this.

Although in her narration Stahl does not deny the calculated image-enhancing aspect of Mrs. Reagan's crusade (after all, how could she deny it and maintain any credibility?), she is still careful not to let Friedan's accusations go unanswered, even if it is by a White House underling who would lose his head if he confirmed the obvious. In fact, it is hard to say whether Stahl's inclusion of the Rosebush response is an obligatory gesture at fairness laced with reportorial irony, or whether Stahl uses Rosebush as a surrogate to voice her own affirmative take on Mrs. Reagan's sincerity.

Friedan appears again near the end of the report when Stahl states that feminists also criticize Mrs. Reagan for not using her power and position to promote women's issues. Friedan's sound bite seems to be directly addressed to the First Lady, although Stahl is the intermediary: "Well, all right, Nancy Reagan, if you're going to be more of a person now, then why don't you stand up a bit for women's rights?" Then, in a bite which immediately follows Friedan's question without any intervening reportorial narration, Mrs. Reagan provides a ringing endorsement of the patriarchal bargain: "I have the best of two worlds. I had a career which I loved, I enjoyed. And then I met the man I wanted to marry and I gave it up. Now that, it seems to me, is the whole idea of women. You have a choice of what you want to do." In this endorsement Mrs. Reagan seems to be saying that women can choose career or family, but not both. Left unstated here is the fact that, in this bargain, men can continue a career they "love and enjoy" after marriage—that they do not have to make the same "choice" confronting women. So what Nancy Reagan celebrates as "freedom of choice" for women is in fact a casting of "the whole idea of women" in terms that naturalize a sexist double standard.

Stahl, however, does not pursue the implications of Mrs. Reagan's position on women. Instead, she moves to the conclusion of her report by defusing criticism of Mrs. Reagan's own well-known troubles with mothering. "Nancy Reagan has even come to terms with family problems," asserts Stahl. "She's made up with her stepson Michael, and embraced Maureen, a staunch feminist, as one of her closest friends. Most of all, she feels more comfortable in her role as First Lady. Nancy Reagan in her second term. More confident. More independent. She says she's a late bloomer." And with that old-fashioned image as the last word, Stahl ends a classic example of a report that operates as a rite of inclusion—a report designed to redeem and mainstream the stigmatized First Lady. From our more skeptical perspective it is precisely the type of coverage that the crusading public relations of Mrs. Reagan's antidrug activism was meant to generate.

Engendering Transgression

If the 1960s counterculture too naively assumed that transcendence was possible through fun, the 1980s have seen the emergence of a notion of transcendence through discipline itself. This has contributed to a number of devastating consequences: e.g., policing women's bodies (and most recently, religious beliefs) in the name of a reconstituted notion of the mother (and her pleasures), not as a nurturing and valued social producer but as a potential criminal *vis-à-vis* the fetus. Pleasure becomes selfish abuse, even as having a child becomes, on the one hand, a social responsibility, and on the other, the only legitimate pleasurable experience for the mother. Similarly, "Just say no"—whether to drugs, alcohol, nicotine, sex or cholesterol—becomes a new moral principle, and addiction (of any sort) becomes that which, above all else, must be surveilled. All of this is done in the name of health, happiness and responsibility.
—Lawrence Grossberg[55]

In the 1980s a particularly sinister type of maternal discourse would capture the imagination of moral crusaders both inside and outside the news establishment. Standing in stark contrast to the composite Reaganite "she" in need of protection, the crack mother was a composite "she-devil" in need of discipline—and yet, like the "emblematic mother" idealized in the New

Right's profamily agenda, the transgressing mother also was reduced to her reproductive identity. In this reduction, the media construction of this she-devil would animate a complex set of meanings that often dovetailed with the fetal rights rhetoric of the antiabortion movement. Like the "murderous mommies" denounced by Operation Rescue, the crack mother was berated in the drug news as an enemy to the innocent life within. "Crack Babies: The Worst Threat Is Mom Herself," announced a *Washington Post* headline that stands as something of a journalistic credo in the late 1980s.[56]

Policing Pregnancy   Our analysis suggests that the journalistic discovery and obsession with the crack baby is most properly understood as part of a larger policing of pregnancy by the medical-industrial complex. Most of the primary locations for the early stories on cocaine and pregnancy are hospital maternity wards. And, in this news discourse, the maternity ward operates as a surveillance setting where the spectacle of birth subjects transgressing mothers to the scrutiny and the stigmatization of medical authority.

The first such story appeared on September 11, 1985. In it, CBS's Susan Spencer announces the results of a preliminary study showing that pregnant women using cocaine suffered spontaneous abortions three times more frequently than the average rate. Although the story did feature the remorseful confession of "Linda," a white woman in Chicago who gave birth to a baby who displayed the "jittery" symptoms of cocaine withdrawal, Spencer (perhaps in part because she is a woman journalist) did not dwell on the sins of the mother. Instead, she structured the package as a general health warning addressed specifically and explicitly to women. Using the second-person singular address, "you," Spencer concludes her report with the direct exhortation: "The message is clear, if you are pregnant and use cocaine—stop."

By December 20, 1985, however, when CBS issued the second network alert about this potential danger, the warning is laced with a righteous indignation that would set the tone for the "crack mother" subplot of subsequent coverage. Perhaps because it is told by a male journalist, Terry Drinkwater, the story is not so much directed at warning pregnant women as it is focused on the damage done to their babies—babies who, according to substitute anchor Charles Kuralt's lead-in, are "victims who aren't even old enough to know better." The shift from warning women to demonizing mothers is captured best in the final moments of the report when Drinkwater considers the case of an eighteen-month-old girl who a social worker predicts may grow up to be a "twenty-one year old with an IQ of perhaps 50, barely able to dress herself, and probably unable to live alone." In a closing line that is saturated with moral disgust, the crusading Drinkwater reports, "The mother told authorities she was just a recreational drug user."

Such journalistic horror stories that cast the mother in the role of the monster would take on racial overtones in 1986. In keeping with the general paradigm shift in the framing of drug news, the "cocaine mothers" of the 1985 warnings were predominantly white, and the "crack mothers" of the coverage in 1986–88 were predominantly women of color. In this shift, the discourse of the crack mother would resonate with the "cultural Moynihanism" of the "new racism"—the "epidemic" of crack babies became yet another example of the "poverty of values" crippling America's largely black inner cities.

The Other Mothers   The harsh journalistic treatment of crack mothers is very much in keeping with a pattern of racial disparity that Rickie Solinger uncovered in her provocative study of the treatment of illegitimacy before *Roe* v. *Wade*. According to her study, in the 1950s transgressing mothers-to-be were the subject of "racially distinct" policies and practices. While all unwed mothers were reduced to their reproductive identities as "breeders," errant white women were viewed as "socially productive" since their babies had value on the adoption market:

> White women in this situation were defined as occupying a state of "shame," a condition that admitted rehabilitation and redemption. The pathway was prescribed: casework treatment in a maternity home, relinquishment of the baby for adoption, and rededication of the offending woman to the marriage market. . . . White illegitimacy was generally not perceived as a "cultural" or racial defect, or as a public expense, so the stigma suffered by the white unwed mother was individual and familial.[57]

As an individual and familial defect, unwed motherhood for white transgressors was treated as a symptom of "mental illness" that could be rectified through therapeutic solutions. Like the white cocaine transgressors of the early 1980s news coverage, the white unwed mother was an "offender" whose psyche could be rehabilitated and whose reproductive capacity could be recovered and legitimated.

Black single pregnancy, in contrast, was defined as "socially unproductive," "the product of uncontrolled sexual indulgence, the product, in fact, of an absense of psyche." Where white transgressors suffered from a

"shame" that could be eliminated by way of the purifying confession and repentance, the pathology of delinquency was literally inscribed on the body
and in the genes of the black unwed mother. Her very life, in other words,
was a punishable offense requiring "punitive, legal sanctions." Again quoting Solinger: "Black women, illegitimately pregnant, were not shamed but
simply blamed, blamed for the population explosion, for escalating welfare
costs, for the existence of unwanted babies, and blamed for the tenacious
grip of poverty on blacks in America. There was no redemption possible
for these women, only sterilization, harassment by welfare officials, and
public policies that threatened to starve them and their babies." [58] In other
words, Solinger's study has discovered virtually the same white/black,
shame/blame, offender/delinquent, therapy/pathology, recovery/discrimi-
nation, inclusion/exclusion oppositions that we have suggested organize
the cocaine narrative of the 1980s.

While the treatment of white unwed pregnancy has changed significantly since the 1950s, the social construction of black illegitimacy as
pathology persists. As was painfully evident during a pivotal moment during the second presidential debate of the 1992 campaign, blaming illegitimacy for the economic distress suffered by people of color remains very
much in vogue. In an interchange with an African American woman who
wanted to know how the candidates "could honestly find a cure for the
economic problems of the common people" if they had no personal experience with poverty, President Bush offered a particularly revealing non
sequitur:

BUSH: Well, listen, you ought to—you ought to be in the White House for a day
    and hear what I hear and see what I see and read the mail I read and touch
    the people I touch from time to time. I was in the Lomaxa AME Church. It's a
    black church just outside of Washington, D.C. And I read in the—in the bul-
    letin about *teenage pregnancies,* about the difficulties that families are having
    to meet ends—make ends meet. [Bush's emphasis]

Bush rambles on. But the church bulletin comment, given the identity of
the questioner and racial makeup of the congregation, is clearly a botched
attempt to blame illegitimacy for black economic distress.

Tomorrow's Delinquents However, four days earlier, during the first
presidential debate, H. Ross Perot—he who claims to have survived an
assassination attempt by Black Panthers—was much more coherent and

effective in mobilizing against the deviancy of crack mothers to justify his
opposition to drug legalization:

> Any time you think you want to legalize drugs, go to a neonatal unit, if you
> can get in. They are 100 and 200 percent capacity up and down the East Coast.
> And the reason is crack babies being born. Baby's in the hospital 42 days.
> Typical cost to you and me is $125,000. Again and again and again the mother
> disappears in three days and the child becomes a ward of the state because he's
> permanently and genetically damaged. Just look at those little children and if
> anybody can think about leave—legalizing drugs, they've lost me. . . .

In this dense response, Perot captures the key features of "common sense"
regarding crack mothers circulated in the news coverage of the late 1980s:
the chemical scapegoating of social ills born of economic deprivation; the
racially coded framing of certain forms of fertility as "socially unproductive," as "subsidized deviancy"; and perhaps most disturbingly, the stigmatization, not only of the mothers, but of "crack babies" themselves as
"permanently and genetically damaged." Even *Rolling Stone,* a publication
once associated with the counterculture, has claimed that these babies are
"like no others, brain damaged in ways yet unknown, oblivious to any
affection."[59] But these demonizing claims seem mild compared to those
made by Boston University President John Silber who, apparently, believes
that crack babies are not even invested with a *soul.* Speaking as "a mere
generalist" in an account published in the *Boston Globe,* Silber (a self-
proclaimed enemy of "political correctness") displayed a rare insight into
the limits of human redemption by stating that "St. Thomas would have a
hard time justifying" priorities that allot insufficient primary health care
to children who could go on and live "to the greater glory of God, while
spending immense amounts on crack babies who won't ever achieve the
intellectual development to have consciousness of God."[60]

But, as pediatricians Barry Zuckerman and Deborah Frank argue, such
divinely inspired public policy statements and widespread demonization of
crack babies have "evolved in the absence of any credible scientific data regarding the sequelae of prenatal exposure to cocaine beyond the newborn
period": "Moreover, this furor over prenatal exposure to cocaine obscures
in the public mind any debate regarding society's responsibility for other
conditions, such as lack of access to prenatal or pediatric care, malnutri-
tion, measles, or lead poisoning, which jeopardize the development of many
impoverished American children, whether substance-exposed or not."[61]

There was, indeed, an "epidemic" of premature and unhealthy babies in the 1980s—as well as a rise in the infant mortality rate. But these health crises had less to do with cocaine than with Reagan's budget cuts and his doctrines of privatization and deregulation. In 1981 alone, cuts in Medicaid and other public assistance programs stripped more than 1 million poor women and their children of medical benefits. Because of the rising cost of health insurance under deregulation, the number of people without coverage soared during the 1980s, as did the percentage of births to mothers with inadequate prenatal care. Faludi's summary of research on the impact of no health insurance and no prenatal care on America's kids speaks to the graveness of the situation:

> A 1989 University of California research team reviewed records of more than 146,000 births between 1982 and 1986 in California, and found that babies born to parents with no health insurance—a group whose numbers had grown by 45 percent in those years—were 30 percent more likely to die, be seriously ill at birth, and suffer low birth weight; uninsured black women were more than twice as likely as insured black women to have sickly newborns. A similar 1985 Florida report tracing the dire effects of lost prenatal care concluded, "In the end, it is safer for the baby to be born to a drug-using, anemic, or diabetic mother who visits the doctor throughout her pregnancy than to be born to a normal woman who does not."[62]

The health insurance crisis, of course, did not result in a moral panic, even though it was damaging to many more children than the so-called cocaine epidemic. But, then again, it did not fit into the "poverty of values" view of social and economic problems. Because the crack mother did reinforce the "new consensus" that what African Americans need is "sexual restraint, marital commitment, and parental discipline," she did provide the basic stuff for the making of an old-fashioned, racially inflected moral panic—both in the news media and in the political arena.

Of the nine news packages in our study devoted to the drug baby epidemic (see CBS, 9/11/85, 12/30/85; NBC, 2/18/86; ABC, 7/11/86; CBS, 8/29/86, 11/12/86; ABC, 10/13/88; NBC, 10/24/88, 10/25/88), the last stands out as an extreme illustration of the panic idiom at work in the drug news. Reported by Michelle Gillen, it is the second segment of a two-part investigative series titled "Cocaine Kids." Calling it a "spotlight" report, Brokaw's lead-in draws attention to its surveillance aspect—and he also sets up the we/they rhetoric that structures the story. According to Brokaw, when these cocaine kids "leave the hospital, they don't leave their problems behind."

Overlaying shots of fragile babies in a maternity ward, Gillen's opening narration also situates the contaminated children as "They": "They are the nation's tiniest drug victims. They are growing into the country's most unwanted children. . . . The babies begin life sick, often premature, suffering withdrawal from the mother's drugs. Their troubles, and ours, are only beginning." After moving from the maternity ward to the streets, where "their mothers often vanish . . . leaving their babies behind for hospitals and overwhelmed foster care systems to cope with," Gillen introduces a genuine specimen of a crack mother whose sexuality and fertility are out of control. Her pseudonym is "Stephanie." The "mother of two crack babies now in foster care," she is now "back on the streets, back on crack and pregnant again." As a part that stands in for the whole, Stephanie's deviancy, according to the logic of Gillen's report, is typical.

The generalization of the worst-case scenario in this type of "trend reporting" is a standard feature of mainstream news discourse. In *personalizing* social problems, such reporting often treats the most visible and extreme instances of a phenomenon (the mentally ill homeless, for example) as paradigmatic of the whole problem.[63] But Gillen's report is a moment in an even larger "trend": the long-standing objectification of women of color as "outsiders within." According to Patricia Hill Collins, as figures in the projection of white fantasies, desires, and fears, this objectification of women of color as the Other has been primarily accomplished by four controlling visions: the mammy, the matriarch, the welfare mother, and the Jezebel.[64] Some as old as slavery, one as fresh as the New Right, these distorted images of black womanhood form the background against which the crack mother would emerge as a hybrid form.

As a composite "she-devil," the crack mother takes the image of the welfare mother, so prominent in the demonology of Reaganism, and fuses it with the sexually aggressive "Jezebel." According to Collins, the Jezebel occupies a central position in the "nexus of elite white male images of Black womanhood because efforts to control Black women's sexuality lie at the heart of Black women's oppression."[65] A particularly menacing image of fertility, the crack mother is the personification of an out-of-control black sexuality that is almost as threatening to the Reaganite imagination as its much older fraternal cousin—the marauding black rapist. Playing on many of the same fears mobilized by the 1988 Willie Horton GOP ads, the news discourse of crack mothers like Stephanie gave the New Right another racially charged code word to deploy in its holy class war against poor people of color.

But Gillen's investigative report is not so much about the sins of a monstrous mother as it is about the demonization of her "They-like" offspring. After visiting foster homes and schools that are burdened by the challenges posed by cocaine kids, Gillen returns to her prophetic opening thesis—that "their problems, and ours, are only beginning." "With crack so cheap and accessible in inner city streets," states Gillen in a stand-up bridge to her closing arguments, "and with the waiting list for drug treatment programs so long, no one is even guessing at how widespread the epidemic of crack babies will be. But the few experts who are paying attention warn that there is a price to be paid—and they say the price will be enormous." Then, in back-to-back sound bites, two experts—Representative George Miller of the House Select Committee on Children and William Gladstone, judge of a Florida juvenile court—provide the approved elite white male spin on cocaine kids that also displays Gillen's We/They language:

REP. MILLER: We are going to have these children, who are the most expensive babies ever born in America, are going to overwhelm every social service delivery system that they come in contact with throughout the rest of their lives.
JUDGE GLADSTONE: These kids have enormous, uh, physical problems, mental problems. They will go into a system that is woefully inadequate, woefully underfunded. They'll grow up to be tomorrow's delinquents.

With these grim predictions, Gillen ends with a close-up shot of a cuddly black baby crawling on a blanket in a foster home. However, the youngster seems positively evil when Gillen's closing narration concludes that "there are tens of thousands of crack babies on the way. A generation born at risk. A generation which may pose an even greater risk."

Gillen's story bears witness to a historical process that is at least as vicious as the so-called cycle of poverty. Where Spencer's package broke the cocaine mother story as a medical warning addressed to women, and Drinkwater's package develops the warning into a horror story that demonized mothers, Gillen goes the next step—the demonization of the cocaine kids themselves. This cycle of "othering" is precisely how the narco-carceral network sustains itself by reproducing deviance. For the modern control culture, cocaine kids have become something of a futures market that can be bid on by the speculating "experts who are paying attention" and know a gold mine of potential deviancy when they see it. Even conscientious members of the medical community are increasingly troubled by the orgy of enterprising research spawned by the crack baby hysteria. According to an editorial by M.D.s Thomas P. Strandjord and W. Alan Hodson published in the *Journal of the American Medical Association,* this crusading research is "fraught with many methodological errors":

> These include sampling bias of the study population, identification and documentation of the actual use of cocaine by mothers (other than by self-reporting), the quantification including time and duration of fetal exposure, other prenatal factors associated with drug-abusing mothers (nutrition, smoking), and the host of social and environmental factors (poverty, neglect, malnutrition, violence, child abuse, lead poisoning) that could adversely affect a developing infant over the first few years of life.[66]

In fact, one of the few studies that attempts to account for some of these intervening factors provides support for a cautious optimism about the long-term developmental consequences of prenatal cocaine exposure. The study (which excluded opium-exposed infants) found no mean developmental differences at two years of age between children labeled as cocaine babies and a control group made up of children from the same social class.[67]

Contrary to the chemical and maternal scapegoating by Perot and Gillen, the worst threat to America's cocaine kids may be, in the long run, the label "crack baby" itself. As Linda C. Mayes, Richard H. Granger, Marc H. Bornstein, and Barry Zuckerman suggest in a commentary also published in the *Journal of the American Medical Association,* the rush to judgment about the "extent and permanency of specific effects of intrauterine cocaine exposure on newborns" is "closely tied to a significant social political issue":

> Why is there today such an urgency to label prenatally cocaine-exposed children as irremediably damaged? What are society's attitudes toward and responsibility for these disadvantaged children? These problems are not explicitly methodologic ones although they are predicated on empirical findings. Moreover, in themselves, they carry significant medical and psychosocial risks for the children. . . . Minimally, expectations for such children are lowered. The attribution of irremediable damage makes it more difficult to find services for these children, and such services may be geared to caretake rather than to challenge children's capacities to remediate effectively. Even more damaging is the difficulty finding adequate homes for such children since potential foster or adoptive parents are often concerned about assuming the care of cocaine-exposed children because of their perceived impairments.[68]

In other words, once attached to a child, the "crack baby" label is something of an albatross, a tag that stigmatizes the subject as "not normal" as the child passes through social service and educational systems.[69] In the words of Dr. Claire Coles, "If a child comes to kindergarten with that label, they're dead. They are very likely to fulfill the worst prophecies."[70] In fact, the stigmatization of such labels often work as self-fulfilling prophecies in a double sense: they justify intense surveillance and tracking procedures that "interpret everything in a negative way" as an effect caused by the damage done in the womb; and they cause the children who are stigmatized by the label to behave, well, like they are hopelessly stigmatized.[71] For cocaine kids, visibility in the narco-carceral network is, indeed, "a trap." In large part because of this visibility, they are preordained to be "tomorrow's delinquents."

The dynamic of a self-fulfilling prophecy was always at the heart of the obsession with crack babies. After the "discovery" of the epidemic, cocaine tended to be blamed for any baby in distress born prematurely to mothers who may or may not have taken the drug but who had unhealthy life-styles that included alcohol, smoking, and drug abuse. As a study published in the *Lancet* found, this prophetic stigmatization was especially apparent in the medical community. According to the study, abstracts on the impact of cocaine use during pregnancy were more likely to be accepted for presentation at the annual meeting of the Society for Pediatric Research if they reported "positive" evidence of impairment—even though rejected papers with "null" or "negative" findings tended to be more methodologically rigorous.[72] Perhaps an even more damaging report on how the racially coded labeling of crack babies has compromised the medical profession appeared in a 1990 issue of the *New England Journal of Medicine*. According to a study by I. J. Chasnoff, H. Landress, and M. Barett, pregnant black women and women on welfare are more likely to be reported for use of illegal drugs to law enforcement agencies by physicians and clinics than are their white and middle-class counterparts.[73] In other words, like Solinger's study of unwed motherhood, Chasnoff and his colleagues discovered the same pernicious discrepancies in the treatment of white "offenders" and black "delinquents"—disparities that we have argued also contaminate television news and its reporting of the 1980s anticocaine crusade.

The Meaning of Family Values

As Coontz observes, at the heart of much of the "hysteria about the 'under-class' and the spread of 'alien' values is what psychologists call projection":

> Instead of facing disturbing tendencies in ourselves, we attribute them to some-thing or someone external—drug dealers, unwed mothers, inner-city teens, or satanist cults. But blaming the "underclass" for drugs, violence, sexual ex-ploitation, materialism, or self-indulgence lets the "over-class" off the hook. It also ignores the amoral, privatistic retreat from social engagement that has been the hallmark of middle-class response to recent social dilemmas.[74]

For us, these projections and evasions provide the master code for decipher-ing the meaning of "family values" as it is deployed in contemporary politi-cal discourse. The subject of much contestation during the 1992 presidential campaign, family values has become a vehicle for constructing a moral veneer on what Coontz terms the "amoral, privatistic retreat from social engagement." As we suggested, the convergence of family values with the cultural Moynihanism of the new racism provides a common grounding for the antiwelfare, anti-affirmative action, antitax, antibusing, and antifemi-nist backlash of the Reagan coalition. Cynically converting the material advantages of the bourgeois nuclear family into a virtue, the family values of the New Right let "the over-class off the hook" by recasting (or "pro-jecting") poverty as a moral transgression and self-righteously reversing economic cause with familial effect. In this warped moral universe, women of color are subject to the multiple jeopardies of a sex/gender system that blames mothers for family problems and a race/class system that blames the pathological black family for economic problems. As Collins observes, the "holier-than-thou" moralism of this standpoint links "gender ideology to explanations of class subordination"—a linkage that diverts attention from the structured inequalities and upward redistribution of wealth of the Reagan-Bush era by blaming black poverty on bad mothering.[75]

9

Denouement:

Second Thoughts

What if there were a drug (I inquired of the mirror) that could chemically in-
duce feelings of upper-middle-classness. It would be attractive to the poor, and
wildly popular among those who had no prayer of ever achieving that comfortable station in life. And it would be despised by people who had worked
long, hard years to obtain the same mental state without resorting to the drug.
It would be popular, cheap, and the cause of anti-social behavior. It would be
a lot like crack. . . . Crack was a parody of Reaganism, I concluded, a brief
high with a bad aftertaste and untold bodily damage.—Jefferson Morley [1]

Refractions on the Crisis Coverage

In preceding chapters we argued that the 1986 crisis coverage represented
nothing less than a paradigm shift in the framing of the cocaine problem. However, the shift from crisis to "postcrisis" coverage is much more
subtle. The siege paradigm continues to prevail in most of 1987–88's aftermath reports—but the crusading enthusiasm is much more restrained, the
reporters more self-conscious about their role in legitimating the war on
drugs. In contrast to the excesses of 1986 coverage, the postcrisis period is
generally more sober, circumspect, critical, and fatalistic.

Acknowledging this modification in the tone and the intensity of network news coverage of the drug crisis, Lloyd Johnston contends: "After the
Omnibus Drug Bill was signed into law, after the Congressional elections
were over, and after a few reporters chastised their colleagues by claiming
that this was all due to a bad case of media hype, there emerged an almost
eerie silence in the media on the subject of drugs. Guilt and withdrawal
for self-assessment among those in the media explains a lot of this 'refrac-

tory period.' " [2] This "refractory period" was ushered in by several reports in 1986 that seemed to anticipate the rush of media criticism that would soon materialize in the partisan and popular press. The first wave of self-reflexive reporting is perhaps best exemplified by ABC's series of five special segments called collectively "Drugs USA." Airing on successive newscasts during the week of September 15–19, 1986, the series includes reports that challenge the conventional wisdom of current drug policies.

On September 18, for example, the fourth installment of the series provides an example of a rare "why" story that makes a good-faith attempt at exploring the history of cocaine pollution. Interestingly, the report acknowledges the long-standing connection between "headlines" and the drug scares. Quoting reporter Bill Blakemore's introduction: "Something like this drug scare has happened before in America. In 1885, pure cocaine hit the market with no laws against it. Doctors extolled it, said it cured almost anything, in wines, cigars, Coca-Cola, even straight. In 1900, the ruined lives began to show though. Headlines rallied people and president." After interviewing David Musto, a Yale scholar whose *The American Disease: Origins of Narcotic Control* [3] is widely recognized as the definitive owlish account of cycles of drug regulation in America, Blakemore presents a thirty-second history lesson that again ends with an acknowledgment of the link between media coverage and drug hysteria:

> In 1906, drugs were controlled. In 1914, made illegal. By 1920, even alcohol was prohibited. Experts call this a national corrections cycle. By the 1930s, any national drug problem was gone for 30 years. Then, in the '60s, Woodstock euphoria and widespread experimentation with drugs. In the '70s, laughter when Woody Allen sneezed at cocaine. But again we're seeing the ruined lives of addiction. Headlines are now rallying people and president.

This, admittedly, surface account is full of holes—but as a minihistory lesson, it still provides much more background on the current drug crisis than any other news package in 1986. At best, other packages consider the crack crisis only in the context of the 1960s' drug culture. Although the account ultimately asserts that the headlines are generated by wasted lives, not moral entrepreneurs, there is still at least some admission that the media play a mobilizing role in rallying people and president.

The second part of Blakemore's report asks "why" questions that distinguish the report as a vintage example of journalism having second thoughts about the root causes of the crisis. In answering the question, "What's caused our current drug problem?" Blakemore includes some of the standard "Just-Say-No" orthodoxy that blames today's problems on the permissive sixties. But what sets Blakemore's report apart is that it also includes the unorthodox idea that the cocaine problem just might have something to do with capitalism. This view is presented in a sound bite from Dr. Ron Siegel: "It appeals to the American ethic and the spirit of capitalism. Look, when someone is under the influence of cocaine, their behavior is really applauded in our society. They work harder, they speak better, they're faster and more alert and reactive and productive." And yet this indictment of capitalist *behavior* is still far from satisfactory because it does not make the connection between drug consumption and the capitalist *values* of a consumer economy that is driven by the destructive highs of short-term profits and instant gratification. The point is that despite its many limitations Blakemore's report still exemplifies a new caution—and a new drive toward complexity—on the part of journalists in their retelling of the cocaine narrative. Other reports in the "Drugs USA" series also display this revisionist impulse (as we shall discuss later in this chapter).

Our analysis also verifies Johnston's observation of an "almost eerie silence in the media on the subject of drugs" in the wake of the crack crisis. In fact, our screening of the *Vanderbilt Television News Index* found not one network news story on cocaine as a social problem in December 1986. Several factors contributed to this refractory period. This was, after all, when revelations regarding the Iran-Contra scandal were commanding a great deal of news space. Furthermore, as we shall see, a kind of generalized journalistic embarrassment about the shenanigans of Geraldo Rivera clearly had a chilling effect. But perhaps the most potent factor was that the crisis no longer qualified as "news." Crises do have life spans, and the passage of the Omnibus Drug Bill provided a sense of closure both for journalists and crusading sources. However, when the cocaine narrative resumed in 1987, reporters were, on the whole, less likely to accept at face value the government's spin on the drug war. In fact, the coverage in 1987–88 included a number of "revisionist" stories that at times even questioned the power of the medical-industrial complex.

Critical Response In chapter 1 we introduced our dialogic approach to interpreting the process by which reporters and anchors rewrite reality in their packaging and anchoring of the news. In that introduction we argued that dialogic analysis was based on the simple proposition that the news is a dynamically open and self-regulating system of communication. Not only oriented toward the package or newscast at hand, journalism's rewriting rituals also take into account other texts (the U.S. Constitution, libel law, NIDA public awareness campaigns, previous network reports, contemporary reports in competing print media, proposed crime bills, etc.) and other contexts (the politics of race, public opinion polls, crime statistics, addiction research, congressional elections, the federal budget deficit, the ratings, etc.).[4] In interrogating the political economy of scandal in chapter 5, we placed the 1980s' reporting of John Belushi's cocaine-related demise in dialogue with the 1920s' coverage of Wallace Reid's death by heroin overdose. Our discussion of the "crime wave dynamic" in chapter 6 argued that the 1986 crisis coverage was fueled, in large part, by "dialogic" or "reciprocal" relations within the community of news workers and between this community and its official sources. In chapter 7 we considered dialogic relations among the "Monitoring the Future" survey, NIDA's "Big Lie" campaign, the opinion polls sponsored by news organizations, the reports of President Reagan's announcement of a new demand-side strategy in his war on drugs, the "Jar Wars" of the 1986 congressional elections, and the passage of the House Omnibus Drug Bill. Chapter 8 placed Nancy Reagan's antidrug activism in dialogue with "traditional family values" and the sinister maternal discourse of the crack mother in dialogue with the cultural Moynihanism of the new racism.

But in this chapter we want to suggest that one of the most powerful contexts regulating journalistic performance is that of the "answering word" of media criticism.[5] Our interrogation of the cocaine narrative treats the climate of media criticism as an integral part of journalism's system of surveillance and spectacle—a self-reflexive surveillance mechanism devoted to generating and publicizing "responsive understandings" that actively contribute to the struggle over the "actual meaning" of a particular newsworthy theme or event. In other words, we do not see the work of AIM, FAIR, *American Journalism Review,* or media columns in the partisan and popular press as standing outside of journalistic discourse. Instead, we see such discourse as part of a self-regulatory system that enforces "professional standards," normalizes journalistic "fairness" and "neutrality," and maintains the modernist and constructed distinctions between information and entertainment, observation and participation, objectivity and bias.

Beginning in the fall of 1986 the campaigning voice of the crisis coverage would itself be the subject of surveillance and criticism by social commentators across the admittedly narrow American political spectrum. Peter Kerr's front-page analysis article in a November 1986 issue of the *New York Times* is a typical middle-of-the-road journalistic acknowledgment of the role extraordinary press attention played in transforming the crack crisis into a political spectacle: "The politics of drugs in Washington was suddenly transformed in the summer by a combination of factors— concern among constituents, the discussion of crack, the athletes' deaths [Len Bias's and Don Rogers's]—but most of all the intense press coverage, according to members of Congress, their aides and White House officials."[6] However, Kerr's analysis essentially stops with descriptive observation.

At about the same time, though, several articles in the partisan press move beyond passive description to a more engaged interpretation of jour- nalistic activism in the war on drugs. Slightly to the left of center, the *New Republic* published Adam Paul Weisman's "I Was a Drug-Hype Junky: 48 hours on Crock Street" in October 1986. Weisman likened journalism's dependence on the cocaine theme to an "addiction":

> For a reporter at a national news organization in 1986, the drug crisis in America is more than a story, it's an addiction—and a dangerous one. Some, but not all, of us feel mighty guilty for having tried to convince readers that practically everyone they know is addicted to crack, and they too are likely to be addicted soon. We know the rush that comes from supporting these claims with a variety of questionable figures, graphs, and charts, and often we enjoy it. Blatant sensationalism is a high.[7]

According to Weisman, this addiction lulled reporters into "a journalistic dream world where the writer is free to write almost anything he chooses because nobody is going to call him on it."

But, for us, an even more provocative response that converges with our analysis in interesting ways was formulated by right-of-center critics who recognized the contradictions—and the threat to individual liberty— of Reaganism's doctrine of corporate privatism. Published in September 1986, William Safire's column on "The Drug Bandwagon" is an early ex- ample of conservative second thoughts about the war on drugs. Safire sug- gested that the news media reaped financial benefits for promoting "narco- mania": "News magazines have been conducting a circulation-building war on drugs for months; television networks are finding prime-time slots for documentaries deploring the crisis; newspapers vie for the most lurid stories on the local angle of the issue now in vogue."[8] By the end of 1986 a few conservative intellectuals even took on the New Right in libertarian attacks on the expansion of police powers attending the crack panic. In back-to-back articles published in the *National Review* in December 1986,

Richard C. Cowan and Richard Vigilante argued that the war on drugs had become "a war against ourselves."

Cowan's piece, "How the Narcs Created Crack" (a prime example of a conservative dove position), contends that the war on drugs has actually "aggravated our society's chronic problems with drugs by mounting a propaganda and enforcement campaign that erodes crucial distinctions between more and less dangerous drugs, makes the marketing of the more dangerous variety the preferred option for dealers, and increases health risks, crime and corruption."[9] For Cowan, these "perversities of drug enforcement encouraged by the crack craze" are the "natural outgrowth of two things: the world view of the anti-drug crusader and the self-interest of the drug-enforcement establishment." According to Cowan, the crusading worldview demonizes drug users as "irrational and self-destructive" when the reality is that most users are "People Like Us." Cowan describes the drug enforcement establishment (what we have termed the medical-industrial complex) as a "narcocracy" that exploits for its own benefit the reductive Us vs. Them rhetoric of antidrug crusaders:

> The anti-drug crusader would suffer a blow to his self-righteous rhetoric if he admitted that drug users and the drugs they use are a varied lot—that many drug users are rationally self-protective and that many of them use mild dosages of not very harmful substances. He could not then depict millions of Americans as either depraved criminals or helpless victims, or paint the country as being in the grip of a major crisis. Similarly, if the narcocracy owned up to the truth, both its self-esteem and its budget would be seriously diminished.[10]

Vigilante's companion article, "Reaganites at Risk," picks up on Cowan's themes by endorsing Representative Phil Crane's decision to oppose the House Omnibus Drug Bill.[11] Noting that Crane was the only Republican congressman to vote against the legislation before it was sent to the Senate, Vigilante presents Crane's explanation for his negative vote as "obvious and sensible."[12] In Vigilante's estimation, other conservative politicians who supported the bill "behaved irresponsibly" by "panicking in the face of a media created crisis." In this panic, the politicians were guilty of "attempting to expand government powers to do for the reckless what they might refuse to do for the poor: protect them from themselves."[13]

In a kind of ahistoric account of the antidrug crusade, Vigilante ignores the Reagan administration's authorship of the war on drugs by scapegoating "the media." For Vigilante, "the media" should be held accountable for popularizing the "corrosive assertion" that drug users have "surrendered their free will" and are therefore "incapable of self-governance":

> The most repulsive instance of this is the tendency of the media to treat millionaire playboy cocaine addicts as victims, as if addiction were a natural catastrophe that befalls the helpless rather than a result of self-indulgence. Similarly, the media tell us that crack is an "epidemic" that "invaded" the ghetto, as if the drug were an insidious agent with a will of its own, ruining otherwise stable and happy lives. Surely it is more likely that crack found a home in the ghetto because it is mostly used by people who take more risks because they have less to lose. It is true that the ghetto poor are in some sense victims, both of racism and of government policies that seek to make them dependents rather than citizens. But to imply that they are utter victims, passive objects of passing plagues, is to surrender to the worst and most elitist fallacies of the welfare state.[14]

Paradoxically, Vigilante treats the media in much the same way that he sees the media treating crack. In other words, for Vigilante, a monolithic media is some kind of "insidious agent with a will of its own" that manipulates otherwise "stable and happy" Reaganites into backing legislation that runs counter to the conservative agenda.

Ultimately, then, in scapegoating the media, Vigilante develops a dove position that manages to dovetail with cultural Moynihanism by arguing that the drug problem is merely a symptom of larger moral crises, such as the breakdown of the family, and the hedonism of a generation "persuaded that self-fulfillment was not only a right but a solemn duty." Cowan and Vigilante, as conscientious objectors in the war on drugs, break with the New Right's hawkish orthodoxy by demonstrating how the neoconservative critique of the "New Class" applies not only to those who would enforce affirmative action, but to those who would regulate morality. Furthermore, between the lines of their critique of the crusading news coverage of 1986 Cowan and Vigilante acknowledge that the crack economy is driven by the same entrepreneurial values that animate Reaganomics: opportunism, risk-taking, competitive individualism, and rational self-interest. For, like Cowan, Vigilante deplores the Us vs. Them rhetoric of antidrug campaigns: "If we think of the drug problem this way, rather than using it as a scapegoat for our larger problems, we will find that we are thinking about ourselves as we really are, not about users and dealers as we imagine them to be, utterly unlike ourselves. That's a lot harder than 'declaring war on crack,' but it's more honest and it will do a lot more good."[15] In other words, Cowan and Vigilante recognize, correctly, that the Reaganites at

risk share more values with the dealer and the addict than they do with the
hawkish administrators of the drug control establishment.

The Rivera Intervention Assessing the impact of Cowan and Vigilante's articles, however, is complicated by a noteworthy coincidence. The
same week that this issue of *National Review* went on sale, a sensational
"television event" would attract widespread condemnation and draw attention to the excesses of crusading antidrug journalism. That event was
Geraldo Rivera's two-hour syndicated special that was telecast live to
141 stations across the country. Titled "American Vice: The Doping of a
Nation," the program featured live police raids that Rivera monitored and
narrated from his command post in a New York studio. A raid in Houston
would precipitate a great deal of controversy. Quoting the *Time* magazine account of the televised raid: "The Texas raid got a big buildup from
Rivera (the authorities, he said, were after 'a pimp and prostitutes . . . this
dude and his ladies who were allegedly dealing dope to truckers'). But the
woman arrested, Terry Rouse, claimed she had been living at the duplex for
only a week, and charges were dismissed by Texas District Judge Donald
Shipley."[16] The *Time* article suggests that the "antics of Rivera's show
highlighted concerns about the increasingly common practice of letting TV
crews tag along on drug raids": "A search warrant, says Judge Shipley,
does not give police 'permission to put the whole nation into somebody's
house with TV cameras.' Some police officials object that the cameras, lights
and onlookers can jeopardize safety. Nor is TV merely an eavesdropper.
During one raid on Rivera's show, an officer could plainly be heard to make
a telling, and disturbing, inquiry: 'We are still live?'"[17] Rouse, by the way,
eventually filed a $30 million lawsuit against Rivera, the arresting officers,
and the program's producers, charging defamation of character, invasion
of privacy, false imprisonment, malicious prosecution, and conspiracy.[18]

The controversy surrounding the Rivera special had a chilling effect on
mainstream television journalism. For many news professionals Rivera's
"participatory journalism" is a constant source of embarrassment. A figure
whose performance violates all of the canons of journalistic objectivity and
neutrality, Rivera is often condemned for blurring distinctions between
news and entertainment. This blurring is evident in the very title of the
special, which refers to a cop action-adventure series popular at the time—
"Miami Vice." But for mainstream journalists the Rivera intervention was
even more distasteful because it forced them to confront their own indulgence in essentially the same kind of "sensationalism." Rivera's deviancy,

then, became a vehicle for drawing attention to how the convergence of the journalistic and the policing points of view in crisis reporting had violated professional norms.

Criticism of Rivera soon spilled over into general condemnations of journalistic performance during the previous year. For the networks, the most devastating response was published in a special *TV Guide* report on February 27, 1987. Written by members of the News Study Group at New York University (Edwin Diamond, Frank Accosta, and Leslie-Jean Thornton), the title of the report asks the provocative question, "Is TV News Hyping America's Cocaine Problem?" [19] The answer provided by the group was "yes." Their research found that the coverage exaggerated the scope of the problem:

> the crack image on the screen in many respects diverged from crack in reality. The CBS preview [to "48 Hours on Crack Street"], for example, was flagged for viewers with a line declaring that, in effect, Crack Street could be "your street." That, we learned, was way off the mark: experts in the field generally agree that crack is a serious problem only in some large metropolitan areas—including the media centers of New York and Los Angeles. And while use of crack could well be on the increase, it may be that the use is mostly among those already using cocaine and other drugs.[20]

Like Peter Kerr in the *New York Times,* the group identified the death of Len Bias as the event that precipitated the flood of exaggerated coverage. But, unlike Kerr, the group questions the integrity of the news media by also identifying another newsworthy event that did not set off a "media feeding frenzy": the release on September 24, 1986, of a DEA report that said the hyperattention being paid to crack might have been "excessive." According to the report, crack was a "secondary rather than primary problem in most areas." "That night [September 24]," the *TV Guide* article charges, "NBC's Tom Brokaw briefly mentioned the DEA story in his lead-in to a report on drug testing in the Boston police force. *ABC's World News Tonight* skipped the DEA story, as did *CBS Evening News*."

Unlike Safire's column and the articles by Weisman, Cowan, and Vigilante, this questioning of the integrity of the network news did not appear as an Op-Ed piece or as a polemical product of the partisan press, which could, like the constant flak from AIM and FAIR, be disregarded as politically motivated and distorted. Instead, because it appeared in the mass-circulation magazine as "objective" scholarly research, it was a challenge that networks could not easily dismiss.

Responses to the charges leveled in the *TV Guide* article varied among the networks. ABC, as we suggest, had already shifted into a more self-conscious "second thoughts" mode in its "Drugs USA" series broadcast in late September 1986. CBS seemed to enter a plea of nolo contendere in a report that treats police sting operations targeting users as "just more hype in the so-called war on drugs." Aired on March 16, 1987, the story opens with reporter Bernard Goldberg narrating archival footage of several police raids from the network's image bank. According to Goldberg, "these were real-life cop shows with more action than *Starsky and Hutch*": "It was the 1986 election-year crack-down on crack. They [the raids] were glitzy. They were flashy. The TV crews were always there. It was great publicity. Your tax dollars at work. But in reality, the name of the show was 'Crime and No Punishment.'" Thus, Goldberg manages to condemn the "hype" of the 1986 coverage without specifically acknowledging CBS's culpability— and without questioning the hawk's view of the war on drugs. In fact, the main thrust of Goldberg's report is that the raids and sting operations give a misleading impression that those arrested would be sufficiently punished for their crimes, which, for Goldberg, was not often enough the case.

In contrast to ABC and CBS, NBC entered a plea of "not guilty." In a report that appeared about two weeks after the publication of the *TV Guide* article, NBC directly answers the charges raised by Diamond and his colleagues. Brokaw sets the stage in his lead-in by referring to "reports that last year's crack epidemic may have been overstated." Reporter Dennis Murphy, then, presents a rebuttal that concludes that "crack was not just hype and hysteria." But to answer the criticism, Murphy turns to familiar sources who were instrumental in authorizing the crack crisis in the first place—Arnold Washton, Lloyd Johnston, and most significantly, Robert Stutman. Stutman's sound bite is especially striking for what it says, and does not say, about the race and class dimensions of the crack crisis: "It [the spread of crack] appears to have somewhat leveled off in the suburbs, *as you and I define the suburbs*. We can't say that certainly for the inner city" (our emphasis). For Stutman, and presumably for Murphy and his viewers, the suburbs ("as you and I define the suburbs") is a code word for white, middle-class America; and the inner city, then, is a code for poor people of color. In this brief sound bite we have a condensation of the generally unstated racial oppositions underlying the news frame of the siege paradigm.

Murphy's reliance on experts who have a vested interest in substantiating and perpetuating the crisis is, ironically, precisely the kind of crusading myopia that fueled the worst excesses of the cracked coverage he sets out to defend. Dissenting opinions from critics of journalism's activism in the war on drugs (Weisman, Safire, Cowan, Vigilante, and the News Study Group) are noticeably absent from Murphy's report. But Murphy is not alone in this oversight. The *Vanderbilt Television News Index and Abstracts* lists only two sound bites from Edwin Diamond in 1987, and neither was related to the "media hype" study. One report (ABC, 3/10/87) dealt with economic problems facing network news divisions; the other (ABC, 3/13/87) is in a "Person of the Week" profile of CBS chairman Lawrence Tisch.[21] Apparently, Diamond is authorized to comment on the economics of the news business, but not on its performance in the war on drugs.

Revisionist Coverage

While the findings of the News Study Group may not have been granted an honest hearing in a network newscast, there can be no doubt that it made a difference, at least for a while, in drug coverage. After the study was made public in February 1987, several revisionist reports began to appear on the networks. This coverage generally employed two corrective strategies that are in keeping with the conventions of routine, mainstream reporting. The first corrective strategy involved focusing journalistic surveillance on the conflicting interests of the medical-industrial complex; the second involved the rehumanization of cocaine transgressors.

Policing the Police   The network news does occasionally run reports about police corruption (see, for example, CBS, 4/08/86). But these reports almost always treat the transgressing police as *individual* renegades who have violated the public trust. As a general rule, the "support the troops" mentality that contaminated all networks' coverage of the Gulf War in 1991 also infected news coverage of the war on drugs. Outside of stray reports, like Bernard Goldberg's cynical questioning of the effectiveness of high-profile police work, the drug news generally treats the police as disinterested and heroic forces in the conduct of the war on drugs—forces that have no economic stakes in the hawkish strategies of punitive retribution. In fact, most coverage takes the current police state as natural and necessary because of the seriousness of the "drug menace." To the best of our knowledge, until Ron Harris's groundbreaking article in the *Los Angeles Times* on April 22, 1990, the racial overtones of the drug war were never seriously investigated by the mainstream news media.[22] Furthermore, until the release of the "counter-police/counter-journalistic" clandestine surveillance of the Rodney King beating, police brutality was never a major issue in news coverage of the war on drugs.

Even so, one NBC "Special Segment" reported by Dennis Murphy is worth mentioning because it *inadvertently* draws attention to the journalistic abuse of raiding footage. Airing November 4, 1987, the report covers a sting operation in which the Miami police set up a storefront for fencing stolen goods. The report is drenched with stigmatizing footage generated by clandestine police cameras of suspects who are fooled by the sham. Murphy uses the police footage, complete with time-code, to introduce a rogue's gallery of sinister characters:

A black "street person called 'Dennis,' " who, according to Murphy, is "an alleged cocaine user with cash-flow problems." Dennis is shown trying to sell a stolen '87 Cadillac DeVille.
"Curt" and "Shawn" (both also black males), who are shown selling a stolen '83 Mercedes SEC "worth $29,000" for a mere $500.
"Rudy," another black male who sold the undercover police a "brand new moving van with a load of furniture aboard" for $1,200. According to Murphy's estimate, it was actually worth $45,000.
"Richie," a white male, who "claims to have killed 14 people." Murphy uses "Richie's" story to conclude his report: "Richie . . . came to the store and offered to kill Dennis for $3,000. He was counting his money when the police arrested him. He was very surprised. Tonight, 49 *others* will be very surprised, too." (our emphasis)

While using clandestine police footage to visualize the deviance of these "others," Murphy employs the standard obtrusive news camera to represent the authority of the law enforcement figures who provide expert sound bites for the story:

Clarence Dickson, Miami's black police chief, who states that "the average [crack] user could spend up to $300 a day."
Rolando Bolano, a Latino agent of the Florida Department of Law Enforcement who uses Us vs. Them language to explain the "nonsense" of taking small sums of money for such expensive property: "For *us,* it's not good business sense. But to *them* it is. Because *they're* thinking in terms of that day. What is going to carry *them* through that day in their addiction." (our emphasis)
Lt. Michael Christopher, a white Miami police officer who delivers the results of the sting operation: "To date we have recovered over a million dollars in stolen

property and identified [or *stigmatized*] approximately 50 criminals in our community."

One police expert, though, received different visual treatment. To protect his identity (and consequently underscore the dangers of clandestine police work), the undercover sergeant who ran the sting operation appears in silhouette. Murphy frames the sergeant's sound bite by revealing that "he bought property from about 60 street people, all of them, he said, hooked on crack":

SERGEANT IN SILHOUETTE: Some of them are very scary. Some of them have extensive criminal pasts. In fact, one of them that we're dealing with now ["Richie" introduced above] has admitted to us that he's committed 14 murders.

In keeping with the narrative and visual oppositions of the segment, the standard obtrusive news camera is used to give body and voice to "commonsense" victims of the property crimes when Murphy interviews a typical white, middle-class couple who come to retrieve their stolen car. "When the owners got their cars back," observes Murphy, "they were delighted—but shocked at how little the fences had paid for them":

WHITE WOMAN: $80.00! Wow!
WHITE MAN: That's less than my monthly payments.

The visual stigmatization of transgressors, the Us vs. Them rhetoric of the policing sound bites, the valuation of white, Middle American, commonsensical property owners—all of these narrative strategies are standard features of drug stories told according to the conventions of the siege paradigm. But this special segment also includes a brief passage that, at least for us, makes strange the familiar racial coding of raiding camerawork.

The passage begins with shots of Chief Clarence Dickson observing the operation in the backroom where the undercover video equipment is located as Murphy's voice-over narration explains that "the cops' cover turned out to be too good": "The day the Miami Police Chief came to check out the sting operation, a bizarre thing happened." Then, over police surveillance footage, Murphy continues: "Miami police were making a buy from 'Dennis' and a nervous new character named 'Michael.' Little did they know that 'Michael' was an undercover county police officer." Michael's "true" identity explains why, although Dennis's face is visible in the cover footage, Michael's is electronically obscured by a "black dot" that not only speaks to the dangers of police work but also to the stigmatization of surveillance footage in the news. In other words, the "black dot" both blots out Michael's identity and rescues him from the realm of "Them."

What follows is a strange rupture in the story's point of view that is explained by Murphy's voice-over: "A county task force had surrounded the store. *These are the pictures they took*." Suddenly, we are treated to the "bizarre" spectacle of the Dade County police raiding the Miami police sting operation. The "pictures they took" even include undercover surveillance footage of Chief Dickson leaving by the backdoor of the store ("The county police didn't recognize him") before showing classic raiding footage of the county police busting the sting operation—only this time the transgressors forced at gunpoint to get face down on the floor include the NBC News crew. "After a few tense moments," confides Murphy, "it dawned on the raiding county police that they hadn't broken up a stolen car ring, they had simply stung the sting." While the ironies of seeing the NBC camera crew suffer the indignities of being "stung" seem to escape Murphy, this is, for us, a defamiliarizing moment in the 1980s cocaine narrative that draws attention to the convergence of the journalistic and policing outlook in the rest of Murphy's report by violating its conventions—a moment that provides perspective by incongruity in its fleeting positioning of the news crew in the stigmatizing place of the transgressor.

Reporting Therapeutic Deviance   TV journalism has, by and large, been more amenable to turning its own entrepreneurial instincts and surveillance mechanisms on the medical-therapeutic arm of the drug control establishment. This was even a feature of some conventional news stories during Phase I coverage. One criticized drug counseling as a business that capitalizes on cocaine addiction (CBS, 2/7/83); another condemned "purported cocaine therapists" and "fly-by-night entrepreneurs" (NBC, 6/3/85). In a more unconventional six-minute report, Meredith Vieira of CBS investigated the hospitalizations of teens with alleged mental, drug, or alcohol problems. In her story, parents protested that the hospitalization had actually exacerbated their children's emotional problems. One therapist warned of misdiagnoses, while another health professional pointed out the economic self-interest of hospitals that report phony symptoms in order to collect insurance monies (CBS, 5/20/85).

During the 1986 crisis, criticism of the drug treatment industry was largely suspended until ABC's "Drugs USA" series (mentioned above). The third special segment in the series is notable for recognizing that drug treatment is big business. Reporter George Strait briefly surveys a range of common treatments from the more conventional twelve-step techniques of Alcoholics Anonymous to the electric shock-aversion therapy favored by Dr. Larry Kroll of the Lifeline Clinic. The main source for the story, though, is Dr. Herb Kleber of Yale University who is highly critical of the industry. "There is a building up of a treatment industry," warns Kleber, "that treats you sometimes for very large sums of money and then discharges you perhaps with a telephone number. I think that it is dreadful practice to do that."

This revisionist impulse again surfaces in Phase III in an exposé of the STRAIGHT drug treatment program (NBC, 11/13/87). In chapter 8 we briefly mentioned the treatment program in connection with the moving confession of Robin Page, a graduate of STRAIGHT, who appeared in ABC's and NBC's coverage of the First Ladies' international drug summit. A thriving "$6 Million, non-profit corporation," STRAIGHT (according to Tom Brokaw's lead-in to Dennis Murphy's exposé) "uses many of the same techniques as military bootcamp." Murphy's investigation discovered a "dark side to STRAIGHT hidden from Nancy Reagan and other admirers, like England's Princess Diana." According to Murphy, STRAIGHT's tough-love technique that involves "kids disciplining kids with little or no adult supervision" resulted in mental and physical abuse that, in many cases, had done more harm than good. Amy Totenberg, a lawyer for one of the STRAIGHT's former juvenile clients, even describes the kids-on-kids discipline of the treatment program as a "*Lord of the Flies* situation."

The most damaging report of "therapeutic deviancy," however, would not appear until 1991: reporter Sylvia Chase's "To the Last Dime" (produced by Stanhope Gould), a segment on ABC's "Prime Time Live" (7/18/91). Almost six years after *Advertising Age* reported ethical problems with 800-COCAINE (which began operations in May 1983), Chase belatedly questions the hotline and other for-profit referral services that funnel potential patients into hospitals run by Psychiatric Institutes of America (PIA), the hotline's parent company (see our discussion of the hotline chronotope in chapter 5). Chase identifies PIA hospitals and therapeutic services that either extend patient stays to collect more insurance or "miraculously cures" and discharges patients on the very day the insurance runs out. Two of the original founders of the cocaine hotline, Mark Gold and A. Carter Pottash, who appear over and over throughout the 1980s as drug experts, would not talk to ABC in this report. However, Chase is able to report that Fair Oaks Hospital and PIA paid a $400,000 fine after an investigation in the early 1990s by the fraud division of the New Jersey Department of Insurance.

As unconventional as this report is, it still suffers from the modern journalistic split between observed and lived experience. In other words, what is not part of the "Prime Time Live" surveillance report is the news turning its critical lens inward. There is no mention in this 1991 report that ABC News and the other networks throughout the 1980s lent creditability and legitimacy to Mark Gold, A. Carter Pottash, and others associated with PIA and the hotline by granting them extensive sound-bite time and allowing them to promote their expertise. Nor does the ABC News program indict its own network for running national advertisements for the hotline that push therapeutic wonders at the same time they mask entrepreneurial motives. A 1990 *Scientific American* article reported, in fact, that one print ad featuring the hotline overinflated (by nearly 7 million) the number of persons using cocaine. Worse yet, the 1989 ad, sponsored by the business consortium Partnership for a Drug Free America (which unapologetically accepts funding from "Anheuser-Busch, R. J. Reynolds Tobacco, and a half-dozen pharmaceutical companies"),[23] stated: "Last year, 15 million Americans used cocaine—and 5 million of those who survived required medical help." This suggests that, in 1988, 10 million people died from cocaine, when the National Institute of Drug Abuse actually reported "62,141 medical emergencies and 3,308 deaths in which cocaine was implicated."[24] In addition, questions also might be raised regarding professionals who use the hotline service to conduct national surveys in order to advance medical-scientific careers since the data from the referral services become the substance of articles published in medical journals.[25]

Rehumanization The rehumanization strategy, strangely enough, tended to graft a therapeutic outlook onto the siege framing of the cocaine problem. In this grafting, the hope of recovery of the standard therapy formula is transformed into a strangely fatalistic *discourse of doom*. As in the siege paradigm's discourse of discrimination, this pronouncement of doom treats drug transgressors as beyond rehabilitation. But, unlike the Us vs. Them rhetoric of the siege paradigm, the fatalistic vision of doom rehumanizes the transgressor as a tragic victim of the "slings and arrows of outrageous fortune" whose "fate" is to make bad choices and pay the consequences.

The ABC "Drugs USA" series pioneered this strategy of rehumanization. The first report in the series was narrated by ABC's Peter Jennings. One of the few stories in 1986 to personalize the cocaine problem, this report looks at the wasted life of a twenty-eight-year-old white man given the pseudonym of "Keith." Described by his sister as "an all-American boy," Keith confesses that at one time he "could have been anything I wanted to be." However, now, because of cocaine, he does not expect to live to see thirty. The report includes interesting footage of Keith watching "Leave It to Beaver" in his bedroom, visiting a former teacher at his high school, and fishing in a pond near his home—but all of the nostalgic images are informed by the despair of Keith's repeated failure at the hands of drug treatment programs. This tragedy is, at once, personal, familial, and societal. It is personal because Keith has given up all hope of being rehabilitated and making something of himself; it is familial because Keith admits to harming his own family (as he confesses, "the easiest people to steal from is your own family"); it is societal because, as Keith's mother puts it, "cocaine destroys the American Dream."

This discourse of doom would even feature a "voice from the grave" in a report aired in the spring of 1988 and narrated by none other than Dan Rather (CBS, 5/12/88). The final installment of a five-part series titled "War on Drugs," this gloomy portrait of the "drug scene" qualifies as one of the most macabre stories in our study. The heart of the report prominently features two experts in death. The first, Dr. Ian Hood, is a former medical examiner: "I've never seen the volume of casual violence before that's been associated with crack cocaine use—especially with young people. They are extremely cavalier in the way in which they will take another person's life." The second, a morgue attendant named Nick Giancana, seconds Dr. Hood's testimony: "Seems like everybody's into drugs for some reason now. I don't know why. It's a shame. We see a lot of lives come through here [the morgue] that are wasted."

Sandwiched between these two expert sound bites is a brief nonsense vignette that, like Keith's in the ABC package, serves to rehumanize drug transgression—only this time the transgressor is a poor, black female named "Lisa" who fears for her life. An excerpt from an interview in her squalid room conducted by an unidentified reporter working at a CBS affiliate, the vignette is overlaid by Rather's morbid grand narration voice-overs:

UNIDENTIFIED LOCAL REPORTER: Did anyone ever threaten to kill you?
LISA: Several times.

RATHER (V.O.): Lisa was an addict who sold bad drugs to support her habit.
LISA: I might be kilt. The fact that I, you know, what I've done to so many people. And I don't remember all of them. Maybe they remember me. I could step down the steps and might get killed.
RATHER (V.O.—after a dissolve to the morgue): Lisa was shot to death at the age of 21. No matter what form it comes in, dope can kill anyone who goes near it.

Lisa's terrified face and despondent words give a fatalistic voice to those who have died—and are destined to die—in the vortex of the crack economy. But, while Lisa is momentarily humanized by this death vision, the report, in the end, does not challenge the moral framing of the siege paradigm. Her death at age twenty-one is only a consequence of her own actions—her choice to get too "near" to "dope." As a voice from the grave, then, Lisa's confession represents an extreme example of the discourse of doom that pervades much of the aftermath coverage—a discourse that in personalizing the drug tragedy essentially treats death as a matter of individual choice.

Boyz n the East Oakland Hood

A Special Segment On April 29, 1988, NBC's John Hart delivered another story that features a voice from the grave. An extremely rare longitudinal study of the inner-city drug problem, the report is notable for its eerie convergences with, and significant divergences from, the tale of two dead brothers told cinematically by writer-director John Singleton in *Boyz N the Hood*. Since, for us, both Hart's report and Singleton's 1991 film represent the outer limits of mainstream popular culture's revision of the cocaine narrative, we have singled out this "special segment" for a special storyboard presentation and comparative analysis.

IMAGE 1: *Brokaw at news desk.*
BROKAW: We began this program tonight with a story of people who beat the odds—the passengers who survived the terrifying mid-air accident when the top of their airliner was ripped away. We can all identify with their fright, and with the elation that they lived through it. Now we ask you to identify with another kind of story of life and death. It's the story of two brothers on the mean streets of East Oakland, a California community that has been all but

abandoned by opportunity. NBC News has been following these brothers and their neighborhood for four years.

Commentary: The exhortative "you" in Brokaw's opening plea is strikingly different from that of the crusading voice common during the crisis coverage. Here, instead of asking viewers to identify with the police on the beat, Brokaw is encouraging viewers to identify with "the policed" on the "mean" East Oakland streets. By placing the report in dialogue with the airplane mishap, Brokaw makes a direct appeal to (white) viewers, who easily identify with the passengers aboard the scalped airplane, to invest that same kind of identification in people who are normally on the nightly news stigmatized as deviant. Put another way, Brokaw in fashioning this unusually long and involved lead-in seems to recognize that rehumanizing black youth requires a fundamental shift in the angle of identification proposed by most news coverage.

In this plea for audience identification, Brokaw also sets up two of the major themes that organize Hart's package: the notion of "beating the odds" and the proposition that East Oakland has been "all but abandoned by opportunity." Both themes are also prominent in *Hood,* except the "California community" abandoned by opportunity in Singleton's story is South Central Los Angeles. Yet, unlike Singleton's film, Brokaw's lead-in frames the situation in East Oakland as a quasi-natural catastrophe compounded by human error (much like a plane crash). The opening segment of Hart's package continues this "naturalization" of a man-made disaster, as well as its refusal to indict broader political and economic forces that have participated in the desertion of this disposable community. Rather than investigate the devastating consequences of Reaganomics in communities like East Oakland, Hart continues the journalistic tradition of blaming the victim by treating the celebration of bad role models linked to the crack economy as the primary reason that "opportunity has all but abandoned" East Oakland.

IMAGE 2: *Horse-drawn caisson carrying Felix Mitchell's casket.*

HART (V.O.): The funeral of Felix Mitchell was a wonder. Bay horses leading. Rolls Royces following. The drug lord of East Oakland who made $20–30,000 a day selling narcotics in a slum, ending up murdered in prison. Some role model.

IMAGE 3: *Black children watching funeral.*

CHILD #1: He cold.

CHILD #2: He really cold.

CHILD #3: I liked the way he died, you know. He died stabbed.

IMAGE 4: *Night arrest scene.*

HART (V.O.): Some neighborhood—where the drug organization was the most visible employer; where a kid could start as a cop spotter for a couple of hundred a week and rise to distributor for thousands.

IMAGE 5: *Wad of money at bust.*

WHITE COP (with mocking cynicism): Oh, my goodness. Look at the money. What do you do for a living?

Commentary: Like John Blackstone's initial report of Felix Mitchell's funeral almost two years earlier (CBS, 8/8/86), John Hart's narration is laced with reportorial disgust (see chapter 7). Also like Blackstone, Hart includes non-sense sound bites, only this time from black children admiring the pomp of the conspicuous funeral procession. With this non-sense framing from "tomorrow's delinquents," Hart establishes East Oakland as a pathological neighborhood that suffers more from a "poverty of values" than from economic deprivation. For Hart, the source of this pathology involves the lack of positive male role models and a drug economy that, according to the logic of the story, has forced "opportunity" to abandon East Oakland. Hart does not consider the alternative explanation for the declining civility in East Oakland—that limited employment opportunities result in the desperate conditions that encourage the thriving drug trade.

While *Hood* is certainly a more sophisticated rendering of these conditions, one that does, in fact, implicate larger political and economic forces in the tragedy of South Central Los Angeles, it also takes a Moynihan-esque position on the absence of strong male role models. But Singleton's treatment of the police is strikingly different from Hart's. During the arrest scene Hart implicitly approves of the white policeman's mockery of the black drug suspect. The kind of overbearing cynicism and self-righteous disgust voiced by the policeman are, after all, also inscribed in the "some role model" and "some neighborhood" pronouncements in Hart's narration. In asking us to identify with the "policed," then, Hart's report does not in any way question how this derisive police presence contributes to the climate of racial hostility, antagonism, and suspicion in East Oakland.

Singleton, on the other hand, does implicate the police as part of the problem. As Michael Eric Dyson correctly observes, the police helicopters that "constantly zoom overhead in *Hood*'s community" are "mobile metaphors" of the "ominous surveillance and scrutiny to which so much of poor black life is increasingly subjected": "The helicopter also signals another tragedy that *Hood* alludes to throughout its narrative: ghetto residents must often flip a coin to distinguish Los Angeles's police from its criminals." Hart's refusal to implicate what Dyson terms "conditions of closely surveilled social anarchy" in the East Oakland tragedy is even more negligent as the story addresses the sad "fates" of Curtis and Gerald Lee—two brothers who bear striking resemblances to characters in *Hood.*[26]

Gerald, the younger, a bespectacled and congenial figure reminiscent of the "Doughboy" character in *Hood,* has a jail record, but unlike Doughboy, Gerald has an older brother in Curtis who provides a "positive role model." In this capacity, Curtis is something of a composite of the Ricky, Tre, and Furious Styles characters in *Hood.* Like Ricky, his athletic talents attract the attention of college recruiters; like Tre, he is still vulnerable to peer pressure; and like Furious Styles, he is not afraid of getting involved in struggles to "clean up" the community.

IMAGE 6: *Gerald Lee on East Oakland Street.*

GERALD LEE: You can't be in East Oakland and stay out of trouble because you see dudes, they get rich. They get rich heck-o-fast. They have flashy cars. Heck-o-big fat [gold] ropes [around their necks]. All the girls. And you go, like, 'Yeah, I want that, too.'

IMAGE 7: *Curtis Lee.*

CURTIS LEE: I could have sold drugs. I could have hooked up with some gang, you know. I could have went to school, which is what I chose to do.

IMAGE 8: *Curtis in school.*

HART (V.O.): This is the story of two brothers from Felix Mitchell's neighborhood. Curtis and Gerald Lee, facing the odds in East Oakland.

IMAGE 9: *Leenell Jennings.*

SCHOOL PRINCIPAL LEENELL JENNINGS: I would think the odds are somewhere around 50–50, and a youngster can go either way.

IMAGE 10: *Evelyn Lee.*

MOTHER EVELYN LEE: And so I teach them, hey, you can get more out of books than you can out in the streets.

IMAGE 11: *Curtis playing basketball.*

HART (V.O.): Curtis' odds appeared to be improved by his basketball. College scouts were looking at him.

IMAGE 12: *Gerald Lee.*

GERALD LEE: Got some scholarships in school and he was like, "Yeah, it's gonna be cool, now."

IMAGE 13: *Gerald Lee with his children.*

HART (v.o.): Gerald, the younger brother, a father at 15 with a jail record by 16. His odds were improved by Curtis' leaning on him.

CURTIS (sound bite): I didn't want to see him make the same mistakes my older brothers made, you know. Neither one of them graduated from high school.

Commentary: Perhaps unintentionally, the sound bites by Jennings and Evelyn Lee implicitly support the moral framing of Hart's package by further demonstrating the absence of strong male authority figures, both in school and at home. But Hart does not explicitly blame either woman for the dismal situation faced by the two brothers. Unfortunately, as Jacquie Jones observes, Singleton's film does come very close to scapegoating single mothers:

> Singleton has said that, "A young boy needs a man to show him how to be a man." Unfortunately, he translates this revelation into a scenario in which women are incapable of raising male children, while men can singularly achieve the feat. Given the fact that virtually no Black men actually raise young boys single-handedly while a large percentage of Black women do, and given the fact that Tre survives while Ricky and Doughboy do not—especially since Ricky is an upright youth with solid values and the promise of a future— Singleton comes dangerously close to blaming Black women for the tragedies currently ransacking Black communities.[27]

While honoring Furious as a heroic single father who (through sheer force of masculine will) saves Tre from almost certain death or imprisonment, Singleton condemns Brenda (Ricky and Doughboy's single mother) as a failed matriarch who "fusses *over* Ricky" and "fusses *at* Doughboy." As Dyson suggests, although Singleton does avoid "flat stereotypical portraits of his female characters," he does not succeed in "challenging the logic that implicitly blames single black women for the plight of black children": "In Singleton's film vision, it is not institutions like the church that save Tre, but a heroic individual—his father Furious. But this leaves far too much out of the picture."[28] While the NBC package suggests that even heroic individualism cannot overcome "the odds" in East Oakland, Hart too leaves far too much out of the picture.

In calculating "the odds in East Oakland," Hart's reportorial disgust is gradually overwhelmed by a debilitating sense of doom. Like Brokaw's invocation of "the odds" in the lead-in, Hart's narration reduces the limited opportunities confronting the Lee brothers to a matter of Darwinian natural selection that seems immune to the interventions of motherhood, school, police, politics, heroic individualism, economics, and history. In other words, "the odds" rhetoric of Hart's account tends to mystify the constrained life chances associated with racial and economic inequality as plain old "bad luck"—and in this mystification the slaughter and criminalization of poor, black youth are transformed into a question of destiny, not discrimination.

IMAGE 14: *Castlemont High School.*

HART (V.O.): But first Castlemont High School needed cleaning up from the gangs and guns and drugs that had the run of it.

IMAGE 15: *Student #1.*

STUDENT #1: How can you come and get your education if you have to duck and dodge bullets?

IMAGE 16: *Student #2.*

STUDENT #2: People down here with .38s, .45s, and Uzis.

IMAGE 17: *Curtis at meeting of the board of education.*

HART (V.O.): Curtis and a friend went to the Board of Education with a petition to get rid of a principal they accused of hiding from the problems.

CURTIS (addressing the school board): Students can't even go to class without the fear of being shot at, jump on by somebody that don't even go to this school.

IMAGE 18: *Principal in school yard.*

HART (V.O.): They won. The high school turned around. The new principal says Curtis was an outstanding student.

PRINCIPAL JENNINGS (sound bite): We were looking forward to his graduation— until this unfortunate situation he was involved in.

IMAGE 19: *Curtis on trial.*

CURTIS (testifying in court): There was a gun there.

LAWYER: And you picked it up?

CURTIS: Yes, I picked it up.

HART (V.O.): Self defense, he said. In a car with the wrong kids; shot at by other wrong kids. Everybody reaching for a gun. A boy died, not by Curtis' gun, but the prosecutor said he shared an intent to kill.

IMAGE 20: *Evelyn Lee.*

EVELYN LEE: I said, "And you fired a gun? All that stuff you went through at Castlemont and you gonna tell me you fired a gun?" It's a bunch to take.

IMAGE 21: *Carole Evans, assistant principal.*

ASSISTANT PRINCIPAL CAROLE EVANS: Got in the wrong car with the wrong guys. Got involved. Wrong decision. And that's not an uncommon situation for some of our kids. They're vulnerable. They make the wrong choices. And they end up with the wrong consequences.

IMAGE 22: *Bailiff reading verdict.*

BAILIFF (in court): We find the defendant, Curtis Lee, Jr., guilty of a felony, to wit, Murder in the Second Degree.

Commentary: The moral ambiguities of Curtis's predicament here are reduced to getting in the "wrong car" with the "wrong guys," making the "wrong choice," and paying the "wrong consequences." Where "getting involved" is celebrated when Curtis petitions the school board to get rid of the old principal, "getting involved" is one of the "wrong decisions" that Curtis makes on this fateful day. The only white source to speak in Hart's package (other than the white police officer's brief ambient comment at the drug bust), assistant principal Carole Evans voices the commonsense outlook underpinning Hart's assessment of the situation in East Oakland. This reportorial privileging of Evans's definition of the situation is evident in the redundancies linking Hart's narration of Curtis in court ("In a car with the wrong kids; shot at by other wrong kids") to Evans's sound bite (". . . wrong car . . . wrong guys . . . wrong decision . . . wrong choices . . . wrong consequences"). With this elaboration on wrongness, the assistant principal, as well as Hart, while humanizing Curtis as a victim of circumstances, still stigmatizes and dehumanizes the "wrong kids" who literally "drive" Curtis to his downfall.

Therefore, where Furious's heroic individualism in *Hood* is redemptive, Curtis's heroic role in cleaning up Castlemont High School is presented as something of an exercise in futility—the futility of a character who tries to cheat his fate. Although we are led to believe that Curtis made a difference at the school, the ultimate fate of his brother Gerald, who we learn was a "straight 'A' student by the time he finished high school," undercuts any sense that improving the school translates into "better odds" for "Boyz N the East Oakland Hood."

IMAGE 23: *Curtis entering Folsom Prison.*

HART (V.O.): Curtis is now serving 17 to life in Folsom Prison. . . .

IMAGE 24: *Mourners at Gerald's casket.*

HART (V.O. cont.): . . . And his younger brother Gerald, who *was* a straight 'A' student by the time he finished high school, is dead.

FEMALE MOURNER (at casket): Oh, my dear God. Dear God!

HART (V.O.): Killed in a car crash while racing from police who thought he was someone else. . . .

IMAGE 25: *Shadowy guard at Folsom Prison.*

HART (V.O. cont.): . . . Curtis will be up for parole in a couple of years. He'll have to survive Folsom Prison where there were 186 stabbings last year. Eight of them fatal. Then, he'll have to survive the old neighborhood to beat the odds. . . .

IMAGE 26: *Judge Benjamin Travis.*

HART (V.O. cont.): . . . His trial judge is not optimistic.

JUDGE BENJAMIN TRAVIS: Well, there's a good possibility of him becoming a victim of the system.

IMAGE 27: *Gerald Lee, alive.*

HART (V.O.): His late brother agreed.

GERALD LEE: If you be around nothing but criminals for 17 years, I think you're gonna be a straight criminal.

Commentary: Gerald Lee's final words, appearing after Hart discloses that he is now dead, is invested with the same kind of poignancy that charges Doughboy's final appearance in *Hood* when he literally "dissolves" into the mean streets of South Central L.A. and we learn he, too, was a casualty of gang violence. As a voice from the grave, Gerald's sound bite provides funereal verification of Judge Travis's prognosis that Curtis stands a good chance of becoming "a victim of the system"—a system in which the judge himself is implicated. It is ironic that Judge Travis, the only strong, black, male role model to speak in Hart's package, presides over the harsh sentencing of Curtis.

Hart and Gerald, correctly, depict the prison for what it is—a dangerous place of exclusion where delinquency is only temporarily contained and where inmates are more likely to be further "criminalized" rather than "rehabilitated." Although Hart does not put it into words, his presentation of the stabbing statistics begs the question: If prison killed a hard case like Felix Mitchell, then what will it do to Curtis Lee? To Hart's credit, the incarceration of people like Curtis is presented as part of the problem.

But to Hart's discredit, he is negligent in connecting the failure of the prison system to the failure of the police surveillance system that is at least partially responsible for taking Gerald's life. Although we learn that Gerald was "killed in a car crash while racing from police who thought he was someone else," Hart fails to link Gerald's decision to run from the police to the kind of demeaning treatment accorded drug suspects that Hart presents, approvingly, in the earlier arrest scenes. Nor does Hart consider how Gerald's fatal flight just might have something to do with the severity of the punishment meted out to his brother for being in the "wrong" place at the "wrong" time and making the "wrong" decision.

IMAGE 28: *Exterior of Folsom Prison.*

CURTIS LEE (over prison scenes): My dream is to be financially stable. Have a family. Two or three kids. Be a dominant effect in their lives, you know. Somebody they can look up to.

IMAGE 29: *School Principal.*

HART (off-screen to principal): It strikes me that we're not just talking about Curtis, but a good part of a whole generation of American young people.
PRINCIPAL JENNINGS: We are indeed talking about a lot of Curtises.

IMAGE 30: *Curtis behind bars.*

CURTIS LEE: I kind of participated in my own fate, you know. But I never thought I would end up here.

Commentary: The conclusion of Hart's report flirts with challenging the "poverty of values" thesis. Curtis, after all, embraces the traditional "family values" of being financially stable, raising 2.5 children, being a strong father figure. Principal Jennings suggests that Curtis is not an isolated case—that a "good part of a whole generation of American young people" who, by and large, believe in the American Dream, suffer fates similar to Curtis's. But Hart, while he presents Curtis's situation as tragic, still treats his imprisonment as a matter of "choice." In Curtis's case, since it is clear that he "knew better," the wrong choice that he makes is even more unforgivable. His knowledge of "right and wrong," then, makes him more culpable as someone who "participates" in his own fate. Ultimately, like Lisa's death in the morbid CBS package, Curtis's prison sentence is cast in terms of individual choice in keeping with the "Just-Say-No" moral framing of the Reagan administration's war on drugs.

Hood, too, is organized around the question of choice. But Singleton's treatment of the choices confronting Doughboy, Ricky, and Tre is much more sophisticated than Hart's rendering of tragic choices made by Gerald and Curtis Lee. To paraphrase Dyson, where Hart's report verges on a "blind fatalism or a mechanistic determinism," Singleton shows how the lives and choices of the central characters are shaped by "personal and impersonal forces":

> Such everyday variations on the question of choice are . . . central to the world Singleton depicts in *Hood*. Singleton obviously understands that people are lodged between social structure and personal fortune, between luck and ambition. He brings a nuanced understanding of choice to each character's large and small acts of valor, courage, and integrity that reveal what contemporary moral philosophers call virtue. But they often miss what Singleton understands: character is not only structured by the choices we make, but by the range of choices we have to choose from—choices for which individuals alone are not responsible.[29]

This "nuanced understanding of choice" is what separates *Hood* from Hart's reports. In failing to implicate the larger *impersonal* forces (i.e., flexible accumulation and the modern police state) in the fates of Gerald and Curtis Lee, Hart makes the people of East Oakland collectively accountable for the severely limited "range of choices" that they confront in the course of their lives in East Oakland.

There is, admittedly, much to admire about Hart's report, just as there is much to admire about *Hood*. Unlike the snapshot views of the color-coded inner city provided by most drug coverage, Hart gives an account enriched by a sense of history, and a sense of place. This is not just a composite view of *the* inner city in the abstract told from the disciplinary point of view of the police. Instead, it provides a richly textured portrait of a particular neighborhood, East Oakland, and the particular life-and-death struggles of two brothers whose experiences, values, and choices have broader significance. But, like *Hood,* this portrait is framed in a way that confirms the prevailing conservative wisdom that sees the simple restoration of patriarchal authority as the solution to the plight of black children. Embracing this definition of the situation is, for us, not just a minor shortcoming, but a fatal flaw.

Singleton's Fatal Flaw   In fact, this flaw animates two very disturbing dialogic responses to *Hood* that, while coming from radically different socioeconomic positions, are linked by a masculinist outlook on the situation in South Central L.A. The first of the responses was displayed on the afternoon that *Hood* began its long first run in a multiplex cinema on the edge of Ypsilanti, Michigan. During the matinee screening, many of the young men in the audience cheered Doughboy's misogyny—a misogyny that Singleton's narrative justifies by establishing Brenda's unfair favoring of Ricky over Doughboy. "Here Singleton is on tough and touchy ground," writes Dyson, "linking the origins of Doughboy's misogyny to maternal mistreatment and neglect."[30] As expressed by laughter when Doughboy says, "Let the ladies eat; ho's gotta eat too," the approval of Doughboy's misogyny by a sizable number of males in the audience was especially apparent during the "welcome home" party celebrating Doughboy's release from jail. After Doughboy's girlfriend objects, declaring that she is not a "ho," there is even more masculine laughter when Doughboy replies, "Oops, I'm sorry, bitch." As Dyson suggests, such terms as "ho," "bitch," "skeezer," and "slut" have, unfortunately, become "the standard linguistic currency that young black males often use to demonstrate their authentic machismo": " 'Bitch' and equally offensive epithets compress womanhood into one indistinguishable whole, so that all women are the negative female, the seductress, temptress, and femme fatale all rolled into one. Hawthorne's scarlet A is demoted one letter and darkened; now an imaginary black B is emblazoned on the forehead of every female."[31] While Singleton tries to distance himself from this misogyny through the characters of Reva (Tre's career-oriented mother) and Brandi (Tre's virtuous Catholic girlfriend), his depiction of Brenda and a stereotypical "crack mother" (who lets her baby wander around the neighborhood in soiled diapers) reinforce the misogynist imagination of certain members of the audience. Therefore, while Singleton's portrayal of womanhood may have been "nuanced," certain members of the audience, who experienced the narrative by identifying primarily with rapper Ice Cube's performance of Doughboy's machismo, were not receptive to Singleton's subtlety. These audience members fabricated what Stuart Hall terms a "negotiated" reading[32] of Hood that was largely immune to the antigang message of the story.

Later in the film, this mutinous audience identification with Doughboy's brutal hypermasculinity was again apparent in the disturbing response to the cold-blooded killing of Ricky's assassins. When Doughboy systematically guns down the "wrong boys," some in the audience cheered his action. That cheer, for us, was infinitely more chilling than Doughboy's horrible black-on-black massacre on the screen. Although Singleton clearly did not intend to provoke such a response in the audience, it does follow the warped logic of a negotiated reading of the film that disregards the sermonizing and glories in the code of male honor that demands revenge.

But this bottom-up masculine response to Hood was not as frightening as a top-down patriarchal response to the film public-ized in the days following the rebellion in South Central L.A. triggered by the Rodney King verdict. About the same time that Vice President Dan Quayle was blaming the uprising on "Murphy Brown" and the "poverty of values" in America's inner cities, California's Republican governor, Pete Wilson, was making the talk show circuit, offering essentially the same explanation for the mass violence in South Central. However, instead of castigating "Murphy Brown," Wilson was celebrating *Hood*.

For instance, consider a lengthy interchange between Wilson and George Will that appeared May 3, 1992, on "This Week with David Brinkley." Will sets up Wilson by presenting the regressive conservative common sense underlying cultural Moynihanism as something "everyone knows." "A lot of people," asserts Will, "despair of government at any level getting purchase on what everyone knows, whether they want to say it or not, is the heart of the problem and that is the fact that too many young black men are fathering children they have no intention of raising." After Will substantiates his moral framing of the rupture in L.A. by citing statistics on illegitimacy ("Eighteen percent of white children in 1988 . . . [were] born to unwed women, 34 percent of Hispanics, 63 percent of blacks"), he formulates a leading question for Wilson: "Now, can you tell us, as governor of a state, anything you can do or recommend that anyone else at any level can do through government to have an effect on this tragedy?"

Wilson's response is a classic study in deflecting attention away from the economic conditions in South Central. In keeping with Bush's "thousand points of light" rhetoric, Wilson suggests that the best thing the government can do is encourage individual voluntarism. In a ringing endorsement of patriarchal power, Wilson commends the efforts of a "nonprofit group" called the "100 Black Men":

> This is a nonprofit group formed by men who are concerned about youngsters who don't have fathers. And what they are concerned about is that black eighth graders in the Los Angeles Unified School System who are B students are dropping out of school and not finishing. That's what they're concerned about, so they have created a mentorship program and what they're doing is not just taking kids to the ball game or to the circus. They're saying, "You keep your little fanny in that chair and study because it's going to affect the rest of your life." But they have to be credible to those children in two ways. It may be the first time in the life of that child that anybody has ever evidenced a real caring for the child, that anybody's ever said, "I care what happens to you." Second, they have to be themselves credible as role models. I think everyone in America should see the movie *Boyz N the Hood*. In that movie, a strong father makes the difference for his teenaged son, a 19-year-old boy about to rush

back out and try to avenge his best friend who has just been gunned down in a mindless, senseless gang war. And this strong father, literally and figuratively, puts his arms around his son and says, "I won't let you go out into the street. I'm not going to let them kill my only son." Now what that movie says is that we need a strong father and welfare is no suitable replacement for that.

In Wilson's countersubversive reading of *Hood*, then, Singleton's condemnation of the belligerent police presence in South Central (which was obviously a decisive factor in generating at least the initial violence) is completely screened out. What remains, after this screening, is a nostalgic treatise on the restoration of masculine power that is not only attractive to the gangster mentality of those who applaud Doughboy's misogyny, but also subject to appropriation by the Pete Wilsons, Daryl Gateses, and Dan Quayleses of this world who would fight machismo with machismo.

In these seemingly diametrically opposed responses to *Hood*, then, we see another instance of how what is condemned as the "poverty of values" in places like South Central L.A. is not all that far removed from the values animating the backlash politics of Reaganism. After all, as the "Tailhook scandal" in the U.S. Navy demonstrates (as well as Anita Hill's charges of sexual harassment during the Senate confirmation hearings on Supreme Court nominee Clarence Thomas), the "bitches-and-hos" view of womanhood is certainly not isolated to inner-city America. This misogyny, in other words, is an aspect of a broader masculinist worldview that has trickled down to the street—a worldview that contaminates not only gangster rap, but also the officer corps in the military, the jock rapists of professional sports, the Senate Judiciary Committee (including the compromised Hill supporter Ted Kennedy), and the patriarchs on the right wing of the U.S. Supreme Court. Although we, too, see this misogyny in terms of impoverished values, they are not values unique to the inner city. In other words, those who share this masculinist outlook in the inner city are not practitioners of an "alien" value system; they are, instead, very much in tune with the gender relations endorsed by the New Right.

Closure

Objectivity requires only that reporters be accountable for *how* they report, not what they report.—Ted Glasser[33]

Journalism ought to be conceived less on the model of information and more on the model of a conversation.—James Carey[34]

A Qualitatively Different Drug Crisis   Lloyd Johnston of the University of Michigan's Institute of Social Research—in an article, "America's Drug Problem in the Media: Is It Real or Is It Memorex?"—argues that the crisis in 1986 was indeed "real":

> It is fair to say that we did have a drug crisis in 1986, once you take into account (a) the addictive potential of cocaine, (b) the increasingly dangerous forms in which it was being used, (c) the rapidly increasing rate of casualties that were occurring, (d) the increasing availability and dropping prices (which made crack, in particular, accessible to younger age groups), and (e) the degree of penetration of cocaine use into youthful populations. *It was admittedly a qualitatively different drug crisis than the one that existed in 1980* [emphasis ours]. Fewer people used illicit drugs in 1986 than in 1980—marijuana use had especially declined, and use of amphetamines and barbiturates was also down —but more people were at risk of addiction and overdose reactions because of what they were using and how they were using it.[35]

According to Johnston's presentation of the evidence for a real crisis in 1986, the lack of success of the war on drugs was probably the most under-rated factor contributing to sensational increases in public concern about drugs. Johnston points out that, although cocaine use did not increase in the 1980s, it did remain steady at peak levels despite dramatic supply-reduction efforts on the part of government:

> As was widely reported in the 1980s, the federal government substantially in-creased its efforts to close the supply of drugs (particularly cocaine) at the U.S. borders. Drug enforcement activities inside the country also continued to increase and, because of their graphic quality, got particular attention in the electronic media. Television coverage made the raids familiar: a battering ram for the door; the heavily armed police; the drugs, money and guns that were confiscated; and the handcuffed suspects being led away. Yet, despite all of this action and the expenditure of billions of dollars on enforcement, there was strong evidence that we were losing the war on drugs. Something was wrong. The public and the media both knew it, and both were increasingly frustrated and alarmed at the inability of their social institutions to handle the problem.[36]

As further evidence that the war on drugs was being lost, Johnston cites DEA figures on the purity and cost of cocaine that indicate increasing avail-ability: the purity rose from about 33 percent in 1982 to about 60 percent in 1986; at the same time, the average wholesale price of a kilogram of cocaine dropped fairly steadily from roughly $60,000 to $35,000.

While we have come to agree with Johnston's assertions that the drug crisis of 1986 did, indeed, reflect dramatically worsening conditions in the "real world," we suggest that the root cause of the crisis was the failure of Reaganomics, not the failure of the war on drugs. As Craig Reinarman and Harry Levine put it: "The spread of cocaine and crack . . . was a godsend to the Right: they used it as an ideological fig leaf to place over unsightly urban ills which had increased markedly under the Reagan administration; they used it as a scapegoat on which to blame many economic and urban problems."[37] From this perspective, the war on drugs was a political coup, not because it reduced demand, but because it was triumphant in framing worsening social problems and economic ills in the moralist terms and culturalist vocabulary of the New Right. As a political spectacle, the crack crisis provided the power bloc with an opportunity to exercise further controls on the core and peripheral labor markets of the post-Fordist economy. Most strikingly, in keeping with what we have identified as the political economy of scandal, after news coverage of the crack crisis succeeded in demonizing an *unemployed* fragment of a disposable labor market (poor, black, urban youth), corporate interests were able to justify, and generate popular support for, a drastic expansion of drug testing of *employed* laborers in the workplace—a development that certainly represents a significant intrusion of employer control over workers' private lives. As we have suggested throughout this study, the only way to grasp the political economy of the war on drugs, in general, and the crack crisis, in particular, is to see it in terms of Reaganism's dual commitment to loosening controls on business interests while tightening margins of illegality on laboring populations.

Furthermore, our analysis of the cocaine narrative demonstrates that this political spectacle was facilitated by "objective" researchers, like Johnston, who attempt to cope with the drug problem, according to Murray Edelman, by "changing the consciousness or behavior of individuals while preserving the institutions that generate consciousness and behavior."[38] Although it is perhaps unfair to single out Johnston because he was, literally, *just* doing his job (and he was, in fact, also mildly courageous in publicly resisting brazen attempts by the Reagan administration to further colonize NIDA),[39] we feel compelled to point out that, perhaps because of a kind of obstinate blindness—or maybe tactical naïveté—Johnston's article on the reality of the crack crisis never once entertains the rather obvious possibility that this "qualitatively different drug crisis" just might have something to do with the politics of race and class in Reagan's America.[40] In other words, by limiting their efforts to controlling "consciousness and behavior," such experts demonstrate how "value-free" research and scientific "objectivity" become ways of "not knowing," even "not seeing." As our study also demonstrates, this "not knowing" and "not seeing" also was promoted by so-called journalistic neutrality during the cocaine narrative.

Naming Names Foucault has forcefully argued that the two poles of the modern control culture—the discipline of the body and regulation of the population—were finally brought together in the nineteenth century's obsession with sex:

> It [sex] fitted in both categories at once, giving rise to infinitesimal surveillances, permanent controls, extremely meticulous orderings of space, indeterminate medical or psychological examinations, to an entire micropower concerned with the body. But it gave rise as well to comprehensive measures, statistical assessments, and interventions aimed at the entire social body or at groups taken as a whole. . . . sexuality was sought out in the smallest details of individual existences; it was tracked down in behavior, pursued in dreams; it was suspected of underlying the least follies. . . . But one also sees it becoming the theme of political operations, economic interventions (through incitements to or curbs on procreation), and ideological campaigns for raising standards of morality and responsibility. . . . Spread out from one pole to the other of this technology of sex was a whole series of different tactics that combined in varying proportions the objective of disciplining the body and that of regulating the population.[41]

Our interrogation of cocaine news of the 1980s has led us to conclude, just as forcefully, that (at least in the United States) this same type of convergence of normalizing technologies also occurred at another site during the decade: the pleasurable and punishable domain of psychoactive drugs. As this study documents, television journalism, in public-izing this convergence, also was implicated deeply in orchestrating and legitimating the "political operations, economic interventions, and ideological campaigns" of both the drug control establishment and the New Right. This crusading journalism not only facilitated the ascendancy of a punitive strategy in national drug control policy, but it has generated a "new consensus" that attributes the many human troubles attending economic distress to "*individual* deviance, immorality, or weakness."[42] Although mainstream television journalism would suffer some second thoughts during its 1986–88 coverage, it was clearly not enough to undermine, to any significant degree, this reactionary consensus.

But, unlike Richard Vigilante in the *National Review,* who blames the excesses of the war on drugs on the anonymous abstraction called "the media," our indictment of the New Right, the drug control establishment, and mainstream journalism has named specific names and organizations that should be held accountable for profiting from drug hysteria. From Ronald and Nancy Reagan, through George Bush and William Bennett, to Daryl Gates and Pete Wilson, the drug warriors on the right have used the cocaine issue to polarize the electorate along color lines while, at the same time, diverting attention away from failed economic policies. As one of the most expansive segments of the service economy during the Reagan era, the drug control industry, in both its hard and soft manifestations, served as a mercenary force that "authorized" this reactionary political operation. The National Institute of Drug Abuse, DARE, Partnership for a Drug Free America, STRAIGHT, Psychiatric Hospitals of America, Mark Gold and Arnold Washton at 800-COCAINE, Richard Miller at COKENDERS, Robert Stutman at the DEA, and free-lancing consultants like Dr. Robert DuPont—these are some of the major players and entrepreneurial interests that have, in Trebach's phrasing, obtained "money and power from playing the ever-popular and ever profitable drug-war game." [43] And to this list of moral entrepreneurs we would add anchors Rather, Brokaw, and Jennings, as well as a host of reporters—from the famous (like Moyers and Stahl) and the not-so-famous (like McLogan and Gillen) to the infamous (like Rivera and Donaldson).

All in all, then, this has been a fairly gloomy assessment of the 1980s drug news. This gloom even infected the final words of an earlier draft of this book that concluded with the dismissive line "And so much for the candidate of change"—a pessimistic note grounded on what we still consider to be unsettling continuities between the rhetorics of Reaganism and Clintonism. Unfortunately, the countersubversive imagination of the Reagan/Bush consensus regarding America's drug/race problems has had such command of the political process that the "air war" of the successful 1992 campaign of Bill Clinton (a member of the contaminated 1960s generation who didn't "inhale") essentially repackaged much of the Reagan backlash. Clinton's gratuitous attack on Sister Souljah, his marginalization of Jesse Jackson, his repudiation of welfare ("welfare is a second chance, not a way of life")—all of these actions were none-too-subtle appeals to Reagan Democrats to come "home" to a newly rejuvenated and "re-centered" party that, under a new covenant, had been freed from the chains of "special interests" and cleansed of the stigmatization of the "liberal" label. In this recovery operation, Clinton, to his credit, managed to condemn the Us vs. Them thinking of the New Right at the same time that he successfully demonized trickle-down economics. But, ironically, in the midst of his condemnation and demonization of the Reagan order, he also reaffirmed a central principle of Reaganomics: Clinton's promise to eliminate 100,000 bureaucrats from the federal payroll and put 100,000 more police on the streets is not only a ringing endorsement of the police state, but it is an encoded pledge to continue the Reagan-Bush administrations' discriminatory program of deregulating entrepreneurial interests while exercising even more discipline on reserve and criminalized labor markets. It was a pledge that should have warmed the cockles of Daryl Gates's heart.

However, since we sincerely hope that this book generates a kind of righteous indignation that translates not only into political awareness, but into real political action, we have had our own second thoughts about concluding with such hopelessness and cynicism. Interestingly, these second thoughts are linked to our own experiences as teachers. Based on our years in the classroom, we believe that intellectuals involved in media analysis tend to devote too much energy toward the construction of doomsday visions of ideological domination that leave many students with no sense that their politics and their actions can make much difference. In short, we have found that critical media studies are generally very good at demonstrating the actuality of domination and not very good at proposing strategies for living that at least tentatively entertain the possibility of freedom, of struggle, of empowerment. Consequently, because we fear the "so-much-for-the-candidate-of-change" punch line might inspire more docility than resistance, we believe it is much more productive to accentuate the positive and the possible in our final words on the 1980s drug news.

A New Covenant for Journalism?   When Horace Greeley started the *New York Tribune* in 1841, he promised a "journal removed from servile partisanship on the one hand and from gagged, mincing neutrality on the other."[44] Journalism history demonstrates that modern reporting accomplished the first half of Greeley's dictum but at the expense of becoming mincingly neutral. By the late nineteenth century most editors and publishers realized that the quickest way to make lots of money was by straddling the middle ground and offending as few folks as possible. Neutrality, in the guise of a moral imperative to objectivity, became modern journalism's ticket to fortune. By the late twentieth century, readers and viewers remain hostage to what James Carey calls "a scientistic journalism devoted to the sanctity of the fact and objectivity. . . ."[45] As Ted Glasser maintains, "objective reporting has transformed journalism into something more technical than intellectual; it has turned the art of story-telling into the technique of report writing." [46]

In indicting television journalism's role in promoting the drug hysteria through much of the 1980s, then, we place much of the blame on modern journalism's own addictions to neutrality and information. In its modern conceptualization, particularly the celebration of data and numbers, the inability of self-scrutiny, and the denial of its own narrative conventions and subjective rituals, daily conventional journalism remains particularly susceptible to aiding and abetting whomever is wielding power. In other words, "neutrality," as it is practiced in modern journalism, is a powerful ideology that routinely masks a top-down perspective on social problems. Glasser argues, "objective journalism" illustrates "a bias in favor of leaders and officials, the prominent and the elite." [47] Journalist Tom Wicker concurs: "Objective journalism almost always favors Establishment positions and exists not least to avoid offense to them." [48]

But most significantly, we worry about modern journalism's foregrounding of front-page, top-of-the-broadcast facts and figures in place of creating provocative public forums for debating significant social issues such as the contemporary drug culture. We agree with James Carey that "Neither journalism nor public life will move forward until we actually rethink, redescribe, and reinterpret what journalism is; not the science of information of our culture but its poetry and conversation." [49] In calling modern journalism back to action, Carey contends: "The public will begin to re-awaken when they are addressed as a conversational partner and are encouraged to join the talk rather than sit passively as spectators before a discussion conducted by journalists and experts." [50]

Joining Carey, we call on journalists to rethink the strictures and rituals that have guided them through much of this century and to embrace a new covenant. For journalism to accomplish its often-stated mission to afflict the comfortable and comfort the afflicted—which the dictates of neutrality, of course, prevent it from doing on most occasions—it must become more self-conscious, more overtly interpretive, more honest, and much less debilitated by its "quasi-objectivity." Contemporary culture awaits a journalism that will provide front-page conversation and top-of-the-broadcast debates about issues of the day. Modern journalism instead has allowed TV and radio talk shows to fill this role for multiple publics who are starved for inclusion, hungry to take part in the collective conversation. As Carey suggests, "The press [and we would add modern TV journalism], by seeing its role as that of informing the public, abandons its role as an agency for carrying on the conversation of our culture."[51] Throughout the 1980s, regarding important issues about drugs, issues about the legal and the illegal, about discipline and punishment, about the police state and the policed, about the mainstream and the marginal, about rich and poor, much of television journalism and many of us as well fell asleep at the wheel, and at some point the crusading moralists of Reaganism, who are not in any way paralyzed by the pretenses of "neutrality" or inhibited by notions of "fairness," happily took control.

Epilogue:

Spin-Offs

Gangster rap reveals the pathology of its creators as well as their skill.
—Jon Pareles of the *New York Times*[1]

I always tell the truth even when I lie.—Scarface of the Geto Boys[2]

Rap music is black America's TV station.—Chuck D of Public Enemy[3]

Rap music is the exemplary postmodern phenomenon. Smoothly packaged for mass consumption in the name of militant opposition to the status quo, the dominant forms of rap music reflect basic contradictions in American life: violent sensibilities alongside pleas for peace, women's degradation beside calls for freedom, technological ingenuity juxtaposed with impoverished visions of the good life and pretentious posturing in the face of pain and suffering. As a cultural genre, rap is as protean as an Emerson essay and as marketable as a Ford automobile. Rap also portends what may await us in the future . . . a global network of markets whose public spheres resemble a huge shopping mall and whose private spheres look like armed camps.—Cornel West[4]

Popularizing the Police State

In this afterword we move beyond the conventional drug news to consider briefly how the cocaine narrative played out in other realms of popular culture that are not subject to journalism's professionalized neutrality and institutionalized blindness. Ranging from the quasi-journalistic discourse of so-called reality programs (like the Fox network's "Cops") to the counterjournalistic discourse of gangster rappers (like Ice-T), these popular spin-offs of the drug news also deserve careful study and analysis.[5] The former, in personalizing and individualizing the police state, celebrates anew the cop as hero. The latter, often mixing in audio from actual TV news stories that cover drugs, offers an oppositional reading of the dominant media portrayals of contemporary urban life with the police state most often portrayed as villain.

Cop Docs When Reagan assumed power in the early 1980s, the most popular television shows in the United States included "Dallas" (#1 in 1980–82, 1983–84) and "Dynasty" (#1 in 1984–85), which both celebrated and criticized the excesses of upper-middle-class America.[6] By the end of the Reagan era, the nonthreatening and assimilationist "Cosby Show," in its four years as America's most-watched program (1985–89), further demonstrated it was okay to be upper middle class, and that such classness was, in fact, color-blind.[7] And in January 1989, as "Cosby" began its decline and as Reagan passed the power to Bush, "Cops" premiered on Rupert Murdoch's Fox network.

Part of a genre tabbed "reali-TV"[8]—which traces its heritage, cheap costs, and affinity to particular Middle American values to CBS's long-running "60 Minutes" (the top-rated show in 1979–80, 1982–83, 1991–92, and again in 1992–93)[9]—verité-style police shows began development during the height of the reportorial drug wars in 1988 and in the middle of a twenty-two-week television writers' strike. As Chad Raphael argues, "existing reali-TV shows [like "60 Minutes," "20/20," "America's Most Wanted," and "Rescue 911"] were largely unaffected by the strike, since they did not depend on writers in the first place."[10]

As crack is to cocaine, then, cop docs are to "Dallas"-"Dynasty"-"Cosby." Like crack, cop docs are cheap to produce, with low overhead, and a big return on minimum investment. Instead of hiring an actor like Ted Danson ("Cheers") at $500,000 per episode, producers and crews just tag along with real police, who often donate the cars, the set, the guns—equipment that the producers of conventional prime-time television usually have to buy and provide.[11] While an hour of conventional prime-time drama, like NBC's "L.A. Law," might run nearly $1.5 million to produce per episode, an hour of verité-style reenactments on CBS's "Top Cops" costs a mere $735,000. While a half hour of "Murphy Brown" costs nearly $800,000, producing a single half hour of Fox's "Cops" costs less than $325,000.[12]

The development of shows such as "America's Most Wanted" (Fox, 1988–), "Cops" (Fox, 1989–), "Top Cops" (CBS, 1990–), "DEA" (Fox, 1990), and "American Detective" (ABC, 1991–) also traces back to NBC's

"Miami Vice," a fictional show that ran from 1984 to 1989 and, like its network news counterparts, celebrated the cocaine narrative with almost every program. Not coincidentally, the only year that "Miami Vice" made the top twenty list of the Nielsen rating system's most popular programs (#9) was the 1985–86 season[13]—the pivotal two years when the news media "discovered" crack cocaine. B. Keith Crew identifies what we argue is the legacy of both "Miami Vice" and network news cocaine stories for the current reali-TV programs: "The myth of crime and punishment." Crew ties this myth and "Miami Vice" directly to Reaganism and "conservative political ideology because it defines the problem of crime strictly as the effect of bad individuals rather than as a product of social and economic conditions."[14] The ties between Reaganism and cop docs are explicit. In November 1990, Reagan made a cameo appearance on CBS's "Top Cops" to honor the Secret Service officer who helped save the former president's life during the 1981 assassination attempt by John Hinckley.

In addition, whereas "Miami Vice" often portrayed individual cops—especially main characters Sonny Crocket and Ricardo Tubbs—as "super-competent" rogue heroes, the police state itself was often represented as an impersonal, lethargic, sometimes sinister, bureaucracy. "Miami Vice" typically focused on conventional two-sided tensions between good and evil and how Crocket and Tubbs negotiated the space in between. Almost always personalized in these drug narratives, good-versus-evil plot lines usually masked the ways in which institutions like the police force were connected to larger social and economic tensions. In similar ways the network news narratives and the cop docs that followed also would focus on individual character pathology rather than on any systemic examination of conditions that made pathology possible.

Cop docs generally have served the two-dimensional we/they rhetoric favored in many of the news narratives that we analyzed. "Top Cops" detective-turned-TV producer Sonny Grosso (of *French Connection* fame) sees the mission of the show in portraying evil individuals against the backdrop of a benign police state: "Every piece of dope, every vial of crack that's sold in a community, somebody knows about. If people can see cops in a more positive light, maybe we can help take back the street from the animals."[15] Grosso's "solution"—to change the symbolic image of police—offers no plan, however, for addressing social and economic inequities. One horrific "use" of cop docs featured Shawn Slater, "a rising star in the Ku Klux Klan," in a recent episode of the CBS News documentary "48 Hours" about the rising tide of racial hatred. Slater is shown at home with some of his Klan friends happily counting the "minorities" who are identified and arrested on the TV show "Cops."[16]

Gangster Rap   The drug news narratives of the 1980s spun off more than dominant replications of mainstream values and celebratory narratives of the police. Countervoices in popular culture (with roots in the 1970s hip-hop tradition that propagated rap music, break-dancing, casual fashion, political graffiti art) developed gangster rap in Los Angeles in 1987 as a direct oppositional response to the network news-inspired drug "epidemic."[17] Rap artists including Ice-T, Ice Cube, the Geto Boys, Public Enemy, and N.W.A., among others, often characterize themselves as "street reporters," offering alternative visions of life in urban America and radical visions of the state (N.W.A.'s 1988 "Fuck Tha Police"). Chuck D of Public Enemy maintains that rap music is the CNN of black America.[18] And Easy-E of N.W.A. argues, "We're like reporters. We give them [our listeners] the truth. People where we come from hear so many lies that the truth stands out like a sore thumb."[19] Certainly, rap offers counterreadings of the war on drugs that are radically different from either cop docs or conventional network news portrayals. Historian Robin D. G. Kelley argues: "In an era when mainstream media, conservative policy specialists and some social scientists are claiming that the increase in street crime can be explained by some pathological culture of violence bereft of the moderating influences of a black middle class, . . . rappers keep returning to the idea that joblessness and crime are directly related."[20]

Perhaps the strongest critique of the contemporary war on drugs emerges in Ice-T's "New Jack Hustler" in which he indicts modern capitalism—the structural model for the illicit drug trade—for turning "the needy into the greedy." Playing a drug dealer with "a capitalist migraine" in the music video version of the song, Ice-T confronts links between capitalism and cocaine: "With cocaine my success came speedy. Got me twisted into a paradox. Every dollar I get another brother drops. Maybe that's the plan and I don't understand." Indicting capitalism's complicity in a kind of urban genocide, the rap singer states the paradox of his character's financial achievement at the expense of his community: "Is this a nightmare? Or the American Dream?" Neuroscientist Michael Gazzaniga puts the paradox another way: "The cocaine-snorting Wall Street broker may vote for Reagan, who inveighs against organized crime, yet that same broker is largely responsible for the existence of the crime structure, with all of its

horrid power beyond the drug business."²¹ Unlike the moralists of the New
Right and the behaviorists of the drug control establishment, Gazzaniga
and Ice-T locate the drug problem in what Cornel West calls the "market
way of life." Quoting West:

> Never before has the seductive market way of life held such sway in nearly
> every sphere of American life. This . . . way of life promotes addictions to
> stimulation and obsessions with comfort and convenience. Addictions and ob-
> sessions—centered primarily around bodily pleasures and status rankings—
> constitute market moralities of various sorts. The common denominator is a
> rugged and ragged individualism and the rapacious hedonism in quest of a
> perennial "high" in body and mind.²²

As we have suggested from the beginning, the murder and mayhem of
the crack economy is not some evil mutation of an "alien value system."
The fault, dear Brutus, lies not in Them, nor in our media stars (like
Candice Bergen on "Murphy Brown"), nor in our scapegoats (like the crack
mother)—the blame lies squarely with Us and the *core* value system that
organized the entrepreneurial market culture of Reagan's America.

The Verdict

> What happened in Los Angeles in April of 1992 was neither a race riot nor a
> class rebellion. Rather, this monumental upheaval was a multi-racial, trans-
> class, and largely male display of justified social rage. For all its ugly, xeno-
> phobic resentment, its air of adolescent carnival, and its downright barbaric
> behavior, it signified the sense of powerlessness in American society. Glib at-
> tempts to reduce its meaning to the pathologies of the black underclass, the
> criminal actions of hoodlums, or the political revolt of the oppressed urban
> masses miss the mark. Of those arrested, only 36 percent were black, more
> than one third had full-time jobs, and most claimed to shun political affiliation.
> What we witnessed in Los Angeles was the consequence of a lethal linkage of
> economic decline, cultural decay, and political lethargy in American life. Race
> was the visible catalyst, not the underlying cause.—Cornel West²³

Certainly in the coverage of drugs and inner-city America during the
Reagan era, conventional journalism supplied us with plenty of description, information, cautionary tales, and horror stories. However, debates
over the racial and class dimensions of drugs, debates over legalization,
and especially debates over the costs of supporting the expanding police

state remained oppressed in a factory of information that succeeded only in manufacturing an apparent national consensus on drug matters: "just say no." [24] With crusading journalists, under the auspices of professional neutrality, leading the consensus-building, it was no great surprise to us when on April 29, 1992, the first verdict came down in the Rodney King beating case.

We also were not surprised when Michael Stone, one of the lawyers for the four indicted police officers, told ABC's "Nightline" that he was able "to put the jurors in the shoes of the police officers." [25] After all, with the surge of stories about crack cocaine in inner-city America, TV news since late 1985 had been doing much the same thing as that L.A. defense lawyer— putting mainstream America into the shoes of the police. Given the routine ritualized visual shot in the coverage of crack and urban America—the hand-held, unstable camera bounding from the back of police vans following gun-toting authorities as they break down the door of yet another crack house—journalism became an agent of the police. In forging an apparent national consensus, the networks put those of us in the comfort of our living room into "the shoes of the police." Unlike many TV crack stories, however, the Rodney King amateur videotape footage that we witnessed, came not from the police (ad)vantage, but from the viewpoint of a witness and bystander. In TV news we do not often see the police and inner-city America from this point of view.

A crack appeared in the consensus.

Indeed, in the aftermath of the King verdict, major TV anchors worldwide descended on Los Angeles to begin soul-searching. At least one conventional journalist got the message—for a moment. Taking a cue from Oprah and Donahue at their best, Ted Koppel and ABC took "Nightline" to South Central L.A. for interviews and panel discussions with Crips, with Bloods, with politicians, and with community and religious leaders. Two rival gang members, Bone and Little Monster, like rap reporters, offered "Nightline" viewers an alternative view of the verdict aftermath, arguing that the white-collar looters involved in the national savings and loan scandal constituted a far more sinister criminal threat to America (and yet most of them never spent time in jail). Koppel also elicited from one gang member an alternative metaphor for the L.A. Police Department: a rival street gang. Koppel also promised to return. Within days of the verdict, Bill Moyers on his PBS "Listening to America" series also disrupted his weekly prerecorded program for a special roundtable discussion on poverty, economics, and racism, which featured rap performer, activist, and Bill Clinton nemesis Sister Souljah, who criticized various Urban League leaders and journalists for using individual pathology as a scapegoat rather than focusing on systemic tears in the economic and political fabric of urban society.

Although Ted Koppel (who, to his credit, did return six months later in November 1992) and the other anchors have long since returned to their sleek, modern, "neutral" surveillance and command centers on the East Coast, critics offer reasons for hope. Cornel West believes that the tragic "chaos" (often the last resort of disenfranchised groups) generated by the L.A. rebellion got some public attention back on social and structural problems and off the ideology of economic individualism that so saturated the Reagan and Bush eras.[26] Pushing his argument a step further, we would add only our own hopeful note—that the fallout from the King incident, the amateur video, and the police acquittals (and subsequent convictions in the federal case) contributed significantly to cracking the so-called new consensus about black urban America, to ending the Reagan-Bush eras, to challenging the brutalities of "rugged and ragged individualism," and to imagining the critical and democratic possibilities of new collective projects.

Appendix A:

Cocaine Stories

Date	Network	Reporter	Subject	Videotape
04/20/81	ABC	Strait, George	John Phillips sentenced	No
01/18/82	NBC	Jamieson, Bob	Cocaine	Yes
01/19/82	NBC	Jamieson, Bob	Cocaine	Yes
02/10/82	ABC	Chase, Rebecca	Cocaine war	Yes
05/28/82	NBC	Sobel, Rebecca	Cocaine deaths	Yes
06/29/82	CBS	Sheppard, Gary	Cocaine	Yes
06/29/82	CBS	Rather, Dan	Narcometer	Yes
06/30/82	ABC	Chase, Rebecca	Football and cocaine	Yes
07/05/82	CBS	Ferrugia, John	Congressional drug scandal	No
07/07/82	ABC	Simpson, Carole	Congressional sex and drug scandal	No
07/08/82	ABC	Strait, George	Look-alike drugs	No
08/25/82	NBC	Jamieson, Bob	NFL drug use	Yes
10/20/82	ABC	Sherr, Lynn	DeLorean arrest	No
10/20/82	CBS	Sheppard, Gary	DeLorean arrest	No
10/20/82	NBC	Ross, Brian	DeLorean arrest	No
10/20/82	NBC	Compton, James	DeLorean arrest	No
10/20/82	NBC	Hart, John	DeLorean arrest	No
10/21/82	NBC	Ross, Brian	DeLorean	No
10/25/82	ABC	Howes, Paul	DeLorean / crackdown	Yes
12/22/82	NBC	Ross, Brian	Hollywood and cocaine	No
01/06/83	NBC	Ross, Brian	Congressional cocaine probe	No
02/07/83	CBS	Bowen, Jerry	Cocaine	Yes
02/15/83	NBC	Stern, Carl	Drug war	Yes

Date	Network	Reporter	Subject	Videotape
03/03/83	NBC	Murphy, Dennis	Richard Dreyfuss	No
03/29/83	NBC	Ross, Brian	Cocaine grannies	Yes
05/10/83	CBS	Wallace, Jane	Cocaine hotline	Yes
05/16/83	ABC	Rose, Judd	Cocaine crisis	Yes
06/27/83	CBS	Dow, David	Baseball player Steve Howe	No
06/27/83	NBC	Matson, Boyd	Baseball player Steve Howe	No
07/15/83	CBS	Bowen, Jerry	Dallas Cowboys' drug use	No
10/17/83	ABC	Spencer, Joe	Baseball and cocaine	No
10/17/83	CBS	Currier, Frank	Baseball and cocaine	No
10/17/83	NBC	Cummins, Jim	Baseball and cocaine	No
12/15/83	ABC	Gandolf, Ray	Sports and drugs	Yes
03/07/84	NBC	Jensen, Mike	Wall Street and drugs	Yes
03/24/84	NBC	Molina, Dan	Drugs and Florida greyhound racing	No
04/17/84	CBS	Graham, Fred	DeLorean trial	No
04/18/84	ABC	Sheppard, Gary	DeLorean trial	No
05/16/84	NBC	Murphy, Dennis	David Kennedy death	No
05/29/84	ABC	Potter, Mark	Cocaine psychosis	Yes
05/29/84	NBC	Murphy, Dennis	Cocaine deaths	Yes
05/30/84	CBS	Peterson, Barry	Cocaine	Yes
05/31/84	CBS	Peterson, Barry	Cocaine	Yes
06/01/84	CBS	Braver, Rita	Cocaine on Long Island	Yes
08/09/84	NBC	Ross, Brian	Cocaine epidemic	Yes
08/10/84	NBC	Orth, Maureen	Cocaine epidemic	Yes
08/29/84	CBS	Graham, Fred	Washington, D.C., Mayor Marion Barry and cocaine allegations	Yes
10/26/84	ABC	Rose, Judd	Los Angeles street gangs	Yes
11/27/84	CBS	Dow, Harold	Cocaine abuse	Yes
11/28/84	CBS	Dow, Harold	Cocaine and teenagers	Yes
12/05/84	NBC	Burrington, David	Drugs in the valley	Yes
12/07/84	NBC	Frazier, Stephen	Stacy Keach sentence	No
12/18/84	NBC	Hurst, Steve	Keach sentence	No
02/11/85	NBC	Bourgholtzer, Frank	Cathy Smith-John Belushi case	No
03/06/85	ABC	Potter, Mark	Congressional report drug abuse	Yes
03/25/85	ABC	Strait, George	Cocaine and heart disease	Yes
05/07/85	NBC	Polk, James	Baseball and drugs	Yes
05/17/85	ABC	Kashiwahara, Ken	Drugs and alcohol poll	Yes
05/31/85	CBS	Dow, Harold	Baseball drug indictments	Yes
05/31/85	CBS	Simon, Bob	Baseball drug indictments	Yes
06/03/85	NBC	Perkins, Jack	Cocaine treatment	Yes
07/16/85	ABC	Gibson, Charles	Cocaine hearings	Yes

Date	Network	Reporter	Subject	Videotape
07/27/85	ABC	Hume, Brit	Congressional hearings stars	Yes
09/03/85	CBS	Dow, Harold	Baseball drug trial	No
09/05/85	CBS	Dow, Harold	Baseball drug trial	No
09/05/85	NBC	Polk, James	Baseball drug trial	No
09/06/85	ABC	King, Jerry	Baseball drug trial	No
09/06/85	CBS	Dow, Harold	Baseball drug trial	No
09/06/85	CBS	Peterson, Barry	Baseball drug trial	No
09/06/85	NBC	Polk, James	Baseball drug trial	No
09/10/85	NBC	Polk, James	Baseball drug trial	No
09/11/85	CBS	Spencer, Susan	Cocaine and pregnant mothers	Yes
09/11/85	NBC	Polk, James	Baseball drug trial	No
09/20/85	CBS	Dow, Harold	Baseball drug trial	Yes
09/20/85	CBS	Stahl, Lesley	Nancy Reagan music video	Yes
09/24/85	NBC	Valeriani, Richard	Baseball drug abuse	Yes
10/02/85	NBC	Polk, James	Football cocaine	Yes
12/01/85	NBC	McLogan, Jennifer	Cocaine crack	Yes
12/04/85	CBS	Young, Steve	Cocaine crack	Yes
12/30/85	CBS	Drinkwater, Terry	Cocaine use	Yes
01/15/86	NBC	Hager, Robert	Ricky Nelson airplane crash follow-up	No
01/21/86	CBS	Teichner, Martha	Nelson crash follow-up	No
03/20/86	ABC	Chase, Rebecca	Children and drugs	Yes
04/08/86	CBS	Van Sant, Peter	Police corruption	Yes
05/23/86	NBC	Murphy, Dennis	Cocaine crack	Yes
05/27/86	ABC	McKenzie, John	Drugs crack	Yes
05/27/86	CBS	Dow, Harold	Drugs crack	Yes
06/19/86	NBC	Hager, Robert	Len Bias death	No
06/20/86	ABC	Greenwood, Bill	Bias death investigation	No
06/20/86	NBC	Hager, Robert	Bias death investigation	No
06/21/86	NBC	McLogan, Jennifer	Drug abuse crack	Yes
06/23/86	CBS	Engberg, Eric	Bias funeral	No
06/23/86	NBC	Hager, Robert	Bias funeral	No
06/24/86	ABC	Greenwood, Bill	Bias death cocaine	No
06/24/86	CBS	Engberg, Eric	Bias death cocaine	No
06/24/86	NBC	Hager, Robert	Bias death cocaine	No
06/27/86	ABC	Jennings, Peter	Bias Person of Week	Yes
06/28/86	ABC	Naverson, Andrea	Don Rogers death	No
06/28/86	NBC	Atkinson, Stan	Rogers death	No
06/29/86	ABC	Kashiwahara, Ken	Rogers cocaine death	No
06/30/86	ABC	Lampley, Jim	Drugs and athletes	Yes

Date	Network	Reporter	Subject	Videotape
06/30/86	CBS	Engberg, Eric	Rogers cocaine death	No
06/30/86	NBC	Bazell, Robert	Cocaine	Yes
07/06/86	CBS	Pearson, Hampton	Bias death investigation	No
07/07/86	ABC	Jennings, Peter	Drugs and college students	Yes
07/07/86	ABC	Claiborne, Ron	Football drug tests	Yes
07/07/86	CBS	Dow, Harold	Cocaine	Yes
07/07/86	NBC	Nykanen, Mark	Cocaine on campus	Yes
07/09/86	CBS	Teichner, Martha	Cocaine	Yes
07/10/86	ABC	Strait, George	Cocaine	Yes
07/10/86	ABC	Threlkeld, Richard	Drugs on the job	Yes
07/11/86	ABC	Quinones, John	Cocaine babies	Yes
07/12/86	ABC	Blakemore, Bill	Drug testing	Yes
07/12/86	ABC	Von Fremd, Mike	Home drug tests	Yes
07/15/86	ABC	Strait, George	Cocaine	Yes
07/15/86	NBC	Valeriani, Richard	Cocaine	Yes
07/16/86	ABC	Serafin, Barry	Drug war	Yes
07/18/86	ABC	Greenwood, Bill	Len Bias investigation	No
07/21/86	ABC	Greenwood, Bill	Bias investigation	No
07/25/86	NBC	Hager, Robert	Bias investigation	No
07/28/86	ABC	Quinones, John	Crack	Yes
08/04/86	ABC	Donaldson, Sam	White House war on drugs	Yes
08/04/86	ABC	Martin, John	White House war on drugs	Yes
08/04/86	CBS	Plante, Bill	White House war on drugs	Yes
08/04/86	CBS	Dow, Harold	White House war on drugs	Yes
08/04/86	NBC	Valeriani, Richard	Cocaine crack	Yes
08/05/86	ABC	Schadler, Jay	Crack	Yes
08/07/86	CBS	Plante, Bill	White House war on drugs	Yes
08/07/86	CBS	Goldberg, Bernard	White House war on drugs	Yes
08/10/86	NBC	Kaiserski, Jim	Florida drug crackdown	Yes
08/13/86	NBC	Elliot, Robert	David Crosby and drugs	No
08/27/86	NBC	Makawa, James	Cocaine	Yes
08/29/86	CBS	Blackstone, John	Drug king funeral	Yes
08/29/86	CBS	Hall, Bruce	Coke baby custody	Yes
08/31/86	CBS	Schieffer, Bob	Drugs and crime	Yes
08/31/86	CBS	Dow, Harold	New York City drug crackdown	Yes
09/01/86	CBS	Rather, Dan	Poll on drugs	Yes
09/01/86	CBS	Arnott, Dr. Bob	Drugs and brain damage	Yes
09/02/86	CBS	Rather, Dan	Cocaine special	Yes
09/04/86	CBS	Potter, Ned	Drug war	Yes
09/04/86	CBS	Rather, Dan	Reagan's TV address on drug abuse	Yes

Date	Network	Reporter	Subject	Videotape
09/10/86	CBS	Corderi, Victoria	New Jersey crack raids	Yes
09/15/86	ABC	Jennings, Peter	Drugs USA I	Yes
09/16/86	ABC	Johnson, Dr. Tim	Drugs USA II	Yes
09/17/86	ABC	Strait, George	Drugs USA III	Yes
09/18/86	ABC	Blakemore, Bill	Drugs USA IV	Yes
10/06/86	ABC	O'Reilly, Bill	Vermont drug trafficking	Yes
10/07/86	CBS	Dow, David	Cocaine and children	Yes
10/21/86	NBC	Cummins, Jim	DeLorean trial	No
10/23/86	NBC	Polk, James	Flying high	Yes
10/27/86	CBS	Plante, Bill	Drug bill	Yes
10/27/86	CBS	Currier, Frank	Drug bill	Yes
10/29/86	ABC	Greenwood, Bill	Bias aftermath	No
10/29/86	CBS	Tucker, Lem	Lefty Driesell and Bias death	No
10/29/86	NBC	Hager, Robert	Driesell	No
11/12/86	CBS	Boros, Karen	Pregnancy and drugs	Yes
11/24/86	CBS	Whitaker, Bill	Miami drug war	Yes
01/10/87	ABC	Magnus, Edie	Detroit murder rate	Yes
02/27/87	NBC	Murphy, Dennis	Crack epidemic	Yes
03/09/87	ABC	Kashiwahara, Ken	California street gangs	Yes
03/12/87	CBS	Dow, Harold	College basketball	Yes
03/16/87	CBS	Goldberg, Bernard	Sting operations	Yes
03/29/87	ABC	Jarriell, Tom	Youth suicide	No
04/01/87	ABC	Martin, John	Dwight Gooden drug abuse	No
04/01/87	CBS	Roth, Richard	Gooden drug abuse	No
04/01/87	NBC	McLogan, Jennifer	Gooden drug abuse	No
04/10/87	ABC	O'Reilly, Bill	Drug abuse program	Yes
04/11/87	NBC	Jones, Kenley	Julian Bond allegations	No
04/14/87	ABC	Dale, Al	Bond allegations	Yes
04/14/87	CBS	Whitaker, Bill	Bond allegations	No
04/14/87	NBC	Hazinski, David	Bond allegations	No
04/16/87	ABC	Bergantino, Joe	Wall Street drug bust	No
04/16/87	CBS	Dow, Harold	Wall Street drug bust	No
04/16/87	NBC	Jensen, Mike	Wall Street drug bust	Yes
04/17/87	CBS	Drinkwater, Terry	Sports drug charges	No
05/14/87	CBS	Smith, Harry	Sports and drugs	Yes
05/17/87	NBC	Lipsyte, Robert	Michael Ray Richardson's addiction story	No
06/17/87	ABC	Potter, Mark	Julian Bond controversy	No
06/17/87	CBS	Whitaker, Bill	Bond scandal	No
06/17/87	NBC	Jones, Kenley	Bond controversy	No

Date	Network	Reporter	Subject	Videotape
06/19/87	CBS	Braver, Rita	Marion Barry controversy	No
06/19/87	NBC	Hager, Robert	Barry controversy	No
08/02/87	CBS	Dow, Harold	Sports and drugs	Yes
08/16/87	CBS	Blackstone, John	Homeless and farm work	Yes
08/28/87	ABC	RT?	Paul Molitor Person of the Week	No
09/07/87	CBS	Dow, David	New York City youth gangs	Yes
10/25/87	ABC	Judd, Jackie	Washington, D.C.,	
			Mayor Marion Barry	No
11/04/87	NBC	Murphy, Dennis	Miami drug bust	Yes
11/13/87	NBC	Murphy, Dennis	Straight kids	Yes
01/13/88	NBC	Hager, Robert	Cocaine use in high schools	Yes
01/26/86	NBC	Lewis, George	Truckers and drugs	Yes
03/01/88	NBC	Brokaw, Tom	Cocaine/war on drugs	Yes
03/01/88	NBC	Chancellor, John	Commentary on the war on drugs	Yes
03/14/88	CBS	Dow, David	Crack cocaine	Yes
04/26/88	CBS	Faw, Bob	War on Drugs II	Yes
04/28/88	CBS	Schlesinger, Richard	War on Drugs III	Yes
05/02/88	CBS	Dow, Harold	War on Drugs IV	Yes
05/12/88	CBS	Rather, Dan	War on Drugs V	Yes
05/16/88	NBC	Nelson, Noah	Deadly Demand I	Yes
05/17/88	NBC	Oliver, Don	Deadly Demand II	Yes
05/18/88	CBS	Plante, Bill	Reagan war on drugs	Yes
05/18/88	CBS	Adams, Jacqueline	War on drugs	Yes
05/18/88	NBC	Bernard, Stan	Deadly Demand III	Yes
05/19/88	NBC	Brokaw, Tom	Lonise Bias and Matthew Byrne	Yes
05/27/88	ABC	Quinones, John	War on drugs (Tampa)	Yes
05/28/88	NBC	Jones, Kenley	War on drugs (Atlanta)	Yes
06/03/88	CBS	Vasquez, Juan	War on drugs (Miami)	Yes
06/15/88	ABC	Compton, Ann	Campaign '88	Yes
06/15/88	ABC	Zelnick, Bob	Military role in drug war	Yes
06/15/88	ABC	Gregory, Bettina	Livermore Lab investigation	Yes
06/30/88	ABC	Kast, Sheilah	White House war on drugs	No
06/30/88	CBS	Plante, Bill	White House war on drugs	No
06/30/88	NBC	Wallace, Chris	White House war on drugs	No
07/11/88	ABC	Nissen, Beth	Drugs in the inner city	Yes
07/13/88	ABC	Blakemore, Bill	Crack cocaine rush	Yes
08/30/88	CBS	Vasquez, Juan	Colombia drug war	Yes
08/30/88	CBS	Potter, Deborah	Campaign '88	Yes
09/02/88	ABC	BS?	Lawrence Taylor Person of the Week	No
10/13/88	ABC	Collins, Peter	Cocaine babies	Yes

Date	Network	Reporter	Subject	Videotape
10/24/88	NBC	Bazell, Robert	Cocaine Kids I	Yes
10/25/88	NBC	Gillen, Michele	Cocaine Kids II	Yes
11/01/88	NBC	Hager, Robert	Washington, D.C., homicides	Yes
11/23/88	CBS	Dow, Harold	Drugs in New York City schools	Yes
11/27/88	NBC	Cummins, Jim	Drugs in Des Moines, Iowa	Yes
12/20/88	NBC	Cummins, Jim	Spotlight Crack, Inc.	Yes
12/26/88	CBS	Morton, Bruce	Marion Barry investigation	No
12/28/88	NBC	Chancellor, John	Commentary on drug treatment	Yes
12/29/88	NBC	Stern, Carl	Washington, D.C., Mayor Barry	Yes

Appendix B:

Noncocaine Stories

Date	Network	Reporter	Subject	Videotape
04/22/81	ABC	Schell, Tom	Hollywood hearings	Yes
05/03/81	ABC	Nunn, Ray	Drugs and youth	Yes
06/30/81	NBC	Nykanen, Mark	T's and Blues	Yes
07/13/81	CBS	Ferrugia, John	Nancy Reagan	Yes
08/17/81	CBS	Peterson, Barry	San Francisco program	Yes
11/05/82	NBC	Hager, Robert	Radio call-in	Yes
04/08/83	CBS	Rabel, Ed	Drug war	Yes
02/29/84	ABC	Schadler, Jay	Arkansas school drug program	Yes
03/11/84	CBS	Kelley, Chris	New York City drug crackdown	Yes
03/28/84	ABC	Sheppard, Gary	Drug bust aftermath	Yes
08/15/84	CBS	Braver, Rita	PCP	Yes
10/11/84	NBC	Nelson, Noah	California marijuana war	Yes
10/22/84	CBS	Moyers, Bill	America's kids	Yes
10/23/84	CBS	Moyers, Bill	America's kids	Yes
02/21/85	NBC	Lewis, George	Crime in America	Yes
04/24/85	ABC	Simpson, Carole	Drug abuse conference	Yes
04/24/85	CBS	Stahl, Lesley	Nancy Reagan conference	Yes
04/24/85	NBC	Mitchell, Andrea	Nancy Reagan conference	Yes
05/03/85	ABC	Dobbs, Greg	Nancy Reagan conference	No
05/25/85	CBS	Schuster, Gary	Charity tennis exhibition	No
06/25/85	CBS	Peterson, Barry	California drug war	Yes
11/12/85	NBC	Valeriani, Richard	High school student drug test	Yes
01/21/86	NBC	Bazell, Robert	Aids and Addicts	Yes

Date	Network	Reporter	Subject	Videotape
01/23/86	CBS	Smith, Harry	Marijuana lacing	Yes
02/18/86	NBC	Nelson, Noah	Drug babies	Yes
03/28/86	NBC	Myers, Lisa	Heroin black tar	Yes
05/22/86	ABC	Jennings, Peter	Drugs and teenagers	Yes
07/03/86	CBS	Burget, Nadine	Drug abuse	Yes
09/19/86	ABC	Threlkeld, Richard	Drugs USA V	Yes
03/05/87	CBS	Young, Steve	TV Commercials	Yes
04/07/87	CBS	Goldberg, Bernard	Heroin addicts	Yes
05/14/87	NBC	Bazell, Robert	AIDS	Yes
08/02/87	CBS	FS?	Marijuana laws	Yes
11/16/87	CBS	Dow, Harold	AIDS-drug link	Yes
11/16/87	NBC	Bazell, Robert	AIDS addicts	Yes
03/10/88	ABC	Greenwood, Bill	Telephone beepers	Yes
03/18/88	CBS	Dow, Harold	New York City drug war	Yes
04/25/88	CBS	Braver, Rita	War on Drugs I	Yes
04/29/88	CBS	Braver, Rita	Health and Human Services Secretary on drug policy	Yes
04/29/88	NBC	Hart, John	Oakland, California, drug war	Yes
05/16/88	CBS	Spencer, Susan	Surgeon General and nicotine addiction	No
05/16/88	NBC	Hager, Robert	Surgeon General and nicotine addiction	Yes

Appendix C:

Chronology of Kernel Events in the Cocaine Narrative

Phase I Coverage

February 28, 1981	*TV Guide* publishes Frank Swerton's exposé of widespread cocaine abuse in Hollywood: "Hollywood's Cocaine Connection."
April 20, 1981	Singer John Phillips of the famous rock group the Mamas & the Papas begins a prison sentence for pushing drugs. See ABC, 4/20/81.
April 1981	A House Select Committee headed by Rep. Leo Zeferetti holds public hearings in Los Angeles about drug abuse in the entertainment industry. See ABC, 4/22/86.
1982	NIDA launches "Just-Say-No" campaign.
March 5, 1982	Actor John Belushi dies.
June–August 1982	Major drug scandal rocks the National Football League's New Orleans Saints. The scandal culminates on August 25, 1982, when Mike Strachan, a former football player, pleads guilty to selling drugs to active players Chuck Muncie and George Rogers. See ABC, 6/30/82, and NBC, 8/25/82.
June 1982–January 1983	Ongoing drug and sex scandal on Capitol Hill. FBI investigating cocaine trafficking by former congressional pages. Allegations of homosexual liaisons involving pages also reported. Former Rep. Barry Goldwater, Jr., is implicated. See CBS, 7/5/82; ABC, 7/7/82; NBC, 1/6/83.

October 14, 1982	President Ronald Reagan, in a speech at the Justice Department, declares war on drugs.
October 20, 1982	John DeLorean arrested. See ABC, CBS, and NBC, 10/20/82; NBC, 10/21/82; ABC, 10/25/82.
May 10, 1983	CBS's Jane Wallace delivers the first network report on the 800-COCAINE hotline.
June–October 1983	First major scandal cycle in professional baseball. Key transgressors include Steve Howe, Vida Blue, Jerry Martin, Willie Wilson, Willie Aikens. See CBS and NBC, 6/27/83; ABC, CBS, and NBC, 10/17/83.
July 1983	A drug scandal taints the NFL's Dallas Cowboys. Players Tony Dorsett, Ron Springs, Harvey Martin, Larry Bethea, and Tony Hill are said to be implicated in a federal investigation of cocaine trafficking. See CBS, 7/15/83.
April 18, 1984	John DeLorean goes on trial. See CBS, 4/17/84, and ABC, 4/18/84.
April 25, 1984	David Kennedy, son of Robert Kennedy, dies of drug overdose. See NBC, 5/16/84.
August 16, 1984	After a twenty-two-week trial, John DeLorean acquitted of all charges.
August 29, 1984	Washington, D.C., Mayor Marion Barry implicated in a cocaine scandal. See CBS, 8/29/84.
October 26, 1984	ABC's Judd Rose makes the first reference to "rock" cocaine in a network newscast.
December 7, 1984	Stacy Keach begins prison sentence in England for smuggling cocaine into the country. See NBC, 12/7/84 and 12/18/84.
February 11, 1985	Cathy Evelyn Smith is charged in connection with the death of John Belushi. See NBC, 2/11/85.
April 24, 1985	Nancy Reagan hosts a drug abuse conference for the world's First Ladies. See ABC, CBS, and NBC, 4/24/85.
July 1985	House hearings on cocaine abuse feature testimony by Stacy Keach and former professional football player Carl Eller. See ABC, 7/16/85 and 7/27/85.
May–September 1985	Centered on the Pittsburgh Pirates, a second major scandal cycle disturbs professional baseball. Key transgressors include Keith Hernandez, Enos Cabell, Jeff Leonard, Tim Raines, Lonnie Smith, Lee Mazzilli, Dale Berra, Al Holland, Rod Scurry, Joaquin Andujar, Willie Stargell, Bill Madlock, and Dave Parker. The scandal climaxes with the trial and conviction of

Curtis Strong, a Pittsburgh caterer. See NBC, 5/7/85;
CBS, 5/31/85; CBS, 9/3/85; CBS and NBC, 9/5/85; ABC,
CBS, and NBC, 9/6/85; NBC, 9/10/85, 9/11/85, and
9/24/85.

Phase II Coverage

November 29, 1985	The *New York Times* prints its first front-page story on crack.
December 1, 1985	NBC's Jennifer McLogan files the first network news report that specifically deals with crack cocaine. She describes crack as the "frightening wave of the future."
January 1986	Investigators link cocaine "free-basing" to the plane crash that killed singer Ricky Nelson. See NBC, 1/15/86, and CBS, 1/21/86.
April 1986	NIDA launches "Cocaine, The Big Lie" campaign.
May 18, 1986	Three of New York City's newspapers (*Times, Daily News,* and *Newsday*) feature substantial articles on the crack trend. The *New York Times*'s Peter Kerr calls this convergence a "turning point" in the news coverage of the crack problem. Within ten days, the three national networks file copycat crack stories. See NBC, 5/23/86; ABC and CBS, 5/27/86.
June 17, 1986	The Boston Celtics select University of Maryland basketball star Len Bias as their top draft pick in the National Basketball Association's annual draft.
June 19, 1986	*The most significant kernel event in the 1980s cocaine narrative—Len Bias dies.* See NBC, 6/19/86; ABC and NBC, 6/20/86; CBS and NBC, 6/23/86; ABC, CBS, and NBC, 6/24/86; ABC, 6/27/86.
June 28, 1986	Cleveland Browns' football player Don Rogers dies of heart failure believed to be triggered by cocaine intoxication. See ABC, CBS, and NBC, 6/28/86; CBS, 6/30/86.
June 1986	Grand jury investigation implicates Terry Long, David Gregg, and Brian Tribble in Bias's death. University of Maryland basketball coach Charles "Lefty" Driesell is also questioned about allegations that he may have interfered with the investigation. See CBS, 7/6/86; ABC, 7/18/86 and 7/21/86; NBC, 7/25/86.

July 28, 1986	ABC's John Quinones files a report that features the first *raiding footage* by a hand-held camera directly covering police in a crack house bust.
August–September 1986	A number of polls find that the American public considers drugs to be the nation's top problem. See ABC, 8/4/86; CBS, 9/1/86; ABC, 9/15/86, 9/16/86, 9/17/86, 9/18/86, and 9/19/86.
August 4, 1986	Responding to the polls, President Reagan calls for a "national crusade against drugs" and volunteers to set an example by submitting to a urine test. See ABC, NBC, and CBS, 8/4/86.
August–November 1986	"Jar Wars" on the campaign trail.
September 2, 1986	CBS airs the prime-time special, "48 Hours on Crack Street."
September 5, 1986	NBC airs the prime-time special, "Cocaine Country."
September 14, 1986	President Reagan and First Lady Nancy Reagan, in a joint television address, charge that drugs are "tearing our country apart."
September 15–19, 1986	ABC airs its five-part series of special reports, "Drugs USA," on five consecutive evening newscasts.
October 27, 1986	President Reagan signs the Drug Free America Act. See CBS, 10/27/86.
October 29, 1986	Driesell resigns as University of Maryland basketball coach. See ABC, CBS, and NBC, 10/29/86.
December 2, 1986	Geraldo Rivera hosts "American Vice: The Doping of America," a two-hour syndicated special carried by 141 stations. The telecast features three live drug busts, one resulting in the arrest of Terry Rouse, who eventually sues Rivera and the police.

Phase III Coverage

February 7, 1987	*TV Guide* publishes "Is TV News Hyping America's Cocaine Problem?" (an article by Edwin Diamond, Frank Accosta, and Leslie-Jean Thornton).
April 1, 1987	New York Mets star pitcher Dwight Gooden enters drug rehabilitation program after testing positive for cocaine use. See ABC, CBS, and NBC, 4/1/87.
April 16, 1987	Police and press converge on Wall Street for a mass arrest of suspected drug dealers. See ABC, CBS, and NBC, 4/16/87.

April 17, 1987	Drug conspiracy and possession charges are filed against Phoenix Suns basketball players Jay Humphries, James Edwards, Grant Gondrezick. Players Walter Davis and William Bedford are also implicated in the scandal. See CBS, 4/17/87.
April–June 1987	Georgia politicians Julian Bond and Andrew Young are implicated in a cocaine scandal. See NBC, 4/11/87; ABC, CBS, and NBC, 4/14/87; ABC and CBS, 6/17/87.
June–October 1987	Washington, D.C., Mayor Barry implicated for second time in a cocaine scandal. See CBS and NBC, 6/19/87; ABC, 10/25/87.
May–November 1988	The war on drugs becomes a major campaign issue. See CBS, 5/18/88; ABC, 6/15/88; CBS, 8/30/88.
June 30, 1988	White House employees (National Security Council secretaries and Secret Service guards) are reportedly under investigation for off-duty drug use. See ABC, NBC, and CBS, 6/30/88.
July 7, 1988	ABC airs prime-time special, "Drugs: Why This Plague?"
October 1988	Willie Horton ads become the centerpiece of the Bush-Quayle campaign.
December 1988	Mayor Barry is caught up in a third drug scandal. See CBS, 12/26/88; NBC, 12/29/88.

Notes

Introduction

1. Michael S. Gazzaniga, *Mind Matters: How Mind and Brain Interact to Create Our Conscious Lives* (Boston: Houghton Mifflin, 1988), p. 141.

2. From *The Natural Mind: An Investigation of Drugs and the Higher Consciousness,* rev. ed. (Boston: Houghton Mifflin, 1986), p. 191.

3. Steve Fore, "Lost in Translation: The Social Uses of Mass Communications Research," *Afterimage* 20 (April 1993): 9.

4. Stuart Hall, "Cultural Studies and Its Theoretical Legacies," in Lawrence Grossberg, Cary Nelson, and Paul Treichler, eds., *Cultural Studies* (New York: Routledge, 1992), p. 286.

5. Edward Said, "Opponents, Audiences, Constituencies and Community," in Hal Foster, ed., *The Anti-Aesthetic* (Port Townsend, Wash.: Bay Press, 1983), p. 57.

6. See *TV: The Most Popular Art* (Garden City, N.Y.: Doubleday Anchor, 1974).

7. In Horace Newcomb, ed., *Television: The Critical View* (New York: Oxford University Press, 1987).

8. In James W. Carey, ed., *Media, Myths, and Narratives: Television and the Press* (Beverly Hills: Sage, 1988), pp. 48–66.

9. See, for example, Herbert Gans, *Deciding What's News: A Study of* CBS Evening News, NBC Nightly News, Newsweek *and* Time (New York: Vintage Books, 1979); Gaye Tuchman, *Making News: A Study in the Construction of Reality* (New York: Free Press, 1978); Michael Schudson, *Discovering the News: A Social History of American Newspapers* (New York: Basic Books, 1978); David Eason, "The New Journalism and the Image-World: Two Modes of Organizing Experience," *Critical Studies in Mass Communication* 1 (March 1984): 51–65; Theodore L. Glasser and James S. Ettema, "Common Sense and the Education of Young Journalists," *Journalism Educator* 44 (Summer 1989): 18–25, 75; Glasser and Ettema, "Investigative Journalism and the Moral Order," *Critical Studies in Mass Communication* 6 (March 1989): 1–20.

10. Carey's most influential papers have been collected in *Communication as Culture: Essays on Media and Society* (Boston: Unwin Hyman, 1989).

11. London: Macmillan, 1978.

12. London: Methuen, 1978.

13. London: BFI, 1978.

14. London: Routledge and Kegan Paul, 1976.

15. London: Routledge and Kegan Paul, 1980.

16. London: Writers and Readers, 1982.

17. For a valuable overview of this work, see Graeme Turner, *British Cultural Studies: An Introduction* (Boston: Unwin Hyman, 1990).

18. See Katerina Clark and Michael Holquist, *Mikhail Bakhtin* (Cambridge, Mass.: Harvard University Press, 1984), p. 11.

19. Consider here the following politically diverse assessments: Max Horkheimer and Theodor Adorno, *The Dialectic of Enlightenment* (New York: Herder and Herder, 1972); Jurgen Habermas, *The Philosophical Discourse of Modernity* (Cambridge, Mass.: MIT Press, 1987); Daniel Bell, *The Cultural Contradictions of Capitalism* (New York: Basic Books, 1976); and Allan Bloom, *The Closing of the American Mind* (New York: Simon and Schuster, 1987).

20. From "Space, Knowledge, and Power," a transcribed interview with Michel Foucault conducted by Paul Rabinow, trans. Christian Hubert, in Rabinow, ed., *The Foucault Reader* (New York: Pantheon, 1984), pp. 248–50.

21. See Stephanie Coontz, *The Way We Never Were: American Families and the Nostalgia Trap* (New York: Basic Books, 1992).

22. See Mikhail Bakhtin, "Discourse in the Novel," in Bakhtin, *The Dialogic Imagination: Four Essays*, ed. Michael Holquist, trans. Caryl Emerson and Holquist (Austin: University of Texas Press, 1981), p. 282.

1   The Cocaine Narrative

1. During this period CNN News was not cataloged at Vanderbilt.

2. Cocaine-related stories became increasingly popular in both print and broadcast news from 1980 through 1986. The *New York Times Index* for 1981 listed only 79 cocaine stories. However, just two years later that number would more than double with 173 articles and editorials indexed. By 1985 the listing had swelled to 204. In 1986 crack received its own category and referred readers to 114 different stories; cocaine that same year referenced nearly 360 articles (an average of nearly one a day). By 1988 these numbers started to level off and drop slightly with 92 citations listed under crack and 225 under cocaine. In each of these years there are also two to four times as many articles cited and summarized under broader categories such as "drug abuse and addiction" and "drug trafficking" (which received its own separate heading for the first time in the 1981 index).

3. For an interpretive analysis of "foreign intrigue" stories based on the same assumptions about journalism's role in the social construction of reality, see Richard Campbell and Jimmie L. Reeves, "TV News Narration and Common Sense: Updating the Soviet Threat," *Journal of Film and Video* 41, no. 2 (Summer 1989): 58–74.

4. Our view of "framing" has been particularly influenced and enriched by the collaborative work of William A. Gamson and André Modigliani, "Media Discourse and Public Opinion on Nuclear Power: A Constructionist Approach," *American Journal of Sociology* 95:1 (July 1989): 1–37; and "The Changing Culture of Affirmative Action," in Margaret M. Braungart, ed., *Research in Political Sociology: A Research Annual* 3 (London: JAI Press, 1987), pp. 137–78.

5. Sarah Ruth Kozloff provides an excellent overview of narrative theory for media scholars in "Narrative Theory and Television," in Robert C. Allen, ed., *Channels of Discourse: Television and Contemporary Criticism* (Chapel Hill: University of North Carolina Press, 1987), pp. 42–73. Also see Shlomith Rimmon-Kenan, *Narrative Fiction: Contemporary Poetics* (London: Methuen, 1983), p. 15.

6. Kozloff, "Narrative Theory," pp. 45–46. Also see Tzvetan Todorov, "The Grammar of Narrative," in *The Poetics of Prose*, trans. Richard Howard (Ithaca, N.Y.: Cornell University Press, 1977), p. 111.

7. For the most influential study of the cultural significance of pollution and purification rituals, see Mary Douglas, *Purity and Danger: An Analysis of the Concepts of Pollution and Taboo* (London: Ark Paperbacks, 1988).

8. Howard S. Becker, *Outsiders: Studies in the Sociology of Deviance* (London: Collier-Macmillan, 1963), p. 157.

9. Kozloff, "Narrative Theory," p. 46; Seymour Chatman, *Story and Discourse: Narrative Structure in Fiction and Film* (Ithaca, N.Y.: Cornell University Press, 1978), pp. 53–56.

10. See, for instance, Steven Wisotsky, *Beyond the War on Drugs: Overcoming a Failed Public Policy* (Buffalo, N.Y.: Prometheus Books, 1990).

11. Steve Fore, "Lost in Translation: The Social Uses of Mass Communications Research," *Afterimage* 20 (April 1993): 11.

12. John E. Merriam, "National Media Coverage of Drug Issues, 1983–1987," in Pamela J. Shoemaker, ed., *Communication Campaigns About Drugs: Government, Media, and the Public* (Hillsdale, N.J.: Lawrence Erlbaum Associates, 1989), pp. 21–28.

13. Donald L. Shaw and Maxwell E. McCombs, "Dealing With Illicit Drugs," in Shoemaker, ed., *Communication Campaigns*, pp. 113–20.

14. Ibid., p. 117.

15. Ibid., p. 119.

16. Again, see Becker, *Outsiders*.

17. For an excellent introduction to Foucault's work, see Paul Rabinow, ed., *The Foucault Reader* (New York: Pantheon, 1984).

18. See Foucault's exhaustive and exhausting archaeological study of the human sciences, *The Order of Things* (New York: Vintage Books, 1973).

19. Pope, quoted in David Harvey, *The Condition of Postmodernity: An Enquiry into the Origins of Cultural Change* (Oxford: Basil Blackwell, 1989), p. 13.

20. See Harvey's discussion of Nietzsche, ibid., pp. 15–16. Also see Foucault, "Nietzsche, Genealogy, History," in *The Foucault Reader*, pp. 76–100.

21. Harvey, *The Condition of Postmodernity*, p. 13.

22. As paraphrased by Rabinow in his Introduction to *The Foucault Reader*,

pp. 26–27. See also Frank Parkin's discussion of "legal-rational domination" in *Max Weber* (London and New York: Tavistock, 1986), pp. 87–89.

23. Introduction, *The Foucault Reader*, p. 13.

24. Lester Grinspoon and James B. Bakalar, *Cocaine: A Drug and Its Social Evolution*, rev. ed. (New York: Basic Books, 1985), p. 215.

25. Bennett, quoted in Daniel Lazare, "The Drug War Is Killing Us," *Village Voice* (January 23, 1990), p. 25.

26. See Ralph Brauer, "The Drug War of Words," *Nation* (May 21, 1990), p. 705.

27. Mikhail Bakhtin, "Discourse in the Novel," in his *The Dialogic Imagination: Four Essays*, ed. Michael Holquist; trans. Caryl Emerson and Michael Holquist (Austin: University of Texas Press, 1981), p. 281.

28. Ibid., p. 282.

29. Bakhtin, "Epic and Novel: Toward a Methodology for the Study of the Novel," in *The Dialogic Imagination*, pp. 30–31.

30. Marshall Berman, *All That Is Solid Melts into Air: The Experience of Modernity* (New York: Simon and Schuster, 1982), p. 15.

31. Mike Wallace and Gary Paul Gates, *Close Encounters* (New York: Berkley Books, 1984), p. 123.

32. Theodore L. Glasser and James S. Ettema, "Common Sense and the Education of Young Journalists," *Journalism Educator* 44 (Summer 1989): 20. See also Gaye Tuchman, *Making News: A Study in the Construction of Reality* (New York: Free Press, 1978).

33. Glasser and Ettema, "Common Sense," pp. 20–21.

34. James W. Carey, "Technology and Ideology: The Case of the Telegraph," in his *Communication as Culture: Essays on Media and Society* (Boston: Unwin Hyman, 1989), p. 211. Kenneth Gergen also notes, "Within the modern ethos it was appropriate that a newspaper reporter like Hemingway could turn novelist. Dialogue was essentially a dispassionate report of 'the facts of the case,'" in *The Saturated Self: Dilemmas of Identity in Contemporary Life* (New York: Basic Books, 1991), p. 36.

35. Carey, "Technology and Ideology," pp. 210–11.

36. See T. J. Jackson Lears, "From Salvation to Self-Realization: Advertising and the Therapeutic Roots of Consumer Culture, 1880–1930," in Robert Wightman Fox and Lears, eds., *The Culture of Consumption: Critical Essays in American History, 1880–1980* (New York: Pantheon, 1983), p. 15. See also Warren I. Susman, *Culture as History: The Transformation of American Society in the Twentieth Century* (New York: Pantheon, 1973/1984), pp. 275–80.

37. Gergen, *The Saturated Self*, p. 38.

38. Ibid., p. 6.

39. Ibid., p. 246.

40. Ibid., p. 19.

41. Philip Bell, "Drugs as News: Defining the Social," *Mass Communication Review Yearbook*, vol. 5, eds. Michael Gurevitch and Mark R. Levy (Beverly Hills: Sage, 1985), p. 311.

42. Ibid., p. 303.

43. For a rare network story that compares both legal and illegal drugs, see ABC News, 9/19/86.

44. Carey, "A Cultural Approach to Communication," in *Communication as Culture*, p. 19.

2   Merchants of Modern Discipline

1. Dwight B. Heath, "U.S. Drug Control Policy: A Cultural Perspective," *Daedalus* 121, no. 3 (Summer 1992): 293.

2. William J. Bennett, "Drugs Damage American Society," in Neal Bernards, ed., *War on Drugs: Opposing Viewpoints* (San Diego: Greenhaven Press, 1990), p. 53.

3. Philippe Bourgois, "In Search of Horatio Alger: Culture and Ideology in the Crack Economy," *Contemporary Drug Problems* (Winter 1989), p. 637.

4. See Peter Reuter, "Hawks Ascendant: The Punitive Trend of American Drug Policy," *Daedalus* 121, no. 3 (Summer 1992): 15–52.

5. Or to say this in another way, Craig Reinarman and Harry G. Levine update the 1985 report and argue, "For every one cocaine-related death in the United States in 1987 there were approximately 300 tobacco-related deaths and 100 alcohol-related deaths." "The Crack Attack: Politics and Media in America's Latest Drug Scare," in *Images of Issues: Typifying Contemporary Social Problems*, ed. Joel Best (New York: Aldine de Gruyter, 1989), p. 120. Also see Ethan Nadelmann, "Drug Prohibition in the United States: Cost, Consequences, and Alternatives," *Science* 945 (September 1, 1989): 943. For a rare network story that compares both legal and illegal drugs, see ABC News, 9/19/86.

6. Daniel Lazare, "The Drug War Is Killing Us," *Village Voice* (January 23, 1990), p. 22.

7. However, in identifying with "the policed," we are careful not to put ourselves in the arrogant position of "speaking for" the surveilled groups in America's inner cities.

8. Murray Edelman, *Constructing the Political Spectacle* (Chicago: University of Chicago Press, 1988), p. 25.

9. These numbers are derived from a graphic presented by Mathea Falco, "Foreign Drugs, Foreign Wars," *Daedalus* 121, no. 3 (Summer 1992): 6.

10. Our view of rites of inclusion and exclusion represents an elaboration of Thomas Schatz's scheme for organizing Hollywood film genres. Schatz argues that these popular genres can be analyzed as "rites of order" (westerns, detective films, gangster films) and "rites of integration" (musicals, family melodramas, screwball comedies). There are certain similarities between Schatz's rites of order and our rites of exclusion as well as between his rites of integration and our rites of inclusion. But since we see all television drug news as preoccupied with "order," we chose to use the terms inclusion and exclusion. See Schatz, *Hollywood Genres* (New York: Random House, 1981).

11. Stanley Cohen, *Visions of Social Control: Crime Punishment and Classification* (Cambridge: Polity Press, 1985), p. 232.

12. Joanne Morreale, *A New Beginning: A Textual Frame Analysis of the Political Campaign Film* (Albany: State University of New York Press, 1991), p. 40.

13. Hubert L. Dreyfus and Paul Rabinow, *Michel Foucault: Beyond Structuralism and Hermeneutics* (Chicago: University of Chicago Press, 1983), p. 175.

14. See "The Repressive Hypothesis," in *The Foucault Reader*, pp. 301–29.

15. Stanton Peele, *Diseasing of America: Addiction Treatment Out of Control* (Boston: Houghton Mifflin, 1989), p. 126.

16. See "To the Last Dime" (produced by Stanhope Gould), ABC's "Prime Time Live" (7/18/91).

17. Michel Foucault, *Discipline and Punish: The Birth of the Prison*, trans. Alan Sheridan (New York: Vintage Books, 1979), p. 251.

18. We are indebted to Robin D. G. Kelley for this insight.

19. Michael Massing, "What Ever Happened to the 'War on Drugs'?" *New York Review of Books* 39 (June 11, 1992): 45.

20. See Mark Tatge, "Drug-user Discrimination Criticized," *Cleveland Plain Dealer* (May 30, 1993), p. 16-A. See also the three-part series in *USA Today*, "Is the Drug War Racist?" July 23–27, 1993. See also Jerome G. Miller, *Search and Destroy: The Plight of African American Males in the Criminal Justice System*, forthcoming. Miller, executive director of the National Center on Institutions and Alternatives, based in Alexandria, Virginia, says such disparities exist in most states and under federal law.

21. See Clarence Lusane, *Pipe Dream Blues: Racism and the War on Drugs* (Boston: South End Press, 1991), p. 44; "Prisoners in 1988," *Bureau of Justice Statistics Bulletin* (Washington, D.C.: U.S. Department of Justice), April 1989; "Profile of State Prison Inmates, 1986," *Special Report, Bureau of Justice Statistics* (Washington, D.C.: U.S. Department of Justice), January 1988.

22. Massing, "What Ever Happened," p. 45.

23. Lusane, *Pipe Dream Blues*, p. 44; Sharon LaFraniere, "U.S. Has Most Prisoners Per Capita in the World," *Washington Post* (January 5, 1991), p. A3.

24. Massing, "What Ever Happened," p. 45.

25. Lusane, *Pipe Dream Blues*, p. 44; Ron Harris, "Blacks Feel Brunt of Drug War," *Los Angeles Times* (April 22, 1990), p. A1; "Young Black Men Most Likely to Be Jailed," *Washington Afro-American* (March 10, 1991), p. A1.

26. See Roger Rouse, "Power in Popular Culture: Transformation, Discimination and the Politics of Flexible Accumulation," pp. 1–2 of paper presented at "Power: Thinking Through the Disciplines," a conference sponsored by the Center for the Study of Social Transformations, University of Michigan, January 24–26, 1992.

27. See Joseph R. Gusfield, *Symbolic Crusade: Status Politics and the American Temperance Movement*, 2d ed. (Urbana: University of Illinois Press, 1986), p. 194.

28. David F. Musto, *The American Disease: Origins of Narcotic Control* (New Haven, Conn.: Yale University Press, 1973), p. 5.

29. See Thomas Szasz, *Ceremonial Chemistry: The Ritual Persecution of Drugs,*

Addicts, and Pushers (Garden City, N.Y.: Doubleday Anchor, 1974), pp. 75–87; Lusane, *Pipe Dream Blues*, p. 31; Musto, *The American Disease*, pp. 5–6.

30. John Helmer, *Drugs and Minority Oppression* (New York: Seabury Press, 1975), pp. 29–33.

31. Furthermore, as Denise Herd documents, racism was also prominent in the prohibition movement: "By the turn of the twentieth century the divergent concerns about black savagery, alcohol disinhibition and social disorder converged in the image of the drunken black beast. Prohibition was urged in order to protect the white populace, particularly females, from the drunken debauches of half-crazed black men." See Herd, "The Paradox of Temperance: Blacks and the Alcohol Question in Nineteenth-Century America," in Susanna Barrows and Robin Room, eds., *Drinking: Behavior and Belief in Modern History* (Berkeley: University of California Press, 1991), pp. 354–75.

32. Lester Grinspoon and James B. Bakalar, *Cocaine: A Drug and Its Social Evolution*, rev. ed. (New York: Basic Books, 1985), pp. 39, 40.

33. Musto, *The American Disease*, pp. 254–55.

34. David E. Smith, *The Coke Book* (New York: Berkeley Books, 1986), p. 82.

35. Ibid.

36. Grinspoon and Bakalar indicate that the move to control patent medicines was facilitated by anti-Semitism: "A 1908 article in the *New York Times*, 'The Growing Menace of Cocaine,' declared that cocaine 'wrecks its victims more swiftly and surely than opium.' It was easily available in patent medicines and popular among Negroes in the South, where 'Jew peddlers' sold it to them," *Cocaine*, p. 38.

37. As we explore in chapter 2, the criminalization of marijuana consumption in 1937 also was inspired by racial antagonisms fanned by lurid stories in the press that linked sex, race, and drugs. According to Lusane, "The Hearst newspaper empire was the chief vehicle for the spread of these racial tales," *Pipe Dream Blues*, p. 37.

38. Craig Reinarman, "Moral Entrepreneurs and Political Economy: Historical and Ethnographic Notes on the Construction of the Cocaine Menace," *Contemporary Crises* 3 (1979): 230.

39. Grinspoon and Bakalar, *Cocaine*, p. 256.

40. As Dreyfus and Rabinow put it: "Political technologies advance by taking what is essentially a political problem, removing it from the realm of political discourse, and recasting it in the neutral language of science. Once this was accomplished, the problems have become technical ones for specialists to debate. In fact, the language of reform is, from the outset, an essential component of these political technologies. Bio-Power [a Foucaultian term referring to the modern control culture's orientation on managing "life" by disciplining the body and regulating the population] spread under the banner of making people healthy and protecting them. When there was resistance, or failure to achieve its stated aims, this was construed as further proof of the need to reinforce and extend the power of the experts. A technical matrix was established. By definition, there ought to be a way of solving any technical problem. Once this matrix was established, the spread of bio-power was assured, for there was nothing else to appeal to; any other standard

could be shown to be abnormal or to present merely technical problems. We are promised normalization and happiness through science and law. When they fail, this only justifies the need for more of the same" (*Michel Foucault*, p. 196).

41. Steve Waldman, Mark Miller, and Richard Sandza, "Turf Wars in the Federal Bureaucracy," *Newsweek* (April 19, 1989), pp. 24–25.

42. The term "carceral network" is associated with Foucault's critique of modern power. According to Barry Smart, Foucault has used this idea to describe "the series of institutions and organizations employing disciplinary techniques of normalization." Smart contends, "Within the carceral network are to be found institutions of penalty such as almshouses for young girls (the innocent and the delinquent), and colonies for vagrant children and minors; and even more removed from mechanisms of penalty, charitable societies, moral improvement associations and organizations offering assistance. Thereby the carceral network effects a linkage between legal forms of punishment and the most minute forms of correction. . . ." See Smart, *Michel Foucault* (London and New York: Tavistock, 1985), p. 92, and Foucault, *Discipline and Punish*, p. 303.

43. Philip Bell, "Drugs as News: Defining the Social," *Mass Communication Review Yearbook*, vol. 5, ed. Michael Gurevitch and Mark R. Levy (Beverly Hills: Sage, 1985), p. 312.

44. Elsewhere, we have called this a "hierarchy of discourse." See Richard Campbell and Jimmie L. Reeves, "Covering the Homeless: The Joyce Brown Story," *Critical Studies in Mass Communication* 6 (March 1989): 21–42. The concept of "hierarchy of credibility" is associated with Howard S. Becker's work and appropriated by many contemporary news researchers analyzing the journalistic construction of deviance. See Becker, "Whose Side Are We On?" *Social Problems* 14 (1967): 239–47; Mark Fishman, *Manufacturing the News* (Austin: University of Texas Press, 1980), p. 94; Richard V. Ericson, Patricia M. Baranek, and Janet B. L. Chan, *Visualizing Deviance: A Study of News Organization* (Toronto: University of Toronto Press, 1987), p. 283.

45. Ibid., p. 308.

46. Ibid.

47. Robert L. DuPont, letter responding to our questionnaire (June 17, 1991).

48. Arnold Trebach, *The Great Drug War* (New York: Macmillan, 1987), p. 5.

3   Visualizing the Drug News

1. From "Writing the News (By Telling the 'Story')," in Robert Karl Manoff and Michael Schudson, eds., *Reading the News* (New York: Pantheon Books, 1986), pp. 228–29.

2. From "News as Entertainment: The Search for Dramatic Unity," in Elie Abel, ed., *What's News: The Media in American Society* (San Francisco: Institute for Contemporary Studies, 1981), pp. 134–35.

3. See Sarah Ruth Kozloff, "Narrative Theory and Television," in Robert C. Allen, ed., *Channels of Discourse: Television and Contemporary Criticism* (Chapel Hill: University of North Carolina Press, 1987), pp. 42–73; and Seymour Chat-

man, *Story and Discourse: Narrative Structure in Fiction and Film* (Ithaca, N.Y.: Cornell University Press, 1978).

4. Kozloff, "Narrative Theory," p. 45.

5. See Barry Smart, *Michel Foucault* (London and New York: Tavistock, 1985), p. 92.

6. Thomas Mathiesen, "The Eagle and the Sun: On Panoptical Systems and Mass Media in Modern Society," in John Lowman, Robert J. Menzies, and T. S. Palys, eds., *Transcarceration: Essays in the Sociology of Social Control* (Aldershot, Eng.: Gower, 1987), p. 59.

7. Michel Foucault, *Discipline and Punish: The Birth of the Prison,* trans. Alan Sheridan (New York: Vintage Books, 1979), pp. 216–17.

8. Mathiesen, "The Eagle and the Sun," p. 65.

9. Vincent Price, *Public Opinion* (Newbury Park, Calif.: Sage, 1992), pp. 38–39.

10. Ibid., p. 78.

11. See Harold Lasswell, "The Structure and Function of Communication in Society," in Lyman Bryson, ed., *The Communication of Ideas* (New York: Institute for Religions and Social Studies, 1948), pp. 37–51.

12. Price, *Public Opinion,* p. 80. In this discussion Price offers three citations: Warren Breed, "Social Control in the Newsroom: A Functional Analysis," *Social Forces* 33 (1955): 326–35; Bernard Roshco, *Newsmaking* (Chicago: University of Chicago Press, 1975); James S. Ettema, D. C. Whitney, and D. B. Wackman, "Professional Mass Communicators," in C. R. Berger and S. H. Chaffee, eds., *Handbook of Communication Science* (Newbury Park, Calif.: Sage, 1987), pp. 747–80.

13. Price, *Public Opinion,* p. 81.

14. Ibid.

15. The idea of "bifocality" is appropriated from George Lipsitz. But the bifocality that Lipsitz speaks of is diametrically opposed to that operating in spectacles of surveillance; it is, instead, a bifocality of resistance. See Lipsitz, *Time Passages* (Minneapolis: University of Minnesota Press, 1990), p. 135.

16. For a provocative feminist analysis of the "domestic gaze," see Lynn Spigel, "Installing the Television Set: Popular Discourses on Television and Domestic Space, 1948–1955," in Spigel and Denise Mann, eds., *Private Screenings: Television and the Female Consumer* (Minneapolis: University of Minnesota Press, 1992), pp. 3–38.

17. Foucault, *Discipline and Punish,* p. 200.

18. Hubert L. Dreyfus and Paul Rabinow, *Michel Foucault: Beyond Structuralism and Hermeneutics* (Chicago: University of Chicago Press, 1983), p. 189.

19. Richard V. Ericson, Patricia M. Baranek, and Janet B. L. Chan, *Negotiating Control: A Study of News Sources* (Toronto: University of Toronto Press, 1989), pp. 3–4; in addition, the authors cite three sources: Peter Berger, *Facing Up to Modernity* (New York: Basic Books, 1977); G. Konrad and I. Szelenyi, *The Intellectual on the Road to Class Power* (New York: Harcourt Brace Jovanovich, 1979); and Alvin Gouldner, *The Future of Intellectuals and the Rise of the New Class* (New York: Macmillan, 1979).

20. Walker Percy, *Lost in the Cosmos: The Last Self-Help Book* (New York: Farrar, Straus and Giroux, 1983), p. 75.

NOTES TO CHAPTER 3 286

21. Alfred Schutz, "The Well-Informed Citizen," in his *Collected Papers, Vol. 2: Studies in Social Theory* (The Hague: Martinus Nijhoff, 1964), pp. 131–32.

22. See Alfred Schutz and Thomas Luckmann, *The Structures of the Life-World*, trans. R. M. Zaner and T. Engelhardt (Evanston, Ill.: Northwestern University Press, 1973), p. 331.

23. See Richard V. Ericson, Patricia M. Baranek, and Janet B. L. Chan, *Visualizing Deviance: A Study of News Organization* (Toronto: University of Toronto Press, 1987), pp. 4–5.

24. See Michel Foucault, *Madness and Civilization*, trans. R. Howard (New York: Vintage, 1965), p. 38.

25. Our discussion of the "horizons of common sense" is heavily indebted to Roger Silverstone, *The Message of Television: Myth and Narrative in Contemporary Culture* (London: Heinemann, 1981).

26. Sherry B. Ortner, *Sherpas Through Their Rituals* (New York: Cambridge University Press, 1978), p. 1.

27. On June 1, 1993, Connie Chung appeared on the *CBS Evening News* as Dan Rather's coanchor, breaking the daily evening patriarchal stranglehold on the anchor position that had been in place on the three major networks since the early 1980s. In the late 1970s on ABC, Barbara Walters had been the only other woman to coanchor the daily network news. With the exception of CNN, women have usually been assigned to anchor duties during the week to replace a vacationing male anchor or on weekends when there is a smaller viewing audience and, by contemporary journalistic standards, less "important" news taking place.

28. In cinema studies this distinction is associated with questions of authorship. According to Robert C. Allen and Douglas Gomery: "As one step in historicizing the concept of authorship in the cinema, we might borrow a distinction made by formalist critic Boris Tomasevskij between authors with biographies and authors without biographies. Most popular products are produced by authors 'without biographies' in the sense that to their audiences they are anonymous," *Film History: Theory and Practice* (New York: Knopf, 1985), pp. 88–89.

29. P. H. Weaver, "Newspaper News and Television News," in D. Cater and R. Adler, eds., *Television as a Social Force: New Approaches to TV Criticism* (New York: Praeger, 1975), p. 84. Also see Shanto Iyengar and Donald R. Kinder's discussion of "authoritative news," in *News That Matters: Television and American Opinion* (Chicago: University of Chicago Press, 1987), p. 126.

30. Narrative theory only recently has begun to grasp the symbolic importance of setting in the telling of a story. One of the most sophisticated theories of setting appears in Mikhail Bakhtin's work on the novel. Bakhtin argues that the novel has had a revolutionary impact on the existential orientation of modern storytelling. To comprehend and document this impact, he relies on an analytic concept meant to describe the temporal and spatial orientation of a story. The word for this concept is "chronotope," a term borrowed from mathematics. Chronotope literally means "time space." Although this term is associated with Einstein's Theory of Relativity, Bakhtin employs it in literary criticism "almost as a metaphor." In his words, "What counts for us is the fact that it expresses the inseparability of space and time (time as the fourth dimension of space)." See "Forms of Time and of

the Chronotope in the Novel," in his *The Dialogic Imagination: Four Essays*, ed. Michael Holquist, trans. Caryl Emerson and Michael Holquist (Austin: University of Texas Press, 1981), p. 84.

31. According to Iyengar and Kinder (*News That Matters*, p. 126), "most Americans, most of the time, seem to find this authoritative pose irresistible. According to various national surveys, Americans believe by a wide margin that television—not magazines, radio, or newspapers—provides the most intelligent, complete, and impartial news coverage. In a national survey carried out in June of 1984, 79 percent of the American public expressed approval of how the networks handle the job of reporting the news. While the public's confidence in many national institutions has been eroding, faith in television news has actually been increasing." See also the following sources cited by Iyengar and Kinder: *ABC News Poll*, Survey 0041 (June 1984); R. T. Bower, *The Changing Television Audience in America* (New York: Columbia University Press, 1985).

32. Iyengar and Kinder, *News That Matters*, p. 126.

33. Gaye Tuchman, *Making News: A Study in the Construction of Reality* (New York: Free Press, 1978), pp. 82–103. See also Robert P. Snow, *Creating Media Culture* (Beverly Hills: Sage, 1983), pp. 47–53.

34. Weaver, "Newspaper News and Television News," p. 90.

35. Steve Fore, "Lost in Translation: The Social Uses of Mass Communications Research," *Afterimage* 20 (April 1993): 10.

36. Ibid. For another discussion of these points, see Richard Campbell and Jimmie L. Reeves, "TV News Narration and Common Sense: Updating the Soviet Threat," *Journal of Film and Video* 41, no. 2 (Summer 1989): 58–74.

37. Stuart Hall, Chas Critcher, Tony Jefferson, John Clarke, and Brian Roberts, *Policing the Crisis: Mugging, the State, and Law and Order* (London: Macmillan, 1978), p. 62.

38. Ibid., p. 61. See also Richard Campbell, 60 Minutes *and the News: A Mythology for Middle America* (Urbana: University of Illinois Press, 1991), pp. 9–16; and Campbell and Reeves, "TV News Narration and Common Sense," pp. 58–74.

39. David Eason, "Telling Stories and Making Sense," *Journal of Popular Culture* 15 (Fall 1981): 125.

40. Hall et al., *Policing the Crisis*, p. 62.

41. From the transcription of an interview with Foucault conducted by Alessandro Fontana and Pasquale Pasquino published in Foucault, *Power/Knowledge: Selected Interviews and Other Writings, 1972–1977*, ed. Colin Gordon; reprinted as "Truth and Power" in Paul Rabinow, ed., *Foucault Reader* (New York: Pantheon Books, 1984), p. 61.

42. According to Robert M. Entman, the success of African American journalists is, unfortunately, a mixed blessing. As we shall discuss in chapter 4, their visibility plays into the racial politics of Reaganism by, in Entman's words, engendering "an impression that racial discrimination is no longer a problem." See "Modern Racism and the Images of Blacks in Local Television News," *Critical Studies in Mass Communication* 7 (December 1990): 332–45.

43. William Hoynes and David Croteau, "All the Usual Suspects: *MacNeil/Lehrer* and *Nightline*," *Extra!* special issue 3, no. 4 (Winter 1990): 2. This article re-

ports on the original "Nightline" study and offers a follow-up study on "Nightline" and "MacNeil/Lehrer," both of which reveal roughly the same gender patterns. See Hoynes and Croteau, "Are You on the *Nightline* Guest List?" *Extra!* 2, no. 4 (January/February 1989): 2–15.

44. We should point out that our interpretation of this disparity is not based on the same assumptions as those informing the FAIR study. FAIR, taking an information-oriented view of the news, sees this disparity in terms of bias and holds the news organizations accountable for what we see as a much larger disparity in the demographic makeup of authorized knowers in the U.S. knowledge culture. Put another way, we believe FAIR unfairly places too much blame on news organizations for the way that power and knowledge is unfairly distributed in a culture still largely dominated by white, male authority. On the other hand, we also want to unmask these power arrangements—but without turning the news media into a scapegoat and the single causal agent.

45. These numbers include all cocaine stories, not just the video sample.

46. According to Clarence Lusane, Anslinger's hatred of people of color was "legendary": "In official memos to this staff, he would refer to a Black person as a 'ginger-colored nigger.' He testified before a southern-controlled Congress the 'coloreds with big lips lured White women with jazz and marijuana.' In the 1940s, he ordered files be kept on all jazz and swing musicians." See Lusane, *Pipe Dream Blues: Racism and the War on Drugs* (Boston: South End Press, 1991), p. 38.

47. Howard S. Becker, *Outsiders: Studies in the Sociology of Deviance* (London: Colliers-Macmillan), p. 141.

48. Quoted in ibid., p. 142. Also see Harry J. Anslinger, with Courtney Ryley Cooper, "Marijuana: Assassin of Youth," *American Magazine* 124 (July 1937): 19, 150.

49. Quoted by Becker, *Outsiders,* pp. 142–43. Also see *Taxation of Marijuana,* Hearings before the Committee on Ways and Means of the House of Representatives, 75th Congress, 1st Session, on H.R. 6385, April 27–30 and May 4, 1937, p. 7.

50. We address this proactive strategy in more detail in the conclusion to chapter 4 in an analysis based on Ericson et al., *Negotiating Control,* pp. 92–93.

51. Michael Massing, "What Ever Happened to the 'War on Drugs'?" *New York Review of Books* 39 (June 11, 1992): 42–46.

52. Massing, "What Ever Happened," p. 44. See also Robert M. Stutman and Richard Esposito, *Dead on Delivery: Inside the Drug Wars, Straight from the Street* (New York: Warner Books, 1992).

4   Reaganism

1. Thomas Byrne Edsall and Mary D. Edsall, *Chain Reaction: The Impact of Race, Rights, and Taxes on American Politics* (New York: W. W. Norton, 1991), p. 174.

2. Robert C. Allen and Douglas Gomery, *Film History: Theory and Practice* (New York: Knopf, 1985), p. 53.

3. For instance, the news coverage of U.S. involvement in the Persian Gulf after the Iraqi invasion of Kuwait began by trying to place the Bush intervention in the messy contexts of neocolonialism, the regional politics of rich and poor, the control of the global oil economy, the long-standing U.S. alliance with Israel, the threat of Islamic fundamentalism, and the rise of Arab nationalism. But, by the time the United States launched its air war on January 16, 1991, this complexity had largely been simplified and personalized as a "showdown" (a CBS News promotion slogan) between great men: Saddam Hussein, a great villain on the order of Hitler— and George Bush, a great statesman on the order of Churchill. The same pattern of "great man" simplification was also apparent in news coverage of the U.S. invasion of Panama in which the CIA's role in drug smuggling operations to support the Contras in Nicaragua was lost in the journalistic infatuation with vilifying Manuel Noriega.

4. George Gilder, *The Spirit of Enterprise* (New York: Simon and Schuster, 1985), p. 147. Also quoted in Robert B. Reich, *Tales of a New America: The Anxious Liberal's Guide to the Future* (New York: Vintage Books, 1988), p. 109.

5. Gilder, *The Spirit of Enterprise*, p. 290. Quoted in Reich, *Tales of a New America*, p. 25.

6. Reich, *Tales of a New America*, p. 108.

7. See Michel Foucault, "What Is an Author?" in Paul Rabinow, ed., *The Foucault Reader* (New York: Pantheon, 1984), pp. 101–20.

8. The view of Reaganism as being engaged in a "war of positions" is informed by Stuart Hall's discussion of Thatcherism, which, in turn, is informed by the writings of Antonio Gramsci. See Stuart Hall, *The Hard Road to Renewal: Thatcherism and the Crisis on the Left* (London: Verso, 1988), p. 3.

9. Rosalind Pollack Petchesky, "Antiabortion, Antifeminism, and the Rise of the New Right," *Feminist Studies* 7, no. 2 (Summer 1981): 208.

10. For another compelling analysis of the rise of the New Right, see Lawrence Grossberg, *We Gotta Get Out of This Place: Popular Conservatism and Postmodern Culture* (New York: Routledge, 1992), pp. 137–70. See also Christopher Lasch, "Right-Wing Populism and the Revolt Against Liberalism," in his *The True and Only Heaven: Progress and Its Critics* (New York: W. W. Norton, 1991), pp. 476–532.

11. See Walter Stafford's single chapter, "Economic Decline and the Rise of the New Conservatism: Twin Threats to Blacks," in Alphonso Pinkney's *The Myth of Black Progress* (Cambridge: Cambridge University Press, 1984), pp. 32–33.

12. See Michael Rogin, *Ronald Reagan, the Movie and Other Essays in Political Demonology* (Berkeley: University of California Press, 1987).

13. Carol F. Karlsen, *The Devil in the Shape of a Woman: Witchcraft in Colonial New England* (New York: W. W. Norton, 1987).

14. See Garry Wills, *Reagan's America: Innocents at Home* (New York: Doubleday, 1987), pp. 249–58.

15. See Jonathan Reider, "The Rise of the Silent Majority," in Steve Fraser and Gary Gerstle, eds., *The Rise and Fall of the New Deal Order, 1930–1980* (Princeton, N.J.: Princeton University Press, 1989), p. 246.

16. Ibid., p. 249.

NOTES TO CHAPTER 4   290

17. See Edsall and Edsall, *Chain Reaction*, p. 41.

18. See Chilton Williamson, Jr., "Country and Western Marxism," *National Review* (June 9, 1978), p. 711.

19. Reider, "Rise of the Silent Majority," p. 252. Most scholars consider Kevin Phillips to be the most significant of the "budding New Right theorists" during this period. Quoting Michael Omi and Howard Winant: "During the [1968] campaign, political analyst Kevin Phillips submitted a lengthy and rather scholarly analysis of U.S. voting trends to Nixon headquarters, arguing that a Republican victory and long-term electoral realignment was possible on racial grounds. Published the following year as the *Emerging Republican Majority*, Phillips's book suggested a turn to the right and the use of 'coded' anti-black campaign rhetoric (e.g., law and order). Wallace's success, the disarray in Democratic ranks caused by the 'Negro socioeconomic revolution,' and polling data from blue-collar districts around the country convinced Phillips that a strategic approach of this kind (dubbed the 'southern strategy') could fundamentally shift political alignments which had been in effect since 1932." See Omi and Winant, *Racial Formation in the United States: From the 1960s to the 1980s* (New York: Routledge and Kegan Paul, 1986), p. 121; also see Kevin Phillips, *The Emerging Republican Majority* (New York: Anchor, 1970).

20. Cornel West, "The Postmodern Crisis of Black Intellectuals," in Lawrence Grossberg, Cary Nelson, and Paula A. Treichler, eds., *Cultural Studies* (New York: Routledge, 1992), pp. 693, 695.

21. Thomas Byrne Edsall, "The Changing Shape of Power: A Realignment in Public Policy," in Fraser and Gerstle, eds., *Rise and Fall of the New Deal Order*, pp. 283–84.

22. This list is derived from Edsall and Edsall, *Chain Reaction*, p. 45.

23. Mike Davis, *Prisoners of the American Dream: Politics and Economy in the History of the U.S. Working Class* (London: Verso, 1986), p. 16.

24. For a study of how Protestant-Catholic conflict animated the temperance movement, see Joseph R. Gusfield, *Symbolic Crusade: Status Politics and the American Temperance Movement* (Champaign: University of Illinois Press, 1963). Or see the section of Gusfield's book reprinted as "Symbolic Crusade: Status Politics and the American Temperance Movement," in Maureen E. Kelleher, Bruce K. Mack Murray, and Thomas M. Shapiro, eds., *Drugs and Society: A Critical Reader*, 2d ed. (Dubuque, Iowa: Kendall/Hunt, 1988), pp. 15–21.

25. For an informative and revealing genealogy of these disputes of fundamentalism written from a learned insider's point of view, see C. Leonard Allen and Richard T. Hughes, *Discovering Our Roots: The Ancestry of Churches of Christ* (Abilene, Tex.: Abilene Christian University Press, 1988).

26. In Gal. 5:19–21 (King James Version) the Apostle Paul catalogs most of these transgressions: "Now the works of the flesh are manifest, which are these; Adultery, fornication, uncleanness, lasciviousness, idolatry, witchcraft, hatred, variance, emulations [jealousies], wrath, strife, seditions, heresies, envyings, murders, drunkenness, revellings, and such like: of the which I tell you before, as I have also told you in time past, that they which do such things shall not inherit the kingdom of God."

27. See Edsall and Edsall, *Chain Reaction*, p. 133.

28. Ibid., p. 132.

29. See ibid., pp. 131–33. Both Viguerie and Billings are quoted in this passage.

30. Wills, *Reagan's America*, p. 198. Wills cites William James's *The Varieties of Religious Experience: A Study in Human Nature* (New York: Collier Books, 1902/1961), pp. 114–42.

31. Ibid. Wills again cites James's *Varieties of Religious Experience*, pp. 78–113.

32. Amos 8:4–7 [King James Version].

33. Greil Marcus, *Lipstick Traces: A Secret History of the Twentieth Century* (Cambridge, Mass.: Harvard University Press, 1989), p. 134.

34. See Edsall and Edsall, *Chain Reaction*, p. 208.

35. Ibid., p. 177.

36. Ibid.

37. Ibid., pp. 145–48.

38. See Hall, *The Hard Road to Renewal*, pp. 123–60.

39. Davis, *Prisoners of the American Dream*, p. 177.

40. David Harvey, *The Condition of Postmodernity: An Enquiry into the Origins of Cultural Change* (Oxford: Basil Blackwell, 1989), pp. 132–33.

41. See Michael Harrington, *The Other America: Poverty in the United States* (New York: Macmillan, 1962).

42. Harvey, *The Condition of Postmodernity*, p. 138.

43. See ibid. pp. 141–72. This new economic order also has been called "post-Fordism," "overconsumptionism," and "flexible specialization." See also Davis, *Prisoners of the American Dream*, p. 206, and Hall, *The Hard Road to Reaction*, pp. 275–76.

44. Harvey, *The Condition of Postmodernity*, p. 171.

45. Davis, *Prisoners of the American Dream*, p. 206; see also Harvey, *The Condition of Postmodernity*, p. 156.

46. Harvey, *The Condition of Postmodernity*, p. 158.

47. For book-length analyses of the decline of big labor, see Davis, *Prisoners of the American Dream*, and Kim Moody, *An Injury to All: The Decline of American Unionism* (London: Verso, 1988).

48. Harvey, *The Condition of Postmodernity*, p. 150.

49. Ibid., p. 149–50.

50. Davis, *Prisoners of the American Dream*, p. 220. Also see Barry Bluestone and Bennett Harrison, *The Deindustrialization of America: Plant Closings, Community Abandonment, and the Dismantling of Basic Industry* (New York: Basic Books, 1982), p. 95.

51. See Hall, *The Hard Road to Renewal*, p. 142.

52. Reich, *Tales of a New America*, pp. 106–7.

53. Marcus, *Lipstick Traces*, pp. 135–36.

54. Ibid.

55. Edsall and Edsall, *Chain Reaction*, p. 178.

56. Consider the findings of a study of Reagan Democrats that was quoted at length by Edsall and Edsall in their much-discussed magazine article on the politics of race: "white Democratic defectors express a profound distaste for blacks, a sentiment that pervades almost everything they think about government and poli-

tics. Blacks constitute the explanation for their [white defectors] vulnerability and for almost everything that has gone wrong in their lives; not being black is what constitutes being middle class; not living with blacks is what makes a neighborhood a decent place to live. . . . These sentiments have important implications for Democrats, as virtually all progressive symbols and themes have been redefined in racial and pejorative terms. . . . The special status of blacks is perceived by almost all of these individuals as a serious obstacle to their personal advancement. Indeed, discrimination against whites has become a well-assimilated and ready explanation for their status, vulnerability, and failure." See Edsall and Edsall, "When the Official Subject Is Presidential Politics, Taxes, Welfare, Crime, Rights, or Values . . . the Real Subject Is Race," *Atlantic Monthly* (May 1991), p. 56.

57. Robert Entman, "Modern Racism and the Images of Blacks in Local Television News," *Critical Studies in Mass Communication* 7 (December 1990): 332–45. For quantitative studies of the new racism, see the following articles published in Phyllis Katz and Dalmas Taylor, eds., *Eliminating Racism: Profiles in Controversy* (New York: Plenum Press, 1988): Lawrence Bobo, "Group Conflict, Prejudice, and the Paradox in Contemporary Racial Attitudes," pp. 85–114; John B. McConahay, "Modern Racism, Ambivalence, and the Modern Racism Scale," pp. 91–125; and David O. Sears, "Symbolic Racism," pp. 53–84. See also Byron M. Roth, "Social Psychology's Racism," *The Public Interest*, no. 96 (Winter 1990): 26–36.

58. Kobena Mercer, " '1968': Periodizing Politics and Identity," in Grossberg et al., eds., *Cultural Studies*, p. 435.

59. Ibid.

60. Consider Merle Black's observation that "Reagan kind of civilized the racial issue. He's taken what Wallace never could do and made it acceptable. It fits in with their [white student's] sense of perceived injustice, with what they see as the status of being a white person not being as it was 15, 20, or 30 years ago." Black's remarks were made in an interview with Haynes Johnson of the *Washington Post* and cited by Omi and Winant, *Racial Formation*, p. 135.

61. Omi and Winant, *Racial Formation*, p. 125.

62. For a genealogy of modern racism, see Cornel West, *Prophesy Deliverance! An Afro-American Revolutionary Christianity* (Philadelphia: Westminster Press, 1982).

63. Omi and Winant, *Racial Formation*, p. 10.

64. Ibid., pp. 14–15.

65. Ibid.

66. E. Franklin Frazier, *The Negro Family in the United States* (New York: Dryden Press, 1939/1948).

67. Gunnar Myrdal, *An American Dilemma: The Negro Problem and Modern Democracy* (New York: Harper and Row, 1944/1962).

68. Rickie Solinger, *Wake Up Little Susie: Single Pregnancy and Race Before Roe v. Wade* (New York: Routledge, 1992), p. 60.

69. Ibid.

70. Patricia Hill Collins, *Black Feminist Thought: Knowledge, Consciousness, and the Politics of Empowerment* (New York: Routledge, 1990), p. 68.

71. Stephanie Coontz, *The Way We Never Were: American Families and the*

Nostalgia Trap (New York: Basic Books, 1992), pp. 237–38. See also the following work cited by Coontz: Joel Perlmann, *Ethnic Differences: Schooling and Social Structure Among the Irish, Italians, Jews, and Blacks in an American City, 1880– 1935* (New York: Cambridge University Press, 1988); Stephen Thernstrom, *A History of the American People*, vol. 2 (New York: Harcourt Brace Jovanovich, 1989), p. 683; Stanley Lieberson, *A Piece of the Pie: Blacks and White Immigrants Since 1880* (Berkeley: University of California Press, 1980); C. Vann Woodward, "The Crisis of Caste," *New Republic* (November 6, 1989), p. 44; and Douglas Massey, "American Apartheid: Segregation and the Making of the Underclass," *American Journal of Sociology* 96 (September 1990): 329–57.

72. Omi and Winant, *Racial Formation*, p. 14.

73. Department of Labor, *The Negro Family: The Case for National Action* (Washington, D.C.: U.S. Government Printing Office, 1965), p. 5.

74. For a summary of this research, see Alphonso Pinkney, *The Myth of Black Progress* (Cambridge: Cambridge University Press, 1984), pp. 7–17. See also Lee Rainwater and William Yancey, eds., *The Moynihan Report and the Politics of Controversy* (Cambridge, Mass.: MIT Press, 1967); Nathan Glazer and Daniel Moynihan, *Beyond the Melting Pot* (Cambridge, Mass.: MIT Press, 1965); Nathan Glazer, *Affirmative Discrimination: Ethnic Inequality and Public Power* (New York: Basic Books, 1975); Ben J. Wattenberg and Richard Scammon, "Black Progress and Liberal Rhetoric," *Commentary* (April 1973); Thomas Sowell, "Are Quotas Good for Blacks?" *Commentary* (June 1978), pp. 39–43; Thomas Sowell, *Affirmative Action Reconsidered: Was It Necessary in Academia?* (Washington, D.C.: American Enterprise Institute, 1975).

75. See Charles Murray, *Losing Ground: American Social Policy* (New York: Basic Books, 1984).

76. See Coontz, *The Way We Never Were*, p. 103.

77. See Craig Reinarman and Harry G. Levine, "The Crack Attack: Politics and Media in America's Latest Drug Scare," in Joel Best, ed., *Images of Issues: Typifying Contemporary Social Problems* (New York: Aldine de Gruyter, 1989), p. 127.

78. Quoted in Edsall and Edsall, "When the Official Subject," p. 80.

79. Quoted in Omi and Winant, *Racial Formation*, p. 134, from an article in the *New York Times* (March 6, 1985).

80. See Mike Davis's discussion of "black lash" in relation to reactionary African American response to the crack panic in *City of Quartz: Excavating the Future in Los Angeles* (London: Verso, 1990), pp. 289–92.

81. Edsall and Edsall state: "The liberal failure to convincingly address increasingly conservative attitudes in the majority electorate, attitudes spurred in part by crime and welfare rates, has damaged the national Democratic party on a variety of counts. Perhaps most importantly, it has signaled a failure to live up to one of the chief obligations of a political party: to secure the safety and well-being of its own constituents, black and white. Secondly, self-imposed Democratic myopia has in no way prevented the majority public from forming 'hard' opinions on crime, drug use, chronic joblessness, and out-of-wedlock births—nor from judging the national Democratic party as excessively 'soft' in its approach to contempo-

rary social issues—nor from voting for politicians whose conservative attitudes on crime and social disorder more completely mirror its own," *Chain Reaction*, pp. 114–15. In coming down on the side of cultural Moynihanism, the Edsalls, unfortunately, tend to take rising crime statistics at face value. For a thorough critique of such statistics, see Kevin N. Wright, *The Great American Crime Myth* (Westport, Conn.: Greenwood Press, 1985).

82. This is derived from Patricia Hill Collins's critique of the Moyers's report. See Collins, "A Comparison of Two Works on Black Family Life," *Signs* 14, no. 4 (1989): 875–84.

83. Quoted in Herman Gray, "Television, Black Americans, and the American Dream," *Critical Studies in Mass Communication* 6 (December 1989): 381.

84. Moynihan, quoted in Steven Mintz and Susan Kellogg, *Domestic Revolutions: A Social History of American Family Life* (New York: Free Press, 1988), p. 211. Also see Daniel Patrick Moynihan, *Family and Nation: The Godkin Lectures, Harvard University* (San Diego, 1986), pp. 27–28.

85. Collins, *Black Feminist Thought*, p. 68.

86. Quoted in Edsall and Edsall, *Chain Reaction*, p. 148.

87. See Coontz, *The Way We Never Were*, p. 235. Also see Ken Auletta, *The Underclass* (New York: Random House, 1982); and "Working Seminar on the Family and American Welfare Policy," *The New Consensus on Family and Welfare: A Community of Self-Reliance* (Washington, D.C.: American Enterprise Institute, 1987).

88. Morton Kondracke, "The Two Black Americas," *New Republic* (February 6, 1989), p. 18.

89. Robb, quoted in Coontz, *The Way We Never Were*, p. 235.

90. Again see Gray, "Television," p. 376. Also quoted by J. Demeter, "Notes on the Media and Race," *Radical America* 20 (1986): 67. In his right-wing attack on affirmative action Dinesh D'Souza also has misappropriated Cosby's success in *Illiberal Education: The Politics of Race and Sex on Campus* (New York: Free Press, 1991), pp. 98–99.

91. Gray, "Television," p. 376. See also Marlon Rigg's documentary, *Color Adjustment*, for a similar discussion (featuring Herman Gray) of Cosby's meaning to white America.

92. Gray, "Television," p. 384.

93. Entman, "Modern Racism," p. 342.

94. Grossberg, *We Gotta Get Out of This Place*, p. 163.

5   The   Trickle-down Paradigm

1. George Lipsitz, *Time Passages: Collective Memory and American Popular Culture* (Minneapolis: University of Minnesota Press, 1990), p. 19.

2. See Oscar Williams, ed., *A Pocket Book of Modern Verse* (New York: Washington Square Press, 1954/1969), p. 210. Significantly, Simon and Garfunkel set these lyrics to music in the 1960s.

3. This final example is a reference to Costner's brief appearance in *Truth or Dare,* a rock documentary of Madonna's controversial 1990 concert tour.

4. Richard Dyer, *Stars* (London: British Film Institute, 1979), p. 111; see also Jimmie L. Reeves, "Television Stardom: A Ritual of Social Typification and Individualization," in James W. Carey, ed., *Media, Myths, and Narratives: Television and the Press* (Newbury Park, Calif.: Sage, 1988), p. 150.

5. This traumatic rupturing of a public persona is perhaps even more apparent in the disclosures of basketball star Earvin "Magic" Johnson's infection by the AIDS virus and Michael Jackson's alleged pedophilia.

6. See David A. Cook, *A History of Narrative Film* (New York: W. W. Norton, 1981), p. 213.

7. Ibid., pp. 212–15, 264–67. Also see Garth Jowett, *Film: The Democratic Art* (Boston: Little, Brown, 1976), pp. 233–59.

8. Stephanie Coontz, *The Way We Never Were: American Families and the Nostalgia Trap* (New York: Basic Books, 1992), p. 107.

9. Mimi White, *Tele-Advising: Therapeutic Discourse in American Television* (Chapel Hill: University of North Carolina Press, 1992), pp. 8, 25.

10. John Cawelti, *Adventure, Mystery, and Romance: Formula Stories as Art and Popular Culture* (Chicago: University of Chicago Press, 1976), p. 35.

11. Hewitt, quoted in Richard Campbell, *60 Minutes and the News* (Urbana: University of Illinois Press, 1991), p. 68.

12. For another discussion of news and the therapeutic, see ibid., pp. 68–92.

13. David Harvey, *The Condition of Postmodernity* (Oxford: Basil Blackwell, 1989), p. 12.

14. Ibid., p. 9; see also Joli Jensen, *Redeeming Modernity: Contradictions in Media Criticism* (Newbury Park, Calif.: Sage, 1990), and Raymond Williams, *The Politics of Modernism* (London: Verso, 1989).

15. Kenneth Gergen, *The Saturated Self: Dilemmas of Identity in Contemporary Life* (New York: Basic Books, 1991), p. 38.

16. Ibid., p. 40.

17. Ibid., p. 41.

18. T. J. Jackson Lears, "From Salvation to Self-Realization: Advertising and the Therapeutic Roots of Consumer Culture, 1880–1930," in Robert Wightman Fox and Lears, eds., *The Culture of Consumption: Critical Essays in American History, 1880–1980* (New York: Pantheon, 1983), p. 15; see also Warren I. Susman, *Culture as History: The Transformation of American Society in the Twentieth Century* (New York: Pantheon, 1973/1984), pp. 275–80.

19. Lears, "From Salvation to Self-Realization," pp. 6–11.

20. Ibid., p. 15.

21. Norman Clark, *Deliver Us from Evil* (New York: W. W. Norton, 1976), p. 12.

22. Susman, *Culture as History*, pp. 271, 276.

23. White, *Tele-Advising*, p. 13.

24. Harry G. Levine, "The Discovery of Addiction: Changing Conceptions of Habitual Drunkenness In America." *Journal of Substance Abuse Treatment* 2, no. 1

(1985): 52; see also Levine, "The Alcohol Problem in America: From Temperance to Alcoholism," *British Journal of Addiction* 79, no. 1 (March 1984): 109–19.

25. Robert N. Bellah, Richard Madsen, William M. Sullivan, Ann Swindler, and Stephen M. Tipton, *Habits of the Heart: Individualism and Commitment in American Life* (Berkeley: University of California Press, 1985), p. 122.

26. See special critical forum on the penny press in *Critical Studies in Mass Communication* 4, no. 4 (December 1987).

27. W. Phillips Davison, "The Third-Person Effect in Communication," *Public Opinion Quarterly* 47 (Spring 1983): 1–15.

28. Michael Schudson, "Trout or Hamburger: Politics and Telemythology," *Tikkun* 6, no. 1 (March-April 1991): 51.

29. See Kevin Phillips, *The Politics of Rich and Poor: Wealth and the American Electorate in the Reagan Aftermath* (New York: Random House, 1990).

30. Susan Mackey-Kallis and Dan F. Hahn, "Questions of Public Will and Private Action: The Power of the Negative in the Reagan's 'Just Say No' Morality Campaign," *Communication Quarterly* 39, no. 1 (Winter 1991): 2.

31. Bellah et al., *Habits of the Heart*, pp. 126–27. In their analysis of modern therapy, Joseph Veroff, Richard A. Kulka, and Elizabeth Douvan assert, "Psychoanalysis (and psychiatry) is the only form of psychic healing that attempts to cure people by detaching them from society and relationships. All other forms—shamanism, faith healing, prayer—bring the community into the healing process . . . ," in *Mental Health in America: Patterns of Help-Seeking from 1957 to 1976* (New York: Basic Books, 1981), pp. 6–7.

32. Nachman Ben-Yehuda, *The Politics and Morality of Deviance* (Albany: State University of New York Press, 1990), p. 70.

33. Ibid., p. 69.

34. For a detailed examination of journalistic values, including individualism, see Herbert Gans, *Deciding What's News: A Study of* CBS Evening News, NBC Nightly News, Newsweek *and* Time (New York: Vintage Books, 1979), pp. 39–69.

35. See Campbell, 60 Minutes *and the News*, pp. 151–56.

36. Gaye Tuchman, *Making News: A Study in the Construction of Reality* (New York: Free Press, 1978), p. 14.

37. Murray Edelman, "The Political Language of the Helping Profession," in his *Political Language: Words That Succeed and Politics That Fail* (New York: Academic Press, 1977), pp. 57–75; see also Stanley Cohen, *Visions of Social Control: Crime, Punishment, and Classification* (Cambridge: Polity Press, 1985), pp. 174–75, 274–76.

38. Ben-Yehuda, *The Politics and Morality of Deviance*, p. 88.

39. Mary Mander, "Narrative Dimensions of the News: Omni-science, Prophecy, and Morality," *Communication* 10 (1987): 63.

40. John Fiske, *Television Culture* (London: Methuen), p. 385.

41. On this point, see James W. Carey, "The Press and Public Discourse," *Kettering Review* (Winter 1992) pp. 9–22, and Christopher Lasch, "Journalism and the Lost Art of Argument," *Gannett Center Journal* 4, no. 2 (Spring 1990): 1–11.

42. Lasch, "Journalism and the Lost Art of Argument," p. 10.

43. Linda Cecere, "Hotline Receives Barrage of Questions on Ethics," *Adver-*

NOTES TO CHAPTER 5

tising Age (October 27, 1986), pp. S-35, 36. The article further investigated claims that "the hotline does not disclose its ownership or relationship to some of the institutions to which it refers patients. Others say it directs those callers unable to pay the stiff prices of many private treatment programs—PIA charges $20,000 for a 28-day treatment—to state-run programs."

44. Hotline data do not appear in our sample after November 1986. To us, this absence suggests that the networks were made aware of the issues raised in the Cecere article and became more cautious about using the hotlines. Even so, we cannot find evidence that the networks reported ethical problems with the hotlines until 1991. See our discussion of this belated report in chapter 9.

45. We are grateful to Craig Reinarman who suggests that much of the drug hysteria flowed from hasty generalizations built on flawed samples; he points out that if the hotline's only data were from users whose habits had become problematic enough to call an emergency hotline, then clearly we are seeing only one perspective of users' complicated and varied relationships to cocaine.

46. See Arnold M. Washton and Mark Gold, "Recent Trends in Cocaine Abuse: A View from the National Hotline, '800-COCAINE,'" *Advances in Alcohol and Substance Abuse* 6, no. 2 (Winter 1986): 31–47.

47. White, *Tele-Advising*, pp. 7–8; see also Michel Foucault, *The History of Sexuality*, vol. 1, trans. Robert Hurley (New York: Pantheon, 1978), pp. 61–62.

48. Peter L. Berger and Thomas Luckmann, *The Social Construction of Reality* (Garden City, N.Y.: Doubleday Anchor, 1967), pp. 112–13.

49. Ibid., p. 114.

50. Joseph R. Gusfield, *Symbolic Crusade: Status Politics and the American Temperance Movement*, 2d ed. (Urbana: University of Illinois Press, 1963, 1986), pp. 66–69, 194.

51. For a discussion of network news anchor salaries, see "Anchorman," a 1983 "60 Minutes" segment, produced by Harry Moses and reported by Mike Wallace.

52. See, for example, Jon Katz, "Can the Media Do the Right Thing?" *Rolling Stone* (May 28, 1992), pp. 30–31.

53. Greider, quoted in Mark Hertsgaard, *On Bended Knee: The Press and the Reagan Presidency* (New York: Farrar, Straus and Giroux, 1988), p. 78.

54. Jon Katz, "AIDS and the Media: Shifting Out of Neutral," *Rolling Stone* (May 27, 1993), p. 32.

55. Yellin, quoted in Hertsgaard, *On Bended Knee*, p. 80.

56. David Eason, "The New Journalism and the Image-World: Two Modes of Organizing Experience," *Critical Studies in Mass Communication* 1 (March 1984): 57.

57. Michael Schudson, *Discovering the News* (New York: Basic Books, 1978), p. 186; see also Gaye Tuchman, "Objectivity as Strategic Ritual: An Examination of Newsmen's Notions of Objectivity," *American Journal of Sociology* 77 (January 1972): 660–79.

58. See Campbell, 60 Minutes *and the News*, pp. 68–92.

NOTES TO CHAPTER 6   298

6 The Siege Paradigm

1. Stanley Cohen, *Visions of Social Control: Crime, Punishment and Classification* (Cambridge: Polity Press, 1985), p. 233.

2. Todd Gitlin, "The War on Drugs and the Enlisted Press," *Columbia Journalism Review* (November–December 1989), p. 17.

3. "Wack" is marijuana laced with heroin or PCP.

4. See Peter Kerr, "Anatomy of the Drug Issue: How, After Years, It Erupted," *New York Times* (November 17, 1986), pp. A1, 12.

5. Quoted by Howard S. Becker, *Outsiders: Studies in the Sociology of Deviance* (London: Collier-Macmillan, 1963), p. 142, and Harry S. Anslinger, with Courtney Ryley Cooper, "Marijuana: Assassin of Youth," *American Magazine* 124 (July 1937): 19, 150.

6. Alter has argued that on a national level the pool of experts, whom he calls the "usual suspects," is often confined to a privileged few: Gloria Steinem on feminism, Carl Sagan on science, Alan Greenspan on economics, John Naisbitt on the future, Henry Kissinger on diplomacy, William Bennett on everything, and so on. See Jonathan Alter, "News Media: Round Up the Usual Suspects," *Newsweek* (March 25, 1985), p. 69.

7. See Peter Stallybrass and Allon White, *The Politics and Poetics of Transgression* (Ithaca, N.Y.: Cornell University Press, 1986).

8. Ibid., pp. 94–96.

9. This description was written by James Howell in the early eighteenth century and quoted by Stallybrass and White, *Politics and Poetics*, p. 97.

10. Stallybrass and White, *Politics and Poetics*, p. 97.

11. Ibid., p. 96.

12. Todd Gitlin, "On Drugs and Mass Media in America's Consumer Society," in Hank Resnick, ed., *Youth and Drugs: Society's Mixed Messages*, Office of Substance Abuse Prevention (OSAP) Prevention Monograph-6, DHHS publication no. (ADM) 90-1689 (Washington, D.C.: U.S. Department of Health and Human Services Prevention, 1990), p. 45.

13. We currently are working with our colleague Hayg Oshagan on this clandestine footage that combines our humanist discourse and more traditional social scientific content analysis.

14. Stutman, quoted in Michael Massing, "What Ever Happened to the 'War on Drugs'?" *New York Review of Books* 39 (June 11, 1992): 44.

15. Richard V. Ericson, Patricia M. Baranek, and Janet B. L. Chan, *Negotiating Control: A Study of News Sources* (Toronto: University of Toronto Press, 1989), pp. 92–93. Also see Suzanne E. Hatty, "Police, Crime and the Media: An Australian Tale," *International Journal of the Sociology of Law* 19 (1991): 171–91.

16. Mike Davis, *City of Quartz: Excavating the Future in Los Angeles* (London: Verso, 1990), p. 267. Davis quotes from a story by Louis Sahagun and Carol McGraw, *Los Angeles Times* (April 7, 1989).

17. Clifford Geertz, "Person, Time and Conduct in Bali," *The Interpretation of Cultures: Selected Essays* (New York: Basic Books, 1973), pp. 443–44.

18. This discussion of subversive aspects of television's sports discourse was originally presented by Jimmie L. Reeves, "TV's World of Sports: Presenting and Playing the Game," in Gary Burns and Robert J. Thompson, eds., *Television Studies: Textual Analysis* (New York: Praeger, 1989), pp. 205–19.

19. Victor Turner, "Social Dramas and Stories About Them," in W. J. T. Mitchell, ed., *On Narrative* (Chicago: University of Chicago Press, 1981), p. 154.

20. James S. Ettema, "Press Rites and Race Relations: A Study of Mass Mediated Ritual," *Critical Studies in Mass Communication* 7, no. 4 (December 1990): 309–31.

21. Ibid., p. 311.

22. For accounts of the controversy surrounding Proposition 42, see William F. Reed, "A New Proposition," *Sports Illustrated* (January 23, 1989), pp. 16–19; and "Race Becomes the Game," *Newsweek* (January 30, 1989), pp. 56–59.

23. Davis, *City of Quartz*, p. 291.

24. Quoted in ibid., p. 292, from an interview by Ken Kelly, *San Francisco Focus* (March 1988), p. 100.

25. Ettema, "Press Rites," p. 511.

26. Turner, "Social Dramas," p. 148.

27. Ettema, "Press Rites," p. 511.

28. For a detailed account of the Tribble trial, see Lewis Cole, *Never Too Young to Die: The Death of Len Bias* (New York: Pantheon, 1989), pp. 183–246.

29. Mark Fishman, "Crime Waves as Ideology," in Stanley Cohen and Jock Young, eds., *The Manufacture of News: Social Problems, Deviance and the Mass Media*, rev. ed. (Beverly Hills: Sage, 1981), p. 98.

30. Terry Williams, *The Cocaine Kids* (Reading, Mass.: Addison-Wesley, 1989).

31. Fishman, "Crime Waves," p. 107.

32. PowerMaster contained an extraordinarily high alcohol content: 5.9 percent by weight, or about 50 percent higher than a regular malt liquor. Heileman is essentially repackaging PowerMaster in a slightly watered-down form (4.5 percent alcohol) as Colt 45 Premium. Other high-alcohol malt liquors targeted at people of color include Olde English 800, Schlitz Malt Liquor's Red Bull, and St. Ives. See "Novello Assails Brewers for Drinks Aimed at Minorities," *Los Angeles Times* (May 20, 1992), p. A16; Jeffrey Scott, "Question of Ethical Marketing Standards Is Still Being Debated," *Atlanta Journal* (July 11, 1991), p. E1. Although not specifically targeted at people of color, two other recent alcoholic beverages also have generated a great deal of controversy: Crazy Horse malt liquor (Hornell Brewing Company) and Black Death Vodka (made in Belgium and imported by Cabo Distributing Company, Inc.). Slash of the popular rock group Guns N' Roses helped pitch Black Death to a youth market. For a compelling editorial on Crazy Horse malt liquor from a Native American point of view, see Michael Dorris, "Noble Savages? We'll Drink to That," *New York Times* (April 21, 1992), p. A23. See also Cynthia Cotts, "Hard Sell in the Drug War," *The Nation* (March 9, 1992), pp. 300–302.

33. Unfortunately, alcohol and drug advertising also has compromised African American publishers. According to Clarence Lusane, "Secretary of Health and Human Services Louis Sullivan, at one point, invited Black magazine owners and

editors to meet and discuss tobacco ads in their publications. All refused to attend the meeting. Some Black newspaper representatives did attend, but no one stopped taking the advertisements." See Lusane, *Pipe Dream Blues: Racism and the War on Drugs* (Boston: South End Press, 1991), pp. 102–5; see also "An Uproar Over Billboards in Poor Areas," *New York Times* (May 1, 1989), p. D10.

34. Gitlin, "The War on Drugs," p. 18.

35. See Pierre Bourdieu, *Distinction: A Social Critique of the Judgment of Taste*, trans. Richard Nice (Cambridge, Mass.: Harvard University Press, 1984).

36. Michel Foucault, *Discipline and Punish: The Birth of the Prison*, trans. Alan Sheridan (New York: Vintage Books, 1979), p. 296.

37. See especially Allan Bloom, *The Closing of the American Mind* (New York: Simon and Schuster, 1987).

38. For overviews of the "political correctness" controversies, see Patricia Aufderheide, ed., *Beyond PC: Toward a Politics of Understanding* (St. Paul, Minn.: Graywolf Press, 1992), and Paul Berman, ed., *Debating P.C.: The Controversy over Political Correctness on College Campuses* (New York: Laurel, 1992).

39. See David F. Musto, *The American Disease: Origins of Narcotic Control*, expanded ed. (New York: Oxford University Press, 1973/1987), p. 3. See also Patricia Erickson et al., *The Steel Drug: Cocaine in Perspective* (Lexington, Mass.: Lexington Books, 1987), p. 7.

40. Susan Sontag, *A Sontag Reader* (New York: Vintage Books, 1983), p. 316.

41. See Stanley Cohen, *Visions of Social Control: Crime, Punishment and Classification* (Cambridge: Polity Press, 1985), pp. 117–18.

42. Ibid., pp. 121–22.

43. For less-than-flattering histories of preindustrial community control in America, see Carol F. Karlsen, *The Devil in the Shape of a Woman: Witchcraft in Colonial New England* (New York: W. W. Norton, 1987); Alexander Saxton, *The Rise and Fall of the White Republic: Class, Politics, and Mass Culture in Nineteenth Century America* (London: Verso, 1990); and Raymond S. Franklin, *Shadows of Race and Class* (Minneapolis: University of Minnesota Press, 1991).

44. Robert B. Reich, *Tales of a New America: The Anxious Liberal's Guide to the Future* (New York: Vintage Books, 1988), p. 171.

45. The lawyer replied, correctly we think, "He that shewed mercy on him." And Jesus told the lawyer to "Go, and do thou likewise." See Luke 10:25–37 (King James Version).

46. Christina Jacqueline Johns, *Power, Ideology, and the War on Drugs: Nothing Succeeds Like Failure* (New York: Praeger, 1992), p. 124. Also see Richard Morin, "Many in Poll Say Bush Plan Is Not Stringent Enough," *Washington Post* (September 8, 1989), p. A18.

47. See "The Informants: In a Drug Program, Some Kids Turn in Their Own Parents," *Wall Street Journal* (April 20, 1992), p. A1. Also see an installment of "Larry King Live" that featured "Jim," a parent turned in by his stepdaughter (a DARE student) for marijuana possession. "The Home Front in the War on Drugs," Transcript #547 (April 24, 1992).

48. In Minnesota alone, 628 public and private schools participate in DARE.

See Mike Kaszuba, "Critics Question Antidrug Program," *Minnesota Star Tribune* (June 7, 1992), p. 18.

49. Jonathan Reider, "The Rise of the Silent Majority," in Steve Fraser and Gary Gerstle, eds., *The Rise and Fall of the New Deal Order, 1930–1980* (Princeton, N.J.: Princeton University Press, 1989), p. 244.

50. Thomas Byrne Edsall and Mary D. Edsall, *Chain Reactions: The Impact of Race, Rights, and Taxes on American Politics* (New York: W. W. Norton, 1991), p. 85.

51. See Ralph Brauer, "The Drug War of Words," *Nation* (May 21, 1990), p. 705.

7   Captivating Public Opinion

1. Barbara Ehrenreich, *Fear of Falling: The Inner Life of the Middle Class* (New York: Pantheon, 1989), p. 247.

2. Actually, we found one previous report that made reference to "rock houses" in Los Angeles (ABC, 10/26/84). Rock is the West Coast term for crack. However, this report was framed as a story about Los Angeles street gangs and violence associated with the drug trade. The specific dangers of rock cocaine are not framed as the "frightening wave of the future."

3. See Stephen D. Reese and Lucig H. Danielian, "Intermedia Influence and the Drug Issue: Converging on Cocaine," in Pamela J. Shoemaker, ed., *Communication Campaigns About Drugs: Government, Media, and the Public* (Hillsdale, N.J.: Lawrence Erlbaum Associates, 1989), pp. 29–45.

4. This metaphor is borrowed from Representative Charles Shumer who was quoted in the *New York Times* as having some second thoughts about the $1.7 billion Omnibus Drug Bill that was passed into law in October 1986. Quoting Shumer: "Maybe we had the wrong solutions but not the wrong problem. . . . What happens is that this occurs in one seismic jump instead of a rational buildup. The down side is that you come up with policies too quickly and that the policies are aimed at looking good rather than solving the problem." See Peter Kerr, "Anatomy of the Drug Issue: How, After Years, It Erupted," *New York Times* (November 17, 1986), p. A12.

5. The *Atlantic, Newsweek, Time,* and *U.S. News and World Report* all featured drugs in cover stories during this period.

6. See Lucig H. Danielian and Stephen D. Reese, "A Closer Look at Intermedia Influences on Agenda Setting: The Cocaine Issue of 1986," in Shoemaker, ed., *Communication Campaigns*, pp. 47–66.

7. For a conventional quantitative analysis of this coverage, see John E. Merriam, "National Media Coverage of Drug Issues, 1983–1987," in Shoemaker, ed., *Communication Campaigns*, pp. 21–28.

8. See Adam Clymer, "Public Found Ready to Sacrifice in Drug Fight," *New York Times* (September 2, 1986), pp. A1, D16.

9. See Stuart Hall, Chas Chritcher, Tony Jefferson, John Clarke, and Brian Roberts, "The Social Production of News: Mugging in the Media," in Stanley

Cohen and Jock Young, eds., *The Manufacture of News: Social Problems, Deviance and the Mass Media,* rev. ed. (Beverly Hills: Sage, 1981), p. 355.

10. See Clymer, "Public Found Ready," p. D16.

11. See Lloyd D. Johnston, "America's Drug Problem in the Media: Is It Real or Is It Memorex?" in Shoemaker, ed., *Communication Campaigns,* p. 99. For another perspective, see also Craig Reinarman and Harry G. Levine, "Crack in Context: Politics and Media in the Making of the Drug Scare," *Contemporary Drug Problems* (Winter 1989), p. 546.

12. John Fiske, *Reading the Popular* (Boston: Unwin Hyman, 1989), p. 194.

13. See Avraham Forman and Susan B. Lachter of the National Institute of Drug Abuse, "The National Institute of Drug Abuse Cocaine Prevention Campaign," in Shoemaker, ed., *Communication Campaigns,* p. 13.

14. See Susan B. Lachter and Avraham Forman, "Drug Abuse in the United States," also in Shoemaker, ed., *Communication Campaigns,* p. 7.

15. Ibid., pp. 9–10.

16. Ibid., p. 11.

17. According to Hubert L. Dreyfus and Paul Rabinow, the isolation of "meticulous rituals of power" is the "conceptual basis for much of Foucault's later work." "In *Discipline and Punish* and *The History of Sexuality* Foucault will identify specific sites in which rituals of power take place—the Panopticon of Bentham and the confessional. He will use these to localize and specify how power works, what it does and how it does it." See *Michel Foucault: Beyond Structuralism and Hermeneutics* (Chicago: University of Chicago Press, 1983), p. 110. Here, we are arguing that, like "the Panopticon of Bentham," the "Monitoring the Future" Survey of ISR and the opinion poll of Gallup are also meticulous rituals of power that actually combine surveillance and confessional technologies.

18. Benjamin Ginsberg, *The Captive Public: How Mass Opinion Promotes State Power* (New York: Basic Books, 1986), p. 32.

19. Johnston, "America's Drug Problem," p. 98.

20. Ginsberg, *The Captive Public,* p. 85.

21. See Clymer, "Public Found Ready," p. D16.

22. Ginsberg, *The Captive Public,* p. 60.

23. See Clymer, "Public Found Ready," p. D16.

24. Ginsberg, *The Captive Public,* pp. 82–83.

25. See Mark Hertsgaard, *On Bended Knee: The Press and the Reagan Presidency* (New York: Farrar, Straus and Giroux, 1988).

26. Shanto Iyengar and Donald R. Kinder, *News that Matters: Television and American Opinion* (Chicago: University of Chicago Press, 1987), p. 131.

27. Iyengar and Kinder's *News that Matters,* which features experimental research methods, represents what is widely recognized as the most sophisticated recent social scientific research on "priming" and "agenda setting."

28. See Michael Schudson, "Trout or Hamburger: Politics and Telemythology," *Tikkun* 6, no. 2 (1991): 50.

29. Hall et al., "The Social Production of News," p. 363.

30. Peter Stallybrass and Allon White, *The Politics and Poetics of Transgression* (Ithaca, N.Y.: Cornell University Press, 1986), p. 191. See also Cynthia Chase, "Re-

view of *Powers of Horror* and *Desire in Language*," *Criticism* 26, no. 2 (1984): 193–201; and Julia Kristeva, *Powers of Horror: An Essay on Abjection,* trans. L. S. Roudiez (New York: Columbia University Press, 1982).

31. See Clymer, "Public Found Ready," p. D16.

32. See Phil Gailey's column "Talking Politics" under the subheading of "Drugs as an Issue," *New York Times* (October 23, 1986), p. B12.

33. Ibid.

34. Johnston, "America's Drug Problem," p. 98.

35. See Reese and Danielian, "Intermedia Influence," pp. 29–45. Also see Danielian and Reese, "A Closer Look," pp. 47–66.

36. See Reinarman and Levine, "Crack in Context," p. 542; and Edwin Diamond, Frank Acosta, and Leslie Jean Thornton, "Is TV News Hyping America's Cocaine Problem?" *TV Guide* (February 7, 1987), pp. 4–10.

8 Family Matters

1. Rosalind Pollack Petchesky, "Antiabortion, Antifeminism, and the Rise of the New Right," *Feminist Studies* 7, no. 2 (Summer 1981): 227.

2. For a summary of feminist critiques of the "functionalist" view of family, see Margaret L. Anderson, "Feminism and the American Family Ideal," *Journal of Comparative Family Studies* 22, no. 2 (Summer 1991): 235–46. Quoting Anderson (p. 237): "Feminists have criticized functionalist theory for its unexamined assumptions about the inevitability of the sexual division of labor and the divisions of male and female roles along instrumental-expressive lines. New concepts of gender in the feminist literature provide a radical departure from a functionalist conception of sex roles and the sexual division of labor in the family. As understood in feminist theory, gender is constructed through specific cultural and historical experiences. Therefore, to understand gender relations in society requires the structural analyses of the social process by which gender is created—not just in the family, but in society as a whole." See also the following sources cited by Anderson: Helene Z. Lopata and Barrie Thorne, "On the Term 'Sex Roles,' " *Signs: Journal of Women in Culture and Society* 3 (1978): 718–21; Judith Stacey and Barrie Thorne, "The Missing Feminist Revolution in Sociology," *Social Problems* 32 (1985): 301–16.

3. Rayna Rapp, "Family and Class in Contemporary America: Notes Toward an Understanding of Ideology," in Barrie Thorne and Marilyn Yalom, eds., *Rethinking Family: Some Feminist Questions* (New York: Longman, 1982), pp. 168–87.

4. Judith Stacey, *Brave New Families: Stories of Domestic Upheaval in Late Twentieth-Century America* (New York: Basic Books, 1990), p. 6.

5. Jackie Byars, *All That Hollywood Allows: Re-Reading Gender in 1950s Melodrama* (Chapel Hill: University of North Carolina Press, 1991), p. 79.

6. See Stacey, *Brave New Families;* Steven Mintz and Susan Kellogg, *Domestic Revolutions: A Social History of American Family Life* (New York: Free Press, 1988); Patricia Hill Collins, *Black Feminist Thought: Knowledge, Consciousness, and Politics of Empowerment* (New York: Routledge, 1990).

7. For an exhaustive examination of these factors, see Elaine Tyler May, *Home-*

ward Bound: American Families in the Cold War Era (New York: Basic Books, 1988).

8. The three-fifths figure is reported in Stacey, *Brave New Families*, p. 10; she is careful to point out that the figure includes childless married couples. Stacey's source for this figure is Kathleen Gerson, *Hard Choices: How Women Decide About Work, Career, and Motherhood* (Berkeley: University of California Press, 1985).

9. See the essays collected in Richard Wightman Fox and T. J. Jackson Lears, eds., *The Culture of Consumption: Critical Essays in American History, 1880–1980* (New York: Pantheon, 1983). See also George Lipsitz, *Time Passages: Collective Memory and American Popular Culture* (New York: Pantheon, 1973/1984.

10. See Lipsitz, *Time Passages*, p. 55.

11. See C. Wright Mills, *White Collar: The American Middle Class* (New York: Oxford University Press, 1956); William H. Whyte, *The Organization Man* (New York: Simon and Schuster, 1956); David Reisman, *The Lonely Crowd: A Study of the Changing American Character* (New Haven, Conn.: Yale University Press, 1956); George Lipsitz, *Class and Culture in Cold War America: "Rainbow at Midnight"* (South Hadley, Mass.: J. F. Bergin, 1982). Interestingly, as Elaine Tyler May notes, because most social observers at the time "considered homemakers to be emancipated and men to be oppressed," no widely read treatise on women's plight during the 1950s would appear until the publication of Betty Friedan's *The Feminine Mystique* (New York: Dell, 1963). See May, "Cold War—Warm Hearth: Politics and the Family in Postwar America," in Steve Fraser and Gary Gerstle, ed., *The Rise and Fall of the New Deal Order, 1930–1980* (Princeton, N.J.: Princeton University Press, 1989), pp. 159, 179.

12. Quoted by Fox and Lears in their introduction to *The Culture of Consumption,* p. x. Also see Richard M. Nixon, *Six Crises* (Garden City, N.Y.: Doubleday, 1962), p. 280.

13. May, "Cold War—Warm Hearth," p. 158.

14. See illustration 1.10 in Michael Rogin, *Ronald Reagan, the Movie and Other Essays in Political Demonology* (Berkeley: University of California Press, 1987).

15. Garry Wills, *Reagan's America: Innocents at Home* (New York: Doubleday, 1987), pp. 279–80.

16. It also should be noted that Reagan's speeches often featured Nixonian reasoning in defending the American free enterprise system. For instance, in a speech delivered on December 8, 1972, at the 77th Congress of American Industry, Reagan implied that free enterprise had actually defeated poverty despite misleading televised reports to the contrary. "Hardly a week passes," complained Reagan, "without some television program sponsored by some of you, but what portrays the horrors of poverty and hunger in the United States—and yet 99 percent of the homes in America have electricity and the basic appliances such as refrigerators and range and 96 percent of the homes have television and telephones; 80 percent have automobiles." As Amos Kiewe and Davis W. Houck point out, there is a major flaw in Reagan's reasoning. Whereas ownership of the standard GE and Ford product line signaled to Reagan that "poverty and hunger had been eradicated, except in the make-believe world of television," the make-believe world of Reagan's speech

305 NOTES TO CHAPTER 8

had been constructed on misleading statistics that were "based solely on homes, hence homeowners." Therefore, as Kiewe and Houck note, "Reagan's proofs that poverty was non-existent begged the question of poverty altogether" because, clearly, "homeowners would tend to fall outside the latitude of the impoverished." See Amos Kiewe and Davis W. Houck, *A Shining City on a Hill: Ronald Reagan's Economic Rhetoric, 1951–1989* (New York: Praeger, 1991), p. 93.

17. See Stephanie Coontz, *The Way We Never Were: American Families and the Nostalgia Trap* (New York: Basic Books, 1992), p. 73.

18. According to Coontz and historian Linda Kerber, Rose Wilder Lane (the daughter of Laura Ingalls Wilder) conceived of the *Little House* stories as "an ideological attack on government programs." Failing as a free-lance writer in the 1930s, Lane returned to her family home in the Ozarks where she "announced that she would no longer write so that she would not have to pay taxes to a New Deal government." However, despite her worst intentions, Lane did continue to pursue writing. In Kerber's words: "she *rewrote* the rough drafts of her mother's memoirs . . . turning them into the *Little House* books in which the isolated family is pitted against the elements and makes it—or doesn't—with no help from the community." See Coontz, *The Way We Never Were*, p. 73, and Linda Kerber, "Women and Individualism in American History," *Massachusetts Review* (Winter 1989), pp. 604–5.

19. According to Coontz, this reliance on federal support is still true of the modern western family: "It would be hard to find a Western family today or at any other time in the past whose land rights, transportation options, economic existence, and even access to water were not dependent on federal funds." Or as western historian Patricia Nelson Limerick puts it: "Territorial experience got Westerners in the habit of federal subsidies, and that habit persisted long after other elements of the Old West had vanished." See Coontz, *The Way We Never Were*, pp. 73–74, and Limerick, *Legacy of Conquest: The Unbroken Past of the American West* (New York: W. W. Norton, 1987), p. 82.

20. Coontz, *The Way We Never Were*, p. 78.

21. For a discussion of "hyperghettoization," see ibid., p. 245.

22. See Judith Stacey's discussion of "Feminism as Midwife to Postindustrial Society," in *Brave New Families*, pp. 12–15. See also the opening chapters of Andrea Press, *Women Watching Television* (Philadelphia: University of Pennsylvania Press, 1991).

23. Susan Householder Van Horn, *Women, Work, and Fertility, 1900–1986* (New York: New York University Press, 1988), p. 152. The figure also appears in Stacey, *Brave New Families*, p. 285, n. 49.

24. Stacey, *Brave New Families*, p. 8.

25. Anderson, "Feminism and the American Family Ideal," p. 237.

26. According to Eli Zaretsky, Charlotte Perkins Gilman was "almost alone in her critique of this position": "Gilman believed that the payment of wages for work outside the home, and not for work within the home, would keep the family as isolated, as inward-turning, and as male-dominated as it had been in the nineteenth century. She called the private economic basis of the family the 'hidden spring' through which the antisocial and antifemale biases of the nineteenth century would

flow into the twentieth." See Zaretsky, "The Place of the Family in the Origins of the Welfare State," in Thorne and Yalom, eds., *Rethinking Family,* pp. 212–14.

27. See Nancy Chodorow and Susan Contratto, "The Fantasy of the Perfect Mother," in Thorne and Yalom, eds., *Rethinking Family,* pp. 54–75.

28. Friedan, quoted in Steven Mintz and Susan Kellogg, *Domestic Revolutions: A Social History of American Family Life* (New York: Free Press, 1988), p. 189.

29. See Sara Ruddick, "Maternal Thinking," in Thorne and Yalom, eds., *Rethinking Family,* pp. 76–94. See also Lipsitz, *Time Passages,* pp. 48–57.

30. David Spiegel, M.D., "Mothering, Fathering, and Mental Illness," in Thorne and Yalom, eds., *Rethinking Family,* p. 95.

31. Rosabeth Moss Kanter, *Work and Family in the United States: A Critical Review and Agenda for Research and Policy* (New York: Sage, 1977). See also Ella Taylor, *Prime-Time Families: Television Culture in Postwar America* (Berkeley: University of California Press, 1989), p. 10.

32. Rogin, *Ronald Reagan,* p. 305.

33. Petchesky, "Antiabortion," p. 224.

34. Despite the profamily veneer of its rhetoric of choice, the Bush education program was essentially an extension of Reaganism's quest to release capitalism from what Walter Karp calls "its republican bondage." "As a matter of course," warns Karp, "the Reaganites hope to turn public education into class education by financing the middle-class exodus from the common schools. When they become schools for a class and not for a commonality, the American republic will have lost the only instrument capable of turning a mass of future jobholders into a plurality of citizens. The common schools of the republic are one of capitalism's fetters, and so of course they must be broken." Karp concludes, and we agree, that Reaganites "want capitalism in America to become what Karl Marx thought it would be by nature—the transcendent force and the measure of all things, the power that reduces free politics to trifling, the citizen to a 'worker,' the public realm to 'the state,' the state to an instrument of repression protecting capitalism from the menace of liberty and equality, with which it grew up as Cain grew up with Abel . . . Marx's description of capitalist society is the Reaganite prescription for America." Quoted in Greil Marcus, *Lipstick Traces: A Secret History of the Twentieth Century* (Cambridge, Mass.: Harvard University Press, 1989), pp. 136–38, from Walter Karp, "Coolidge Redux," *Harper's* (October 1981), pp. 31–32.

35. Judith Stacey, as usual, cuts straight to the heart of the matter: "When 'profamily' forces helped elect Reagan to his first term in 1980, 20 percent of American children lived with a single parent, and 41 percent of the mothers with children under the age of three had joined the paid labor market. When Reagan completed his second term eight years later, these figures had climbed to 24 and 54 percent respectively. The year of Reagan's landslide re-election, 1984, was the first year that more working mothers placed their children in public group child care than in family day care. Reaganites too hastily applauded a modest decline in divorce rates during the 1980s—to a level at which more than half of first marriages still were expected to dissolve. But demographers who studied marital separations as well as divorce found the years from 1980 to 1985 to show 'the highest level of marital

disruption yet recorded for the U.S.' Likewise, birth rates remained low, marriage rates fell, and homeownership rates, which had been rising for decades, declined throughout the Reagan years." See Stacey, *Brave New Families*, p. 15. Stacey cites the following sources in this passage: Larry Liebert, "Gloomy Statistics on the Future of Poor Children," *San Francisco Chronicle* (October 2, 1989), p. A2; Sar A. Levitan, Richard S. Belous, and Frank Gallo, *What's Happening to the American Family? Tensions, Hopes, Realities*, rev. ed. (New York: Pilgrim Press, 1986); Larry Bumpass and Teresa Castro, "Recent Trends and Differentials in Marital Disruption," *Working Paper 87–20*, Center for Demography and Ecology, University of Wisconsin, Madison (June 1987); Karen Skold, "The Interests of Feminists and Children in Child Care," in Sanford M. Dornbusch and Mayra H. Strober, eds., *Feminism, Children, and the New Families* (New York: Guilford Press, 1988).

36. Petchesky, "Antiabortion," p. 224.

37. The distinction between "hard" and "settled" living is borrowed from Stacey, *Brave New Families*.

38. The profiles of Schlafly, Marshner, and LaHaye are derived from Susan Faludi, *Backlash: The Undeclared War Against Women* (New York: Crown, 1991), pp. 239–56. Also see the following sources cited by Faludi: Carol Felsenthal, *The Sweetheart of the Silent Majority: The Biography of Phyllis Schlafly* (New York: Doubleday, 1981); Rebecca E. Klatch, *Women of the New Right* (Philadelphia: Temple University Press, 1987); Connaught C. Marshner, *The New Traditional Woman* (Washington, D.C.: Free Congress Research and Education Foundation, 1982); Beverly LaHaye, *The Spirit-Controlled Woman* (Eugene, Ore.: Harvest House, 1976).

39. Faludi, *Backlash*, p. 256.

40. Stuart Hall, *The Hard Road to Renewal: Thatcherism and the Crisis on the Left* (London: Verso, 1988), p. 145.

41. See Phyllis Schlafly, "ERA Means Unisex Society," *Conservative Digest* 4, no. 7 (July 1978): 14–16. Quoted in Petchesky, "Antiabortion," p. 226.

42. See Stacey, *Brave New Families*, p. 259. Stacey credits Deniz Kandiyoti with coining the term "patriarchal bargain." See Kandiyoti, "Bargaining with Patriarchy," *Gender & Society* 2, no. 3 (September 1988): 274–90.

43. Faludi, *Backlash*, p. 256.

44. Petchesky, citing an article in *Ladies' Home Journal*, reports that Bryant now claims to be more sympathetic to gay and feminist anger: "She sees 'a male chauvinist attitude' in 'the kind of sermon [she] always heard' growing up in the Bible belt—'*wife submit to your husband even if he's wrong*'; and thinks 'that her church has not addressed itself to women's problems'": "Fundamentalists have had their heads in the sand. The church is sick right now and I have to say I'm even part of that sickness. I have often had to stay in pastors' homes and their wives talk to me. Some pastors are so hard-nosed about submission and insensitive to their wives' needs that they don't recognize the frustration—even hatred—within their own households." See Petchesky, "Antiabortion," p. 238. Also see Cliff Jahn, "Anita Bryant's Startling Reversal," *Ladies' Home Journal* (December 1980), pp. 62–68.

45. Stacey, *Brave New Families*, p. 14. Also see, Delia Ephron, "The Teflon Daddy," *New York Times Book Review* (March 26, 1989), p. 8., and "The Daughter Who Begs to Differ," *San Francisco Chronicle* (September 22, 1989), p. A10.

46. For accounts of the Nofziger and Sears dismissals, see Wills, *Reagan's America*.

47. Philip Wylie, *Generation of Vipers* (New York: Farrar, Rinehart, 1942); Christopher Lasch, *Haven in a Heartless World: The Family Besieged* (New York: Basic Books, 1977), p. 153. For a discussion of the myth of the all-powerful mother, and how it even contaminates some feminist writing, see Chodorow and Contratto, "The Fantasy of the Perfect Mother."

48. Wills, *Reagan's America*, p. 185.

49. Ibid., p. 190.

50. Ibid., p. 191.

51. "It messes up your hair."

52. See Richard Campbell, 60 Minutes *and the News: A Mythology for Middle America* (Urbana: University of Illinois Press, 1991).

53. For another description of the centering function of television news and its ties to Middle America, see Campbell, 60 Minutes *and the News*, pp. 137–57.

54. See Wills, *Reagan's America*, p. 190. For other accounts of the incident, see *New York Times* (August 15 and 19, 1984) and *Washington Post* (August 22, 1984).

55. Lawrence Grossberg, *We Gotta Get Out of This Place: Popular Conservativism and Postmodern Culture* (New York: Routledge, 1992), p. 165.

56. See Douglas J. Baharov, "Crack Babies: The Worst Threat Is Mom Herself," *Washington Post* (August 6, 1989), Outlook, p. 1.

57. Rickie Solinger, *Wake Up Little Susie: Single Pregnancy and Race Before* Roe v. Wade (New York: Routledge, 1992), pp. 24–25.

58. Ibid.

59. Ellen Hopkins, "Childhood's End," *Rolling Stone* (October 18, 1990), pp. 66–72, 108, 110.

60. See Richard Saltus, "Silber Attacks Health System," *Boston Globe* (April 30, 1991), Metro/Region, p. 15.

61. Barry Zuckerman and Deborah A. Frank, " 'Crack Kids': Not Broken," *Pediatrics* 89 (February 1992): 337–39.

62. See Faludi, *Backlash*, p. 428. Faludi cites Paula Braverman, Geraldine Oliva, Marie Grisham Miller, Randy Reiter, and Susan Egerter, "Adverse Outcomes and Lack of Health Insurance Among Newborns in an Eight-County Area of California, 1982–1986," *New England Journal of Medicine* 321, no. 8 (August 24, 1989): 508–14; Lynn M. Paltrow, "When Becoming Pregnant Is a Crime," *Criminal Justice Ethics* 9, no. 1 (Winter–Spring 1990): 2–3, 14.

63. See Richard Campbell and Jimmie L. Reeves, "Covering the Homeless: The Joyce Brown Story," *Critical Studies in Mass Communication* 6, no. 1 (March 1989): 21–42.

64. Collins, *Black Feminist Thought*, pp. 68–78.

65. Ibid., p. 77.

66. Thomas P. Strandjord and W. Alan Hodson, "Neonatology," *Journal of the American Medical Association* 268 (July 15, 1992): 377–78.

67. See I. J. Chasnoff, D. R. Griffith, C. Frier, and J. Murray, "Cocaine/Polydrug Use in Pregnancy: Two Year Follow-Up," *Pediatrics* 89 (1992): 284–89.

68. Linda C. Mayes, Richard H. Granger, Marc H. Bornstein, and Barry Zuckerman, "The Problem of Prenatal Cocaine Exposure: A Rush to Judgment," *Journal of the American Medical Association* 267 (January 15, 1992): 406–7.

69. See Al Dale's report on "ABC's World News Tonight" (July 2, 1992), and Dan Rutz, "Crack Babies May Have Better Future than Thought," CNN (August 8, 1992).

70. Coles, quoted in Ellen Goodman, "The Myth of the 'Crack Babies,' " *Boston Globe* (January 12, 1992), Op-Ed, p. 69.

71. See Ellen Whitford, "Cocaine Babies' Problems Overstated? Emory Team's Conclusions Rock the Boat," *Atlanta Constitution* (April 10, 1992), p. H-7.

72. See G. Koren, K. Graham, H. Shear, and T. Einarson, "Bias Against the Null Hypothesis: The Reproductive Hazards of Cocaine," *Lancet* 2 (1989): 1440–42.

73. See I. J. Chasnoff, H. Landress, and M. Barett, "The Prevalence of Illicit-Drug or Alcohol Use During Pregnancy and Discrepancies in Mandatory Reporting in Pinellas County, Florida," *New England Journal of Medicine* 322 (1990): 1202–6.

74. Coontz, *The Way We Never Were*, pp. 270–71.

75. Collins, *Black Feminist Thought*, p. 74.

9   Denouement

1. Jefferson Morley, "What Crack Is Like," *New Republic* (October 2, 1989), pp. 12–13.

2. Lloyd Johnston, "America's Drug Problem in the Media: Is It Real or Is It Memorex?" in Pamela J. Shoemaker, ed., *Communication Campaigns About Drugs: Government, Media, and the Public* (Hillsdale, N.J.: Lawrence Erlbaum Associates, 1989), p. 98.

3. See David F. Musto, *The American Disease: Origins of Narcotic Control*, expanded ed. (New York: Oxford University Press, 1973/1987).

4. This openness can also be conceived of in temporal terms. On one temporal horizon, journalistic discourse is oriented toward the past, or what Bakhtin terms *the already uttered*. According to Bakhtin, "discourse—in any of its forms, quotidian, rhetorical, scholarly—cannot fail but be oriented toward the 'already uttered,' the 'already known,' the 'common opinion,' and so forth." This orientation toward the "already uttered" is especially apparent in news reports that attempt to place the 1980s cocaine crisis in dialogue with the drug culture of the 1960s, or in reports that show clips of *Annie Hall* or *Desperately Seeking Susan* as examples of popular movies that endorse the use of drugs. However, there is another horizon to the temporal openness of news discourse—and that is its orientation on the future, or what Bakhtin terms *the answering word*: "Every word is directed toward an *answer* and cannot escape the profound influence of the answering word that it anticipates." In the case of news discourse the answering word can take several forms. In print journalism the letters to the editor section is a standard

form of the answering word. At times, public opinion polls are a type of answering word that often stands in for the will of the people. The ratings of competing newscasts can even be conceived of in terms of an answering word that has important consequences in the economics of the news business. See Mikhail Bakhtin, "Discourse in the Novel," in Michael Holquist, ed., *The Dialogic Imagination: Four Essays,* trans. Caryl Emerson and Michael Holquist (Austin: University of Texas Press, 1981), pp. 279, 280.

5. Because of the gravity of the "already uttered" and promise of the "answering word," the news report lives (in Bakhtin's words) "on the boundary between its own context and another, alien, context." As most introductory newswriting textbooks confirm, the news report, whether in print or on television, is written with a readership or viewership "in mind." This internalized "sense of audience," in Bakhtin's estimation, is a "fundamental force" in the meaning-making process that actually "participates in the formulation of discourse." As we have shown in previous chapters, this internalized sense of audience response is especially apparent in the Us vs. Them rhetoric of the siege paradigm and the "exhortative 'You'" of the crusading voice. In both the We and the You modes of address, the report speaks to an "active understanding" in the audience, one that the journalist "senses as resistance or support enriching the discourse." As we demonstrate in this chapter, Bakhtin's notion of the "answering word" also has several implications for those interested in studying media criticism as a cultural enterprise. For this study, perhaps the most important of these implications involves seeing the climate of media criticism as a system for generating and publicizing "responsive understandings" that contribute to the struggle over the "actual meaning" of a particular newsworthy theme or event. See Bakhtin, "Discourse in the Novel," p. 284.

6. Peter Kerr, "Anatomy of the Drug Issue: How, After Years, It Erupted," *New York Times* (November 13, 1986), p. B6.

7. Adam Paul Weisman, "48 Hours on Crock Street: I Was a Drug-hype Junky," *New Republic* (October 6, 1986), pp. 14–17.

8. William Safire, "The Drug Bandwagon," *New York Times* (September 11, 1986), p. A27.

9. Richard C. Cowan, "How the Narcs Created Crack," *New Republic* (December 5, 1986), pp. 26–31.

10. Ibid., p. 28.

11. Richard Vigilante, "Reaganites at Risk," *New Republic* (December 5, 1986), pp. 31–34.

12. Vigilante, ibid. (p. 31), provides the following paraphrase of Crane's objections to the bill: "Since drug abuse 'cannot be eliminated from society,' it is bad policy to 'spend $6 billion, cut corners on civil liberties, and expand the power of the government in ways that we might regret later,' in pursuit of an unattainable goal."

13. It should be noted that Vigilante, ibid., is careful to distance his position from that taken by "left-wing intellectuals": "The drug war as it is being waged now violates not merely the abstract symbols of civil liberties beloved of left-wing intellectuals; it violates genuinely important rights in ways Americans have not in the past been disposed to tolerate." Later in the article, Vigilante expands (p. 34) on his concerns about the enterprising moralism of the New Right by suggesting

that "embracing the drug hysteria requires a rejection of essential conservative principles": "In a democratic society, the crucial psychological step in justifying coercion is to become convinced that those who are to be coerced lack the judgment to make minimally sensible decisions about their own welfare. Since this is an ability nearly all people feel they have, those who are to be coerced must be alleged to be unlike us, by virtue either of their abilities or their circumstances. . . ."

14. Ibid.

15. Ibid.

16. See "Live on the Vice Beat," *Time* (December 22, 1986), p. 60.

17. Ibid.

18. See "Woman Sues Over TV Arrest," *New York Times* (December 31, 1986), p. D16.

19. Edwin Diamond, Frank Accosta, and Leslie-Jean Thornton, "Is TV News Hyping America's Cocaine Problem?" *TV Guide* (February 7, 1987), pp. 4–10.

20. Ibid., pp. 5–6.

21. One report (ABC, 3/10/87) dealt with economic problems facing network news divisions; the other (ABC, 3/13/87) is in a "Person of the Week" profile of CBS chairman Lawrence Tisch.

22. Ron Harris, "Blacks Feel Brunt of Drug War," *Los Angeles Times* (April 22, 1990), p. A1.

23. Quoting Tom Hetrick, president of Partnership for a Drug Free America, "We've been called a front for the alcohol industry. We take contributions from cigarette and alcohol companies. I have no apologies whatsoever. . . . Anheuser-Busch, RJR Reynolds are top 100 advertisers. Now, am I going to turn down $100,000 or $50,000 from them? Absolutely not! Are they impacting what I'm doing? Absolutely not! What is it W. C. Fields said? 'No good deed goes unpunished.' " See Larry McShane, "Did the Sizzling Ads Work? Madison Avenue Knows Something About Drugs, After All," *Ann Arbor News* (June 15, 1992), Connection, p. 1. See also Johan Carlisle, "Drug War Propaganda," *Propaganda Review* (Winter 1990), pp. 6–10, 43–44, and Cynthia Cotts, "Hard Sell in the Drug War," *The Nation* (March 9, 1992), pp. 300–302.

24. See "An Antidrug Message Gets Its Facts Wrong," *Scientific American* 262, no. 5 (May 1990): 36.

25. For example, see Peter Herridge and Mark S. Gold, "The New User of Cocaine: Evidence from 800-COCAINE," *Psychiatric Annals* 18, no. 9 (September 1988): 521–22; and Arnold Washton and Mark Gold, "Recent Trends in Cocaine Abuse: A View from the National Hotline, '800-COCAINE,' " *Advances in Alcohol and Substance Abuse* 6, no. 2 (Winter 1986): 31–47.

26. Michael Eric Dyson, "Between Apocalypse and Redemption: John Singleton's *Boyz N the Hood*," *Cultural Critique* (Spring 1992), pp. 129–30.

27. Jacquie Jones, "The New Ghetto Aesthetic," *Wide Angle* 13, nos. 3 and 4 (July–October 1991): 41.

28. Dyson, "Between Apocalypse and Redemption," p. 133.

29. Ibid., p. 135.

30. Ibid., p. 131.

31. Ibid., p. 132.

32. See Stuart Hall, "Encoding/Decoding," in Hall, Dorothy Hobson, Andrew Lowe, and Paul Willis, eds., *Culture, Media, Language* (London: Hutchinson, 1980), pp. 128–38.

33. Theodore L. Glasser, "Objectivity Precludes Responsibility," *The Quill* (February 1984), p. 15.

34. James Carey, "The Press and the Public Discourse," *Kettering Review* (Winter 1992), p. 20.

35. Johnston, "America's Drug Problem," p. 109.

36. Ibid., p. 103.

37. Craig Reinarman and Harry G. Levine, "The Crack Attack: Politics and Media in America's Latest Drug Scare," in Joel Best, ed., *Images of Issues: Typifying Contemporary Social Problems* (New York: Aldine de Gruyter, 1989), p. 127.

38. Murray Edelman, *Constructing the Political Spectacle* (Chicago: University of Chicago Press, 1988), p. 25.

39. See William Booth, "War Breaks Out Over Drug Research Agency," *Science* 241 (August 5, 1988): 648–50.

40. We also feel compelled to point out that Johnston, like most other expert sources, did not choose to defend his record by responding to our questionnaire.

41. Michel Foucault, "Right of Death and Power Over Life," in *The Foucault Reader,* pp. 267–68.

42. See Reinarman and Levine, "The Crack Attack," p. 127.

43. Arnold Trebach, *The Great Drug War* (New York: Macmillan, 1987), p. 5.

44. Greeley, quoted in Christopher Lasch, "Journalism, Publicity and the Lost Art of Argument," *Gannett Center Journal* 4 (Spring 1990): 2.

45. Carey, "The Press and the Public Discourse," p. 15.

46. Glasser, "Objectivity," p. 15.

47. Ibid.

48. Wicker, quoted in ibid., p. 15.

49. Carey, "The Press and the Public Discourse," p. 22.

50. Ibid., p. 21.

51. Ibid., pp. 20–21.

Epilogue

1. Jon Pareles, "Gangster Rap: Life and Music in the Combat Zone," *New York Times* (October 7, 1990), Arts and Leisure, p. 29.

2. Scarface, quoted in Catherine Chriss, "For Houston's Geto Boys, Anything Goes in the World of Gangsta Rap," *Houston Chronicle* (April 5, 1992), Texas magazine, pp. 10ff.

3. Chuck D, quoted in Jay Cocks, "The Empire Strikes Black," *Time* (November 11, 1991), p. 98.

4. Cornel West, "Rap It Up," *Village Voice Literary Supplement* (October 1991), p. 21.

5. In fact, we are currently considering a sequel to this study (*Accounting for Mayhem: Drugs, Sex, Race, and Popular Culture Under the Reagan/Bush Order*)

that investigates the dialogue between these accounts and the mainstream news treatment of the cocaine problem.

6. See Alex McNeil, *Total Television: A Comprehensive Guide to Programming from 1948 to the Present*, 3d ed. (New York: Penguin, 1991), pp. 1058–61.

7. For a critique of "The Cosby Show" and a compelling history of African American images on television, see the documentary by Marlon Riggs, *Color Adjustment* (San Francisco: California Newsreel, 1991).

8. See Ed Siegel, "It's Not Fiction, It's Not News, It's Not Reality, It's Reali-TV," *Boston Sunday Globe* (May 26, 1991), p. A1.

9. See Richard Campbell, 60 Minutes *and the News: A Mythology for Middle America* (Urbana: University of Illinois Press, 1991), and Campbell, "Don Hewitt's Durable Hour," *Columbia Journalism Review* (September/October 1993): 25–28.

10. Chad Raphael, "The Political Economy of Reali-TV," paper presented at the annual meeting of the Association for Education in Journalism and Mass Communication (August 1992), Montreal, p. 13.

11. For information on producers who convince law enforcement to donate equipment, see Scott A. Nelson, "Crime-Time Television," *FBI Law Enforcement Bulletin* (August 1989), pp. 1–9; and Raphael, "The Political Economy of Reali-TV," p. 13.

12. "Fall TV Preview: 1991–92 Primetime at a Glance," *Variety* (August 31, 1992), pp. 29–36.

13. McNeil, *Total Television*, p. 1060–61.

14. B. Keith Crew, "Acting Like Cops: The Social Reality of Crime and Law on TV Police Dramas," in Clinton R. Sanders, ed., *Marginal Conventions: Popular Culture, Mass Media and Social Deviance* (Bowling Green, Ohio: Bowling Green State University Press, 1990), p. 132.

15. Grosso, quoted in Jane Hall, "CBS' 'Top Cops' Patrol TV's Toughest Turf," *Los Angeles Times* (October 25, 1990), p. F1.

16. See John Koch, "A Deadpan 'Feud'; Flimsy '48 Hours,'" *Boston Globe* (February 26, 1992), Living, p. 39.

17. For an overview of the rise of gangster rap, see Robin D. G. Kelley, "Straight from Underground," *Nation* (June 8, 1992): 793–96; and Kelley, "Kickin' Reality, Kickin' Ballistics: The Cultural Politics of Gangsta Rap in Postindustrial Los Angeles," unpublished paper, Department of History and the Center for African and Afro-American Studies, University of Michigan, 1992. We are grateful to Kelley for access to this paper and his research on rap.

18. See John Katz, "Rock, Rap and the Movies Bring You the News," *Rolling Stone* (March 5, 1992), pp. 33–40, 78.

19. Quoted in Dennis Hunt, "The Rap Reality: Truth and Money," *Los Angeles Times* (April 2, 1989), Calendar, p. 80.

20. Kelley, "Straight from Underground," pp. 794–95.

21. Michael S. Gazzaniga, *Mind Matters: How Mind and Brain Interact to Create Our Conscious Lives* (Boston: Houghton Mifflin, 1988), p. 159.

22. Cornel West, *Race Matters* (Boston: Beacon Press, 1993), pp. 29–30.

23. Ibid., p. 1.

24. Again, for the record, for every one death related to cocaine or crack, there

are, respectively, three hundred and one hundred deaths related to nicotine and alcohol.

25. Michael Stone, quoted on "Nightline," ABC News, April 29, 1992.

26. Our interpretation of West here comes from a public lecture he gave for the Center for African and Afro-American Studies at the University of Michigan, May 17, 1992. See also Cornel West, "Learning to Talk of Race," *New York Times Magazine* (August 2, 1992), pp. 24–26.

Index

General Electric, 188
Gergen, Kenneth, 30, 280 n.34
Geriatric drug epidemic, 151
Geto Boys, 255, 258
Giancana, Nick, 234
Gilder, George, 74, 96
Gillen, Michelle, 212–215, 251, 269
Gilman, Charlotte Perkins, 305–306 n.26
Ginsberg, Benjamin, 167–168
Gitlin, Todd, 129, 133, 150
Gladstone, William, 214
Glasser, Theodore, 7, 29, 247, 253
Gold, Mark, 68, 151, 232–233, 251
Goldberg, Bernard, 178–180, 227–228, 266–267, 271
Goldwater, Barry, 77–78
Goldwater, Barry, Jr., 272
Gomery, Douglas, 286 n.28
Gooden, Dwight, 70, 138, 267, 275
Gorbachev, Raisa, 199
Gould, Stanhope, 232
Graham, Phil, 110
Gramsci, Antonio, 5, 8, 85, 289 n.8
Granger, Richard H., 215
Gray, Herman, 100–101, 294 n.91
Great Depression, 187
Greeley, Horace, 252
Greenberg, Stanley B., 182
Greenwood, Bill, 65
Gregg, David, 68, 274
Greider, William, 126–127
Grinspoon, Lester, 25, 45, 283 n.36
Griswold v. Connecticut (1965), 80, 195
Grossberg, Lawrence, 103, 207, 289 n.10
Grosso, Sonny, 257
Gulf War, 228, 289 n.3
Gusfield, Joseph R., 42, 125, 290 n.24

Habermas, Jurgen, 9, 24
Hager, Robert, 65
Hahn, Dan F., 119
Hall, Bruce, 177, 266
Hall, Stuart, 4–5, 8, 62–63, 78, 89, 163–164, 175, 196–197, 245, 289 n.8
Harrington, Michael, 86
Harris, Ron, 229–230

Harrison Act, 43–44
Hart, John, 235–244, 271
Harvey, David, 24, 86–88, 116
Hays, Will, 112
Health insurance crisis, 212
Heath, Dwight B., 35
Helmer, John, 43
Helms, Jesse, 164
Helmsley, Leona, 199
Hemingway, Ernest, 30, 280 n.34
Henry III, William A., 48
Herd, Denise, 283 n.31
Heritage Foundation, 76, 195–196
Hernandez, Keith, 67
Heroin, 18, 112, 129, 201, 221, 271
Herschman, Philip, 125
Hertsgaard, Mark, 169
Hetrick, William Morgan, 67
Hewitt, Don, 115
Hill, Anita, 247
Hinckley, John, 257
Hodson, W. Alan, 215
Hollywood: and anticommunist witch hunts, 77; as chronotope, 19, 109–113, 131, 136; as journalistic beat, 65; and Nancy Reagan, 199–200; and Ronald Reagan, 77, 82; and scandal, 10, 20, 67, 110–114, 128, 137, 141, 263, 272
Hood, Ian, 234
Horton, Willie, 26, 213–214, 276
Hotlines, 68–69, 71, 124, 130, 151, 264, 273, 296–97 n.43–46; 800-COCAINE, 122, 232–233, 251, 273, 311 n.25; and informant society, 159; as postmodern chronotope, 122–123
Houck, Davis H., 304–305 n.16
House Committee on Un-American Activities (HUAC), 77
Housman, A. E., 143
Howe, Steve, 67, 138, 264, 273
Hudson, Rock, 33, 111–113
Hype, 1, 11, 21, 182, 218, 222, 226–228, 275

Iacocca, Lee, 2, 74, 109
Ice Cube, 245, 258

About the Authors

Jimmie L. Reeves is Assistant Professor, Department of Communication, University of Michigan.

Richard Campbell is Assistant Professor, Department of Communication, University of Michigan. He is the author of 60 Minutes *and the News: A Mythology for Middle America*. Both authors are also faculty associates in Michigan's Program in American Culture.